Secure Communication for the 6G and the Internet of Things Networks

This book will serve as a comprehensive resource for readers interested in utilizing different technologies like fog computing, federated learning, cloud computing, and blockchain in 6G-enabled systems for addressing security concerns.

Key Features:

- Discusses technologies such as fog computing, federated learning, and blockchain in 6G-enabled networks to address security threats.
- Provides in-depth coverage of fundamental models, architectures, and security frameworks to help readers understand the complex landscape of 6G and the Internet of Things (IoT) security.
- Covers diverse areas of security frameworks and software-defined fixes for security problems in 6G and the IoT networks.
- Presents protocols for authentication and key exchange in 6G and the IoT networks.
- Highlights the use of the IoT for enhanced biometric authentication in security.

This book is primarily written for senior undergraduates, graduate students, and academic researchers in the fields of electrical engineering, electrical and communications engineering, telecommunications, and computer science and engineering.

Communication Networks and Smart Techniques For 5G and Beyond
Edited by: Nisha Gupta

Secure Communication for the 6G and the Internet of Things Networks: Advances and Applications
Edited by Gaganpreet Kaur, A.V. Senthil Kumar, G. Prasanna Lakshmi, and Rakesh Nyayapati

Secure Communication for the 6G and the Internet of Things Networks

Advances and Applications

Edited by

Gaganpreet Kaur, A.V. Senthil Kumar,
G. Prasanna Lakshmi, and Rakesh Nyayapati

CRC Press
Taylor & Francis Group
Boca Raton London New York

CRC Press is an imprint of the
Taylor & Francis Group, an **informa** business

Designed cover image: Shutterstock

First edition published 2026
by CRC Press
2385 NW Executive Center Drive, Suite 320, Boca Raton FL 33431

and by CRC Press
4 Park Square, Milton Park, Abingdon, Oxon, OX14 4RN

CRC Press is an imprint of Taylor & Francis Group, LLC

ISBN: 978-1-032-95765-4 (hbk)
ISBN: 978-1-032-95766-1 (pbk)
ISBN: 978-1-003-58645-6 (ebk)

DOI: 10.1201/9781003586456

Typeset in Times
by codeMantra

Contents

Preface

Welcome to *Secure Communication for the 6G and the Internet of Things Networks: Advances and Applications*," a forward-looking exploration into the rapidly evolving intersection of next-generation communication technologies and cybersecurity. As editors, we are excited to present this comprehensive work that examines the transformative impact of 6G networks and IoT ecosystems on the global communication landscape, with a specific focus on safeguarding data, infrastructure, and user trust.

This book takes readers on an in-depth journey through the foundational and emerging technologies driving secure communication in ultra-connected environments. With the advent of 6G – a paradigm shift promising ultra-low latency, massive connectivity, and unprecedented data speeds – security and privacy concerns are becoming more critical than ever. Similarly, the IoT, with its billions of interconnected devices and entities spanning physical and cyber-physical domains, presents a complex landscape of both opportunities and vulnerabilities. This book aims to illuminate how these two revolutionary technologies can be harmonized securely and effectively.

As these platforms grow in scale and sophistication, so too do the challenges related to protecting the confidentiality, integrity, and availability of data. Security must no longer be an afterthought – it must be embedded as a core design principle. Modern threats demand a proactive, adaptive, and holistic approach, especially when dealing with heterogeneous and decentralized networks characteristic of both 6G and IoT.

Our primary objective is to provide a rich and accessible resource that bridges theoretical underpinnings with practical implementation. Through a careful blend of technical depth and real-world applicability, we explore the latest advancements in cryptographic techniques, AI-enhanced security protocols, quantum-safe encryption, network slicing, secure edge computing, privacy-preserving mechanisms, and risk assessment frameworks. This book also addresses the crucial balance between stringent security requirements and quality of service: a fundamental consideration in communication systems design.

This volume brings together high-quality, original contributions that tackle pressing open problems, ranging from secure protocol development to risk mitigation strategies in dynamic environments. Both theoretical explorations and experimental results are represented, offering a balanced perspective on the current state and future trajectory of secure communications in 6G and IoT systems.

Designed to serve a wide-ranging audience – including researchers, engineers, students, policymakers, and technology strategists – this book offers hands-on insights, case studies, and implementation frameworks. Whether you are a seasoned cybersecurity professional looking to understand next-generation threat models or a student entering the field with curiosity and ambition, this book provides the foundational and advanced knowledge necessary to thrive in this space.

Through this work, we aim not only to inform but also to inspire innovation in secure communication. The future of global connectivity depends on our ability to create secure, scalable, and intelligent systems. We hope this book will contribute to that mission by empowering readers with the tools and understanding needed to design and implement robust security solutions for tomorrow's hyper-connected world.

We invite you to explore the chapters ahead and engage with the ideas and innovations that are shaping the secure communication frameworks of the future.

As we conclude this preface, we emphasize the importance of secure communication at the intersection of 6G and IoT technologies. This book brings together essential knowledge and innovative strategies to address the unique challenges posed by these rapidly evolving systems.

Our goal is to offer a resource that not only educates but also encourages practical application and forward-thinking solutions. Whether you're a researcher, professional, or student, we believe the insights shared here will support your efforts in building secure, intelligent, and resilient communication networks.

Thank you for joining us on this journey. We look forward to the contributions you will make in shaping the future of secure 6G and IoT ecosystems.

Acknowledgments

We would like to express our deepest gratitude to all those who have contributed to the development of this book.

First and foremost, we extend our sincere appreciation to the authors and researchers whose valuable contributions form the foundation of this work. Their expertise, dedication, and innovative perspectives have made this book a rich and insightful resource.

We are also grateful to the peer reviewers and subject matter experts who generously offered their time and knowledge to ensure the quality and relevance of the content. Their constructive feedback played a crucial role in refining each chapter.

Special thanks go to our academic and professional institutions for their ongoing support and encouragement throughout the preparation of this volume. Their commitment to advancing research and knowledge in emerging technologies has been a constant source of inspiration.

We would also like to acknowledge the technical and editorial teams who worked behind the scenes to bring this book to life. Their meticulous attention to detail and unwavering professionalism made this complex process a smooth and successful journey.

Finally, we are thankful to our families and colleagues for their patience, understanding, and continuous encouragement during the completion of this project.

We hope that this book will serve as a valuable contribution to the field and inspire further innovation and collaboration in the area of secure communication for 6G and IoT networks.

About the Editors

Dr. Gaganpreet Kaur is a Professor in the Department of Computer Science and Engineering at Chitkara University, Punjab, India. She has teaching experience of 20 years and has published more than a hundred research articles in various peer-reviewed, SCI, Scopus journals, and international and national conferences in the field of computer science. She has published more than 17 Indian patents. Her research interests include biometrics, image processing, forensic science, cloud computing, and healthcare.

A.V. Senthil Kumar is presently working as a Professor and Director of PG and Research Department of Computer Applications at Hindusthan College of Arts & Science, Coimbatore, India. To his credit, he has industrial experience of 5 years and teaching experience of 27 years. He has published more than 200 papers in international and national journals and nearly 60 papers in international conferences. His areas of research interest include data mining and networks.

G. Prasanna Lakshmi is working as a Professor in the Department of Computer Science and Engineering and as R&D Coordinator at Sandip University, Nashik, India. She has 23 years of teaching experience and 7 years of industrial experience. Her field of expertise includes data mining, pattern recognition techniques, artificial intelligence, machine learning, and deep learning. She has published research papers in journals and conferences of national and international reputation.

Rakesh Nyayapati, based in Weehawken, New Jersey, United States, is currently an Associate Director at UBS, bringing experience from previous roles at Cognizant and UBS. He holds an ITIL Intermediate – SO in Service Management @ APMG International. With a robust skill set that includes process excellence, vendor management, transition management, IT service management, service delivery, and more, he contributes valuable insights to the industry.

Contributors

Abarna Sundaramurthy
IT/Cyber Security
Rashtriya Raksha University
Puducherry, India

Amitesh Aggarwal
Chitkara University Institute of Engineering and Technology
Chitkara University
Rajpura, India

Abhimanyu Agnihotri
Chitkara College of Pharmacy
Chitkara University
Rajpura, India

Gautam Bansal
Department of Computer Science and Engineering
CGC - College of Engineering
Mohali, India

Anupam Bonkra
Department of Computer Science and Engineering (CSE)
CGC College of Engineering
Landran, Mohali

Neeraj Choudhary
Department of Pharmacognosy
GNA School of Pharmacy
GNA University
Phagwara, India

Francisca Rosique Contreras
Polytechnic University of Cartagena
Cartagena, Spain

Priyanka Datta
Chitkara University Institute of Engineering and Technology
Chitkara University
Rajpura, India

Pummy Dhiman
Chitkara University Institute of Engineering and Technology
Chitkara University
Rajpura, India

Janvi Garg
Chitkara University Institute of Engineering & Technology
Rajpura, India

Neha Garg
Chitkara University Institute of Engineering and Technology
Chitkara University
Rajpura, India

Anjuli Goel
Chitkara University Institute of Engineering and Technology
Chitkara University
Rajpura, India

Raghav Goel
Chitkara University Institute of Engineering and Technology
Chitkara University
Rajpura, India

Yonis Gulzar
Department of Management Information Systems
College of Business Administration
King Faisal University
Al Ahsa, Saudi Arabia

Aryan Gupta
Chitkara University Institute of Engineering and Technology
Chitkara University
Rajpura, India

Jeyakarthic Mohan
Department of Computer and Information Science
Annamalai University
Annamalai Nagar, India

Sushil Kamboj
Department of Computer Science and Engineering
CGC - College of Engineering
Mohali, India

Kanchana Kavaratty Raju
Department of Computer Science
Government Arts College
Udhagamandalam, India

Amandeep Kaur
Chitkara University Institute of
Engineering and Technology
Chitkara University
Rajpura, India

Amanpreet Kaur
Chitkara University Institute of Engineering and Technology
Chitkara University
Rajpura, India

Archana Kaushik
Chitkara College of Pharmacy
Chitkara University
Rajpura, India

Ujjwal Kaushik
Chitkara College of Pharmacy
Chitkara University
Rajpura, India

Kavitha Kalyanasundaram
Department of Electrical & Electronics Engineering
Annamalai University
Annamalai Nagar, India

Keshav
Chitkara University Institute of Engineering & Technology
Rajpura, India

Neha Kukreti
Chitkara College of Pharmacy
Chitkara University
Rajpura, India

Shalini Kumari
Department of Computer Science & Engineering
Chitkara University Institute of Engineering and Technology
Chitkara University
Punjab, India
Mohali, India

Vansh Kwatra
Chitkara University Institute of Engineering and Technology
Chitkara University
Rajpura, India

Pooja Mahajan
Chitkara University Institute of Engineering & Technology
Rajpura, India

Khushi Mahendru
Chitkara University Institute of Engineering & Technology
Rajpura, India

Megha Bhushan
Department of Computer Languages and Systems
University of Seville
Sevilla, Spain

Jivansh Midha
Chitkara University Institute of Engineering and Technology
Chitkara University
Rajpura, India

Hitkar Miglani
Chitkara University Institute of Engineering and Technology
Rajpura, India

Akshay Mudgal
Department of Computer Science
Amity University
India

Aditya Pathania
Chitkara University Institute of Engineering and Technology
Chitkara University
Rajpura, India

Chander Prabha
Chitkara University Institute of Engineering and Technology
Chitkara University
Rajpura, India

Dolly R. Prajapati
Ganpat University
Mehsana, India

Shofia Priyadharshini Dayalan
Department of Electronics and Communication Engineering
Rajalakshmi Institute of Technology
Tamil Nadu, India

Rajasri Kathavarayan
Artificial Intelligence and Machine learning
Christ College of Engineering and Technology
Puducherry, India

Sahil Rana
Department of Computer Science and Engineering
CGC - College of Engineering
Mohali, Punjab

Neetu Rani
Department of Computer Science & Engineering
Chandigarh University
Mohali, India

Sujoy Ray
Department of Computer Science & Engineering
Chandigarh University
Mohali, India

Ketan Sarvakar
Ganpat University
Mehsana, India

Mohd Asif Shah
Kardan University
Kabul, Afghanistan

Chetna Sharma
Chitkara University Institute of Engineering and Technology
Chitkara University
Mohali, India

Dikshant Sharma
Department of Computer Science and Engineering
CGC - College of Engineering
Mohali, India

Anand Singh
Illinois Institute of Technology
Chicago, IL USA

Damandeep Singh
Vidya Jyoti Eduversity
Mohali, India

Ishdeep Singla
Department of Computer Science & Engineering
Chandigarh University
Mohali, India

Vijayabaskaran Pavadaisamy Sivakolundu
Tata Consultancy Services
London, United Kingdom

Gaurav Yadav
Department of Computer Science and Engineering
CGC - College of Engineering
Mohali, Punjab

Narinder Yadav
Department of Computer Science & Engineering
Chandigarh University
Mohali, India

1 Next-Generation Authentication and Key Exchange Protocols for 6G and IoT

Gautam Bansal, Dikshant Sharma, Sahil Rana, Gaurav Yadav, Sushil Kamboj, Anupam Bonkra, and Megha Bhushan

1.1 INTRODUCTION

The advancement in wireless communications called for the 6G networks, which are expected to be much faster, with extremely low latency, and very much automated through the help of AI and edge computing. As more smart devices are being developed and utilized in our society, IoT [1] has also become an essential element of contemporary society owing to its capability of connecting trillions of sensors and devices to other applications and systems across healthcare, transportation, smart cities, and industry, among others. Although 6G is expected to mark monumental enhancements in network possibilities, security still poses a significant problem in recognizing and intercepting the identities of the users and the process of managing key exchanges. Since IoT [2] devices are inherently constrained in terms of resources and can be deployed in open and dynamic environments, conventional security mechanisms may not be effective in handling security threats. However, the 6G networks come with emerging risks, including integration of quantum computing, use of THz communication, AI control, and orchestration of the networks that require stronger security frameworks to counter them. This is for the purpose of user and device identification to ensure that only authorized users and devices gain access to the server, while key exchange is essential for the creation of secure channels for data exchange to counteract unauthorized access and various forms of cyberattacks. This chapter aims to discuss a number of authentication and key exchange protocols with respect to existing security concerns in 6G and IoT networks to meet their requirements and roadmap for development.

1.1.1 Importance of Security in Next-Generation Networks

Security is a vital part of modern networks and 6G and IoT networks in particular, since these networks intend to offer reliable, smart, and fast connections in numerous

DOI: 10.1201/9781003586456-1

applications. As the use of smartphones, the Internet, and the web grows every day, virtual threats are increasing at a rather fast pace as well, which is why security has become the main concern when it comes to sharing data, maintaining networks, and building up trust with users. While the fifth generation networks are mainly focused on various improvements of the new technique assembled in B5G networks, the 6G networks [3] will contain new key technologies that represent both opportunities and threats. As most of the IoT devices are resource constraint, and many of them are not very secure, there are risks of unauthorized access, data compromise, and disrupted service in the form of Distributed Denial of Service (DDoS) attack or even Man-in-the-Middle (MITM) attack. Some of the services include autonomous vehicles, telemedicine services, smart cities, and industrial automation, among others; this means that any violation of the security of the 6G network can lead to loss of lives or any other harm. Authorised access, sharing of secure keys, identifying encrypted data from other information, and other such measures are important in protecting data from being accessed by unauthorized people. However, distributed system security requires solutions based on Zero-Trust modelling, secure multi-party computation, and post-quantum cryptography, among others, to mitigate emerging forms of cyber threats. This means that the implementation of effective security solutions in the next-generation networks is again entirely dependent on the success of 6G and IoT network deployment, as the organizations and users of the world are demanding more security, reliability, and resilient networks from their prospective service providers.

1.1.2 Challenges in Authentication and Key Exchange

The first general problem is scalability, which is going to be a significant problem at 6G due to support for billions of IoT devices. This is especially so, considering that many IoT devices present in the system today are resource constrained; hence, the computation capability is limited. Further, due to the nature of 6G networks being largely distributed, lightweight and distributed authentication models will be preferred as opposed to centralized models, which may hinder the network performance and invite security threats. There is also the issue of quantum computing, which poses a threat to classical cryptographic algorithms in terms of key exchange, and this requires one to employ quantum-resistant cryptographic methods. The privacy issue is also a problem as it has to ensure that the identities of the users are confirmed without exposing their information to unauthorized persons. Besides, new generation threats like AI-stimulated cyberattacks, impersonation, and malware intrusion are threats to authentication schemes [4]. To overcome these obstacles, 6G and IoT networks require deploying new marvellous security solutions, including anomaly, blockchain-based authentication, and post-quantum cryptography to establish a reliable communication channel.

1.2 SECURITY THREATS AND REQUIREMENTS IN 6G AND IOT

Security threats in 6G and IoT networks pose significant challenges, requiring advanced solutions to ensure data integrity, privacy, and resilient authentication.

1.2.1 Emerging Security Threats in 6G Networks

With the advancement of 6G networks, the networks adopt AI, quantum computing, and ultrahigh-speed communications, thus creating new security threats. Table 1.1 lists the new security threats that will arise in 6G networks, their effects, and protection measures.

TABLE 1.1
Emerging Security Threats and Mitigation Strategies in 6G and IoT Networks

Threat Category	Description	Potential Impact	Mitigation Strategies
Quantum computing attacks	Quantum computers can break traditional cryptographic algorithms, making current encryption methods vulnerable.	Compromised confidentiality, unauthorized access to sensitive data.	Adoption of post-quantum cryptography and quantum-safe encryption techniques.
AI-driven cyberattacks	Attackers use AI to develop sophisticated malware, automate attacks, and evade detection systems.	Enhanced attack efficiency, faster and more adaptive threats.	AI-based threat detection, anomaly detection systems, and adversarial AI defences.
MITM attacks	Attackers intercept and manipulate communication between legitimate users.	Data tampering, eavesdropping, and session hijacking.	End-to-end encryption, secure key exchange protocols, and AI-driven intrusion detection.
DDoS attacks	Massive-scale botnets flood network resources, causing service disruptions.	Network downtime, degraded performance, and financial losses.	AI-driven traffic analysis, blockchain-based distributed security, and rate-limiting techniques.
IoT device vulnerabilities	Weak security in IoT devices leads to unauthorized access and exploitation.	Device takeovers, data breaches, and botnet formation.	Lightweight encryption, secure boot mechanisms, and firmware updates.
Identity spoofing & unauthorized access	Attackers impersonate legitimate users or devices to gain access to network resources.	Data theft, privacy breaches, and unauthorized control of network services.	Strong authentication mechanisms, biometric verification, and blockchain-based identity management.
THz communication Vulnerabilities	THz waves have different propagation characteristics, making them vulnerable to new interception techniques.	Eavesdropping and signal manipulation.	Secure THz transmission protocols and advanced physical-layer security techniques.
Side-channel attacks	Exploiting hardware vulnerabilities to extract cryptographic keys or sensitive information.	Unauthorized decryption, system compromise.	Secure hardware design, AI-driven anomaly detection, and shielding techniques.

(Continued)

TABLE 1.1 (*Continued*)
Emerging Security Threats and Mitigation Strategies in 6G and IoT Networks

Threat Category	Description	Potential Impact	Mitigation Strategies
Supply chain attacks	Compromising hardware or software during manufacturing or distribution.	Hardware backdoors, malware injection, large-scale espionage.	Blockchain-based supply chain security, strict hardware validation, and continuous monitoring.
Zero-day exploits	Attacks exploiting previously unknown software or hardware vulnerabilities.	Widespread network compromise, data breaches.	AI-based vulnerability detection, regular security patches, and bug bounty programmes.

1.2.2 IoT-Specific Security Challenges

The use of IoT devices has become a challenge to security due to the following reasons: massive use of these devices, limited resource availability, and the variation of devices. A huge number of IoT devices are primitive devices wherein constrained computational power, memory capacity, and battery power restrict complex security protocols, including encryption and authentication, and therefore, open the devices to cyber threats. Third, the current IoT manufacturers do not comply with central security standards due to the absence of a unified framework for security adoption, which exposes them to more risks of hacking, data leaks, and malware attacks. IoT networks are also diverse and deployed in open and dynamic contexts, so rather easy to attack, for example, via device cloning, unauthorized control, or MITM attacks. However, cyber attackers are even able to use them to recruit hosts for botnets to attack several other targets and even shut down essential services through DDoS attacks [5]. Another risk factor is a privacy problem since IoT devices gather a considerable amount of users' data, and if that data is vulnerable to attacks, then exposure to identity theft occurs. Mitigating these challenges requires employing lightweight cryptography, proper authentication mechanisms, artificial intelligence for anomaly detection, and frequent firmware updates to make IoT's protection efficient against emerging threats.

1.2.3 Requirements for Secure Authentication and Key Exchange

Secure authentication and key exchange are critical for protecting communication channels from eavesdropping and attacks. These processes ensure that entities verify each other's identities and establish encryption keys securely. The requirements are outlined in Table 1.2, which highlights key aspects such as mutual authentication, forward secrecy, and resistance to replay attacks.

1.3 AUTHENTICATION MECHANISMS IN 6G AND IOT

Authentication mechanisms in 6G and IoT must ensure secure, efficient, and scalable identity verification to support massive connectivity and low latency communication.

TABLE 1.2
Requirements for Secure Authentication and Key Exchange

Requirement	Description	Importance
Scalability	The authentication and key exchange mechanisms must support billions of IoT devices and users in 6G networks.	Ensures seamless connectivity without bottlenecks as the number of connected devices grows.
Low latency	Authentication and key exchange must occur in real time with minimal delay to support ultra-fast 6G applications.	Critical for latency-sensitive applications like autonomous vehicles and industrial automation.
Lightweight computation	Security protocols should be optimized for resource-constrained IoT devices with limited processing power and battery life.	Ensures that even low-power devices can participate in secure communications without excessive overhead.
Strong authentication	Devices and users must be accurately verified using multi-factor, biometric, or blockchain-based authentication.	Prevents unauthorized access and impersonation attacks.
Quantum-resistant cryptography	Cryptographic algorithms must be resilient against attacks from quantum computers.	Future-proofs authentication and key exchange against quantum-based threats.
Decentralization	Security mechanisms should minimize reliance on centralized authentication servers.	Enhances reliability, reduces single points of failure, and prevents large-scale breaches.
Privacy preservation	User and device identity information should be protected from exposure during authentication.	Prevents unauthorized tracking, identity theft, and data breaches.
Mutual authentication	Both communicating parties must verify each other's identity before establishing a secure connection.	Ensures trust and prevents MITM attacks.
Forward and backward secrecy	Ensures past and future communications remain secure even if a key is compromised.	Protects sensitive data from being decrypted retroactively.
AI-driven threat detection	Machine learning models should be integrated to detect anomalies in authentication attempts.	Helps identify suspicious activities and prevent unauthorized access in real time.
Interoperability	Security protocols should be compatible across various IoT devices, manufacturers, and network infrastructures.	Ensures seamless integration and secure communication across different platforms.

1.3.1 Traditional Authentication Protocols

Conventional methods have been applied in the field of authentication to ensure the validity of users and devices in the networks. These protocols are based on password authentication, PKI, challenge response mechanism, and application of symmetric as well as asymmetric cryptographic techniques [6]. This has made the Password Authentication Protocol (PAP) common since it is easy to implement, but it is vulnerable to eavesdropping and replay attacks. The Challenge Handshake Authentication protocol (CHAP) [7] also helps improve on the security since it uses the three-way handshake to authenticate the users without using the passwords in plaintext. Another type is certificate based, where entities have to trust each other through the use of a certificate issued by a given CA using PKI. The discussed authentication mechanisms have drawbacks in the contemporary network, especially for IoT and 6G, which are characterized by scalability, low latency, and resource constrain. The Internet has proposed legacy protocols, which are not capable of offering sufficient protection from the new age threats such as phishing, MITM, and quantum threats. For this reason, the subsequent-generation authentication technologies, for example, blockchain-based authentication, biometric authentication, and AI inventive, are being considered for improving security in 6G and IoT networks.

1.3.2 Lightweight Authentication for IoT Devices

Authentication control for IoT devices should be lightweight because of the restricted resources, such as their computing ability, memory, and energy [8]. In an IoT environment, traditional credentials-based procedures that involve computations with combined cryptographic algorithms are not effective in most cases. Figure 1.1 illustrates the secure authentication process in 6G networks, highlighting mutual verification and secure key establishment to protect against potential threats.

This is a challenge that has been solved by designing lightweight authentication protocols with the aim of offering security without having to compromise on the resources needed by the network. One of them is Hash-Based Authentication, which employs cryptographic hash functions instead of consuming key activities carried out by Public-Key. Similarly, there is some other protocol known as the Lightweight Authentication and Key Exchange (LAKE) [9] protocol, which also limits our processing overhead by using symmetric keys for the identification of the nodes. Elliptic curve cryptography (ECC) is also considered because it can enhance the security level, using lower keys, therefore perfect for IoT devices with restrictive requirements on energy and storage. Additionally, challenge response authentication mechanisms, which involve a one-time challenge to verify identity without storing sensitive credentials, further enhance security while reducing computational complexity. To improve scalability and decentralization, blockchain-based authentication and federated identity management systems are also gaining popularity in IoT networks. These lightweight authentication solutions ensure secure communication while maintaining efficiency, making them critical for the success of IoT deployments in 6G networks.

1.3.3 AI and Blockchain-Based Authentication in 6G

AI and blockchain-based authentication in 6G networks provide a secure, intelligent, and decentralized approach to identity verification and data protection. AI [10] leverages machine learning and deep learning algorithms to analyse user behaviour, detect anomalies, and predict potential security threats in real-time, enabling proactive defence mechanisms. Blockchain, with its decentralized and immutable ledger, ensures transparent and tamper-proof authentication, eliminating risks associated with centralized security models and single points of failure. Smart contracts

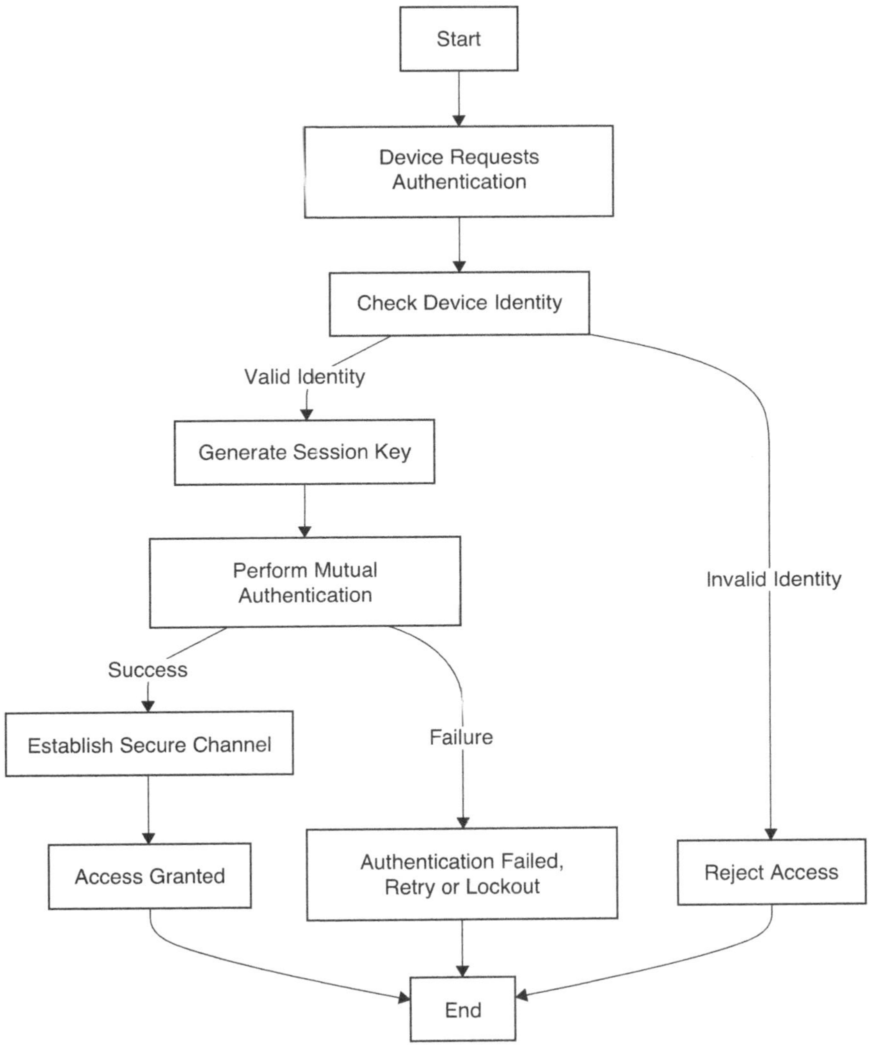

FIGURE 1.1 Secure authentication process in 6G networks, ensuring mutual verification and secure communication.

can further automate authentication protocols, reducing the need for intermediaries while enhancing trust. By combining AI's adaptability with blockchain's security, 6G networks can achieve robust, scalable, and efficient authentication, ensuring secure communication across IoT devices, autonomous systems, smart cities, and industrial automation. This integration not only mitigates cyber threats but also supports seamless and trustworthy connectivity in next-generation wireless networks. Figure 1.2 illustrates a blockchain-based authentication system, ensuring secure access control and identity verification in IoT environments.

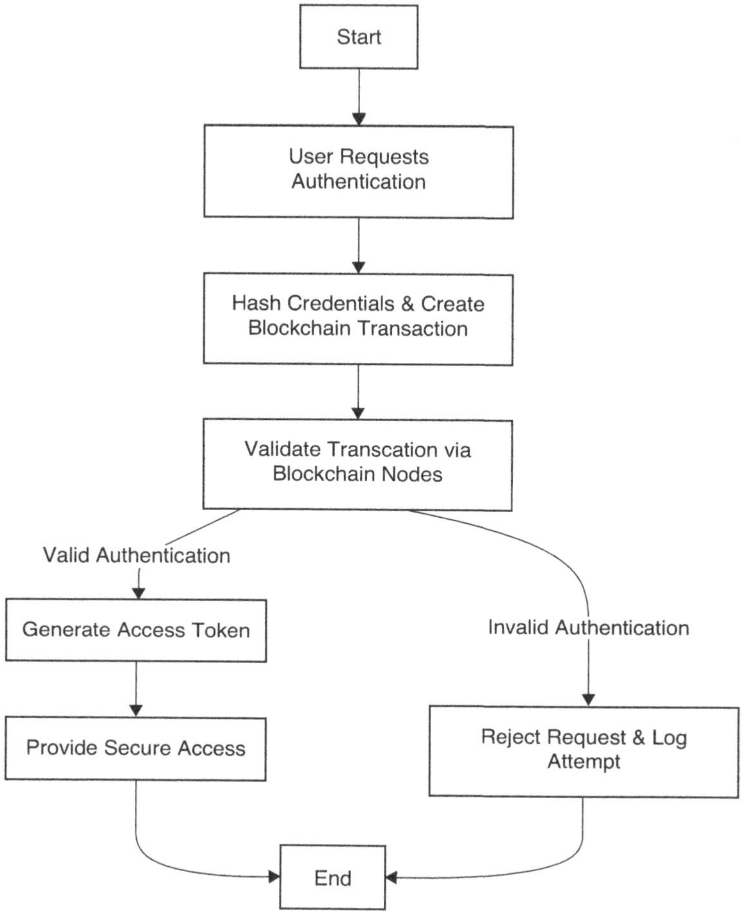

FIGURE 1.2 Decentralized blockchain-based authentication system ensuring secure access in IoT networks.

1.4 KEY EXCHANGE PROTOCOLS IN 6G AND IOT

Key Exchange Protocols in 6G and IoT are cryptographic methods used to securely share encryption keys between devices in next-generation wireless networks and the Internet of Things. These protocols ensure confidentiality, authentication, and resistance against quantum and cyber threats in ultra-connected environments.

1.4.1 Traditional Key Exchange Methods (Diffie-Hellman, RSA, ECC)

Traditional techniques like DH, RSA, and ECC are very important in connection establishment because they facilitate key exchange over an insecure channel [11]. DH enables two parties to establish a secret key even if the key is never transferred between the two, hence protecting it from any form of wiretapping. RSA is based

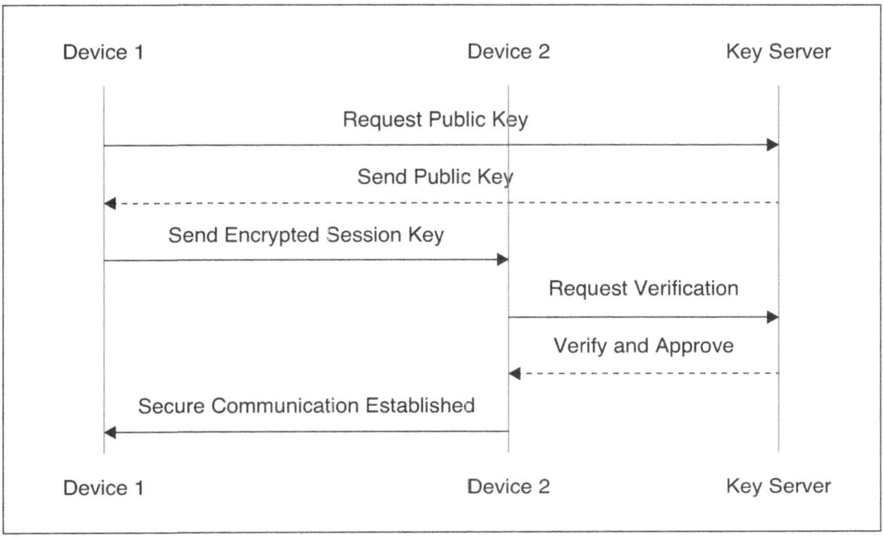

FIGURE 1.3 Secure key exchange process in 6G networks ensuring confidentiality and mutual verification.

on the notion of factoring large prime numbers, which makes it one of the most popular methods for key exchange in SSL/TLS. Figure 1.3 illustrates the secure key exchange process in 6G networks, ensuring both confidentiality and mutual authentication between communicating entities.

ECC is similar to RSA in terms of security level but enjoys more efficiency with a smaller key size, and faster computation to suit resource-limited devices like those of IoT. Although these methods are useful in cryptographic security, new threats and risks, such as those from quantum computers, call for sophisticated post-quantum cryptography.

1.4.2 Post-Quantum Cryptographic Approaches

Post-quantum cryptographic approaches aim to develop encryption methods resistant to quantum computing threats [12]. These techniques ensure long-term security by replacing traditional algorithms with quantum-safe alternatives. Table 1.3 summarizes various post-quantum cryptographic algorithms, their key features, and the underlying security assumptions. Figure 1.4 illustrates the post-quantum key exchange process in 6G, ensuring secure and resilient authentication against emerging quantum threats.

1.4.3 Secure Key Management in Resource-Constrained IoT Devices

Key management is a considerable problem in resource-constrained IoT devices because of their low computational capability, memory, and energy storage. Mainstream key management techniques, as discussed under PKI, are not efficient

TABLE 1.3
Post-Quantum Cryptographic Approaches and Their Security Foundations

Algorithm Type	Examples	Key Features	Security Basis
Lattice-based [13]	NTRU, Kyber, FrodoKEM	Highly secure, efficient, resistant to quantum attacks	Hardness of lattice problems (e.g., Learning With Errors)
Code-based [14]	McEliece	Longstanding security, large key sizes	Hardness of decoding random linear codes
Multivariate-based [15]	Rainbow	Fast signature verification, suitable for lightweight devices	Hardness of solving multivariate polynomial equations
Hash-based [16]	SPHINCS+, XMSS	Stateless, strong security, efficient key generation	Security of cryptographic hash functions
Isogeny-based [17]	SIDH, SIKE	Small key sizes, efficient operations	Hardness of computing isogenies between elliptic curves

for IoT, hence the need for lightweight methods. Lightweight key exchange protocols, ECC, symmetric key techniques like AES [18], and the lightweight block cipher also provide the best way to secure the Internet of Things while consuming the least resources. Also, key storage and authentication are supported by specific hardware elements like TPM and PUFs. To enhance security, key renewal, booting methods, and distributed control of information using blockchain are under consideration. It is a way of making sure that even the constrained IoT devices will be able to manage a cryptographic secure key, and this will prevent attacks and the invasion of cyber threats as well.

1.5 EMERGING TECHNOLOGIES FOR SECURE COMMUNICATION

This has led to the necessity of the use of enhanced secure communication systems due to the increasing development of cyber threats. Conventional cryptography techniques, moreover, currently encounter new problems due to state-of-the-art computers and quantum computer technologies. Now, the major trends that are defining the new environment for protected communication are post-quantum cryptography, blockchain security, homomorphic computations, and AI for threat detection. These innovations will increase data protection, accuracy, and security in domains such as IoT, cloud technology, 6G, and others. Technologically, these emerging systems apply the best use of cryptographic algorithms with the latest network architectures to offer reliable mechanisms to protect such important information in today's interconnected world.

1.5.1 Zero-Trust Architecture (ZTA) in 6G and IoT

ZTA [19] in 6G and IoT is a security model that does not presuppose any trustworthiness of a network's core and requires confirmation for each user intending to access

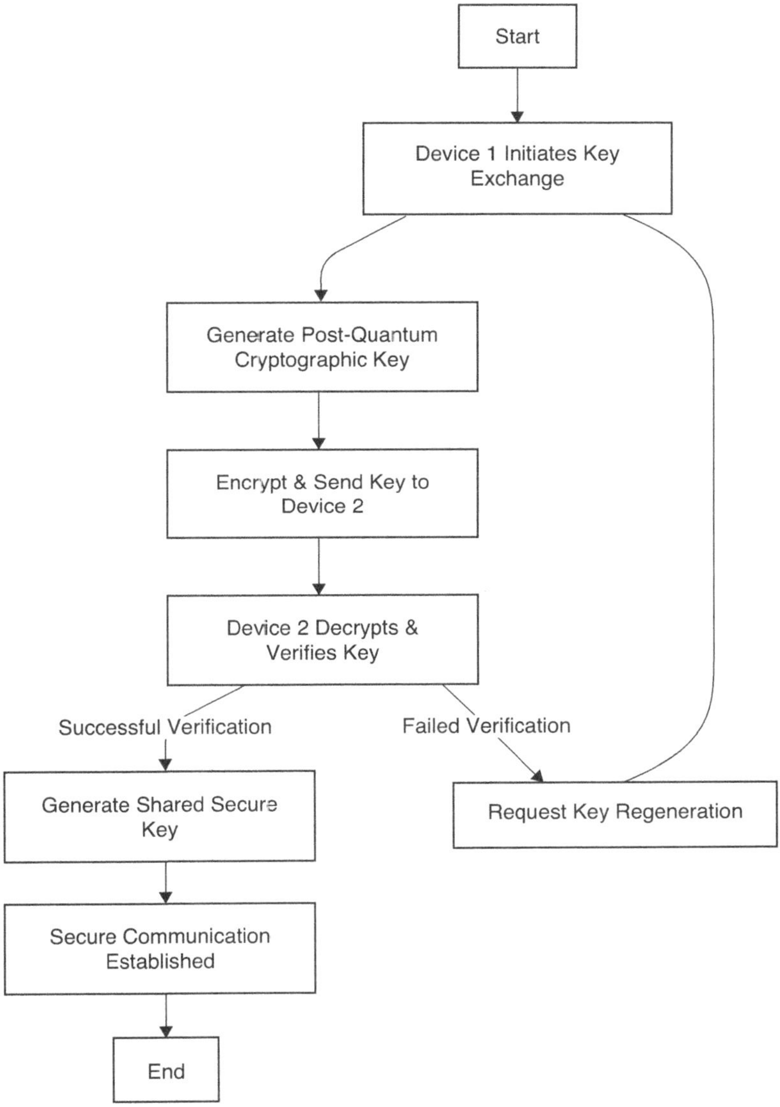

FIGURE 1.4 Post-quantum key exchange process in 6G, ensuring resistance against quantum-based attacks.

a resource. As opposed to most security models that are more or less focused on the network perimeter, ZTA is developed under the premise that threats are internal and external. It becomes vital in 6G networks and IoT systems where trillions of devices, various endpoints, and distributed systems lack sufficient security solutions. In 6G networks, ZTA applies other features of emerging intelligent communication technology, including AI, ML [20], and blockchain, in a bid to boost security features.

Machine learning-based security architectures provide the ability to observe the system's behaviour and shield against attacks, or put in place the capability that would allow only the right entity to operate on the system. Furthermore, it shall also be noted that, through the implementation of blockchain-based identity management, it is possible to avoid such issues as centralized storage of credentials. In IoT environments, devices are generally characterized by limited computational and storage resources, and as such, the Zero-Trust principles should be supported with minimal authentication requirements, including physically unclonable functions, token authentication, and secure enclaves. Edge computing also addresses the issue of security in IoT and the authentication of requests by performing its duty nearer to the site of operation without causing delays in the process. In addition, specifications such as SDN and slicing in 6G networks allow for the dynamic policy update in a way that the devices are only granted the permissions required to perform their functions. Thus, ZTA lets 6G and IoT [21] infrastructures decrease attack vectors to a minimum, eliminate threats' lateral movement, and implement a novel & adaptable security mechanism. As these solutions become more intricate, ZTA provides solutions that are adaptive, reliable, and anticipatory to provide proper protection to the information exchanged and communicated through various and sometimes changing platforms.

1.5.2 5G-AKA and Its Evolution for 6G Networks

The 5G Authentication and Key Agreement, also known as 5G-AKA, is a security process put in place whereby the user equipment and the network can authenticate themselves towards each other when setting up the secure communication links. This is based on 4G AKA but brings more advanced features to ward off the mentioned vulnerabilities, such as identity exposure and replay attacks in 5G-AKA; it uses Subscription Concealed Identifier (SUCI) to hide the IMSI number, preventing those who could track the subscribers. Besides, it has enhanced capability in terms of encryption, key utilization, and protection against MITM attacks that make the protocol suitable for protecting modern mobile networks. When transitioning to the 6G network, there are new security threats and mechanisms, such as quantum cryptography, artificial intelligence-assisted authentication, and decentralized identities for the 5G-AKA to develop. Consequently, 6G may incorporate blockchain-based authentication and smart contracts to increase the agent's transparency and reliability in identification. Furthermore, additional measures of private key management will also apply private set intersection (PSI), blinding techniques, and zero-knowledge proofs (ZKPs) to reinforce the privacy preserving of the identities of the users. The continuation of 5G-AKA for 6G networks will also result in the implementation of edge computing and federated learning (FL) to attain a real-time authentication with low latency for providing a secure connection in highly dense networks. By refining and extending 5G-AKA capabilities, 6G networks will achieve a highly secure and efficient authentication framework that supports emerging technologies such as holographic communications, intelligent IoT devices, and autonomous systems. The continuous evolution of authentication protocols will play a vital role in securing next-generation mobile networks against increasingly sophisticated cyber threats.

1.5.3 FL for Authentication and Privacy

FL can be described as a progressive machine learning method that allows training models in distributed devices without the exchange of data between the devices and a central server [22]. It can be useful in the functioning of the new approaches that are being developed, such as 6G or IoT networks, where the scale of connected devices may involve a vast number of devices; thus, the efficiency and secure methods of authentication are critical while keeping the data private. The method of training the model at the user site in FL also increases security by maintaining users' biometric data, behavioural profiles, or credentials on the device and preventing them from being transmitted over the network. FL does not require the centralization of data as it only sends the updates of the model, which minimizes the threats of data theft, identity theft, and other unlawful access. It would be most beneficial in mobile networks within M2M communications and IoT, smart applications, and other applications that require privacy for the authentication process. To enhance the privacy of the users of the FL, FL incorporates aspects such as differential privacy and homomorphic encryption to obscure the actual updates of the model to prevent adversaries from deducing the original data. It is especially important for personal data protection in applications like mobile payment, health care, or more generally in a smart city framework. Figure 1.5 illustrates an AI-driven threat detection mechanism in 6G networks, ensuring real-time cybersecurity threat mitigation through adaptive learning and automated response systems.

In turn, FL eliminates dependencies on the cloud-based authentication servers as well, thereby eradicating threats such as data interception and unauthorized access. In the context of 6G and IoT, FL is also a globally efficient approach for farmers and provides solutions for real-time authentications of users while maintaining the user's privacy intact. It is, therefore, evident that AI-driven adaptive authentication is pivotal to FL now and to the next-generation authentication framework as it provides a secure approach to dealing with highly dynamic and decentralised networks.

1.6 COMPARATIVE ANALYSIS OF AUTHENTICATION AND KEY EXCHANGE PROTOCOLS

Comparing authentication and key exchange protocols provides an understanding of how secure and efficient various methods are used in ascertaining the identity and securing an environment for communication. This allows for better identification of the efficacy of these protocols, their ability to withstand attacks, and their compatibility with different networks such as 5G/6G and IoT.

1.6.1 Performance and Security Analysis

Security analysis of authentication and key exchange protocols is crucial as it helps determine the efficiency and scalability, as well as identify vulnerabilities that may be prone to cyber threats. Hence, that definition also evaluates their computational complexity, the time taken to authenticate, energy consumption, and their capability to address attacks such as replay, MITM, and quantum attacks when used in certain

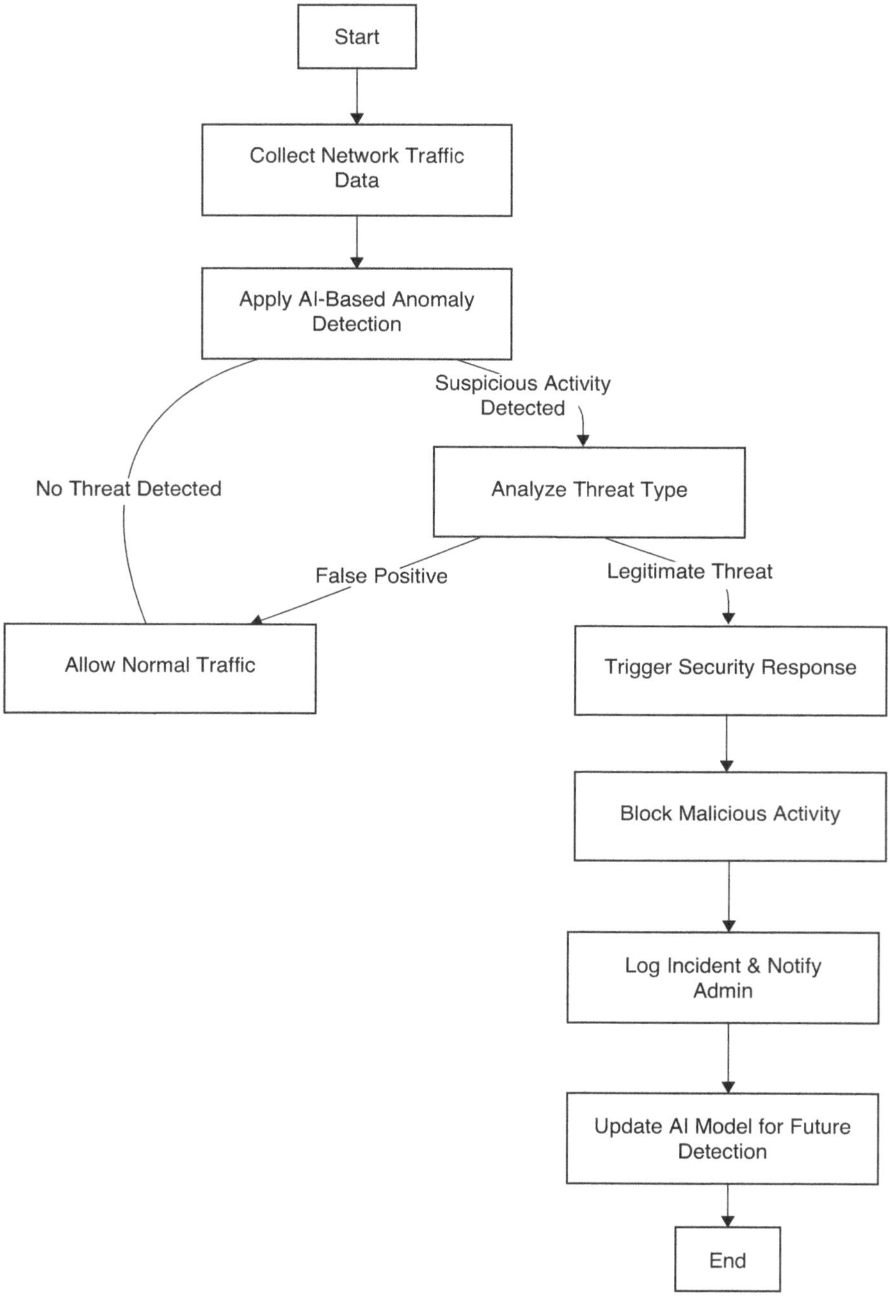

FIGURE 1.5 AI-driven threat detection mechanism in 6G networks for real-time cybersecurity threat mitigation.

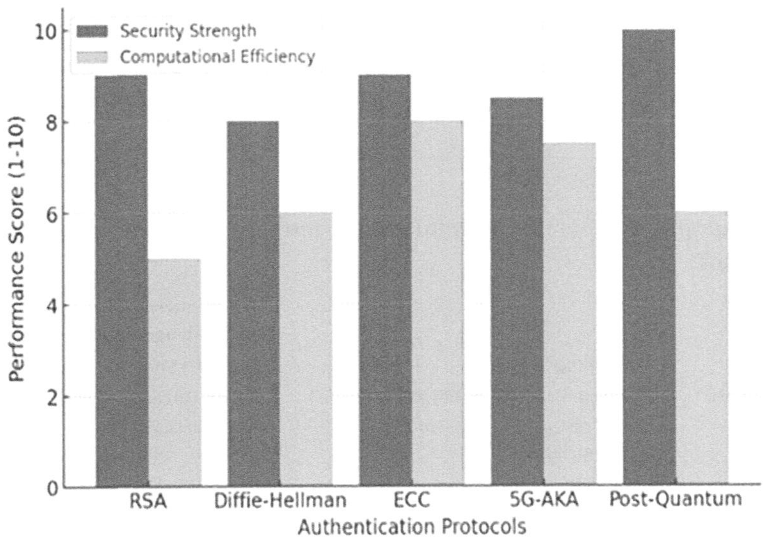

FIGURE 1.6 Performance analyses of authentication protocols in 6G networks.

network settings. While RSA and Diffie-Hellman are secure to implement, they are time-consuming, and on the other hand, the lightweight cryptographic methods, along with the AI-based authentication in the case of 6G networks, are more efficient and flexible. It is likely getting necessary for the security and performance of integration between various mandatory and elective emerging technologies for secure and continuous communication. Figure 1.6 presents a performance analysis of authentication protocols in 6G networks, comparing their security strength and computational efficiency to guide the selection of optimal authentication mechanisms.

Here is a comparative analysis graph showing the security strength and computational efficiency of different authentication protocols. Higher security scores indicate stronger protection against attacks, while higher computational efficiency scores suggest better performance in resource-constrained environments.

1.6.2 Trade-Offs between Security, Latency, and Scalability

In the fields of authentication and key exchange, it is a common problem to identify the correct balance between such characteristics as security, latency, and scalability. High-security measures like post-quantum cryptography and multi-factor authentication are likely to have higher time and computational complexity, which results in improved response time for real-time applications. On the other hand, protocols that are suitable for low latency may involve little security, thus making a system prone to hacking. Versatility intensifies the question, as protocols have to serve the increasing number of users and devices in networks, for example, in 6G and IoT. This is a complicated process because things like adaptive security models, efficient cryptographic algorithms, as well as appropriate resource management, and other tweaks should be employed in order to come up with wider, faster, and scalable ways of authentication.

1.6.3 Case Studies in 6G and IoT Deployments

Table 1.4 includes additional parameters such as Use Case, Security Challenges Addressed, and Deployment Status for 6G and IoT deployments in different countries.

TABLE 1.4
Global Adoption of Secure Authentication and Key Exchange Technologies in 6G and IoT

Country	Technology Used	Use Case	Security Challenges Addressed	Deployment Status
United States [23]	AI-driven Zero-Trust Authentication	6G Networks, IoT Security	Insider Threats, Unauthorized Access	Pilot Testing
China [24]	Blockchain-based Key Exchange	Smart Cities, IoT Devices	MITM Attacks, Data Tampering	Large-Scale Deployment
Germany [25]	Post-Quantum Cryptography for 6G	Secure Communications, Data Privacy	Quantum Computing Threats	Research Phase
Japan [26]	6G Edge Computing Security	Smart Factories, Autonomous Vehicles	Edge Node Vulnerabilities	Field Trials
South Korea [27]	FL for IoT Authentication	Smart Home Devices, Wearables	Data Privacy, Centralized Attack Risks	Commercial Rollout
United Kingdom [28]	Quantum Key Distribution (QKD)	Government Networks, Defence	Eavesdropping, Interception	Experimental Stage
India [29]	Lightweight Cryptographic Authentication	Low-Power IoT Devices	Resource Constraints, Scalability Issues	Proof of Concept
Canada [30]	Biometric-based 6G Authentication	Mobile Payments, Smart Devices	Identity Theft, Spoofing	Pilot Testing
Australia [31]	Secure IoT Device Identity Management	Healthcare IoT, Smart Agriculture	Unauthorized Device Access	Ongoing Research
France [32]	AI-enhanced Multi-Factor Authentication	Banking, Cloud Security	Phishing, Credential Theft	Implementation Stage
Brazil [33]	Privacy-Preserving IoT Communication	Public Wi-Fi Security, IoT Networks	Data Privacy, Anonymity	Testbed Deployment
Russia [34]	Homomorphic Encryption for IoT Networks	Critical Infrastructure Security	Data Confidentiality	Research & Development

(Continued)

TABLE 1.4 (*Continued*)
Global Adoption of Secure Authentication and Key Exchange Technologies in 6G and IoT

Country	Technology Used	Use Case	Security Challenges Addressed	Deployment Status
Singapore [35]	Smart City IoT Security with 6G Integration	Traffic Monitoring, Smart Grids	Cyberattacks on Public Infrastructure	Large-Scale Deployment
Sweden [36]	Blockchain-enabled Device Authentication	Industrial IoT, Supply Chains	Identity Spoofing, Data Integrity	Field Trials
Netherlands [37]	6G-enabled Secure Cloud Authentication	Cloud Computing, Enterprise Security	Cloud Data Breaches	Commercial Pilot
UAE [38]	AI-powered Authentication in Smart Homes	Smart Cities, Home Security	Unauthorized Access, IoT Botnets	Proof of Concept
Italy [39]	Decentralized Identity Management in IoT	Secure Digital Identity, E-Governance	Identity Fraud, Centralized Breaches	Early Development
South Africa [40]	Adaptive 6G Security for Rural Networks	Rural Connectivity, Remote Services	Infrastructure Limitations, Scalability	Research & Testing
Mexico [41]	Secure Edge AI Authentication in IoT	Wearable Devices, Smart Healthcare	AI-based Cyber Threats	Pilot Implementation
Spain [42]	Privacy-Focused IoT Key Exchange	Consumer IoT, Smart Devices	Data Privacy, Unauthorized Sharing	Deployment Phase

1.7 CONCLUSION

6G and IoT are the newer generations of wireless communications and smart connectivity that are rapidly growing at the moment. In more detail, 6G is to become an evolution or a direct successor of the 5G that is designed to deliver ultra-high speeds, very low latency, and highly secure connectivity for different IoT applications in a variety of areas such as healthcare, smart cities, autonomous vehicles, and industrial systems. The use of 6G and IoT will complement each other to offer aspects such as M2M communication, real-time data processing, and intelligence for automating processes in a connected environment. This paper looks at the various enablers that are used for this integration, namely artificial intelligence, edge computing, THz communication, and blockchain, that improve efficiency, security, and system performance. Besides, it highlights some major concerns that are crucial to consider as new opportunities for the development of 6G Internet of Things systems with scalability, energy, security, and spectrum. The paper also described the possibilities of employing this advanced technology in options such as remote healthcare monitoring, smart transportation systems, industrial automation, and AR and VR. Last but not least, the research questions for future research are provided with regard to the standardization, energy consumption, security advancements, and human interfacing in the context of networked robotic systems. Altogether, the results indicate that the integration of 6G and IoT can transform the communications networks, have great impacts on numerous fields, and foster the emergence of a new, intelligent, self-sufficient, and highly efficient digital world.

REFERENCES

1. Bonkra, A., & Dhiman, P. (2021, December). IoT security challenges in cloud environment. In *2021 2nd international Conference on Computational Methods in Science & Technology (ICCMST)* (pp. 30–34). IEEE.
2. Bonkra, A., Dhiman, P., & Sonker, J. (2025). Machine learning techniques for IoT applications in healthcare. In A. K. Rana, S. K. Rana, V. Sharma, & R. Dewan (Eds.), *Intelligent Data-Driven Techniques for Security of Digital Assets* (p. 191). Boca Raton, FL: CRC Press (Taylor & Francis).
3. Dogra, A., Jha, R. K., & Jain, S. (2020). A survey on beyond 5G network with the advent of 6G: Architecture and emerging technologies. *IEEE Access*, *9*, 67512–67547.
4. Pooyandeh, M., Han, K. J., & Sohn, I. (2022). Cybersecurity in the AI-based metaverse: A survey. *Applied Sciences*, *12*(24), 12993.
5. Gupta, B. B., & Dahiya, A. (2021). Distributed Denial of Service (DDoS) Attacks: *Classification, attacks, challenges and countermeasures*. Boca Raton, FL: CRC Press.
6. Goswami, S., Misra, S., & Mukesh, M. (2014, December). A replay attack resilient system for PKI based authentication in challenge-response mode for online application. In *2014 3rd International Conference on Eco-Friendly Computing and Communication Systems* (pp. 144–148). IEEE.
7. Elumogo, J. T. C. (2023). *Exploring a Cyber Security Strategy for the TCP/IP 3-Way Handshake* (Doctoral dissertation, Northcentral University).
8. Tiwana, P. S., & Singh, J. (2024, May). Enhancing multimedia forwarding in software-defined networks: An optimal flow mechanism approach. In *2024 2nd International Conference on Advancement in Computation & Computer Technologies (InCACCT)* (pp. 888–893). IEEE.
9. Tiwana, P. S., & Singh, J. (2023, December). Software-defined networking's adaptive load optimisation technique for multimedia applications with multi-controller utilisation. In *2023 9th International Conference on Signal Processing and Communication (ICSC)* (pp. 167–172). IEEE.
10. Bonkra, A., Bhatt, P. K., Rosak-Szyrocka, J., Muduli, K., Pilař, L., Kaur, A., … & Rana, A. K. (2023). Apple leave disease detection using collaborative ml/dl and artificial intelligence methods: Scientometric analysis. *International Journal of Environmental Research and Public Health*, *20*(4), 3222.
11. Ramakrishna, D., & Shaik, M. A. (2024). A comprehensive analysis of cryptographic algorithms: Evaluating security, efficiency, and future challenges. *IEEE Access*, PP(99), 1–11.
12. Mohammed, A. (2018). Quantum-resistant cryptography: Developing encryption against quantum attacks. *Journal of Innovative Technologies*, 1(1), 1–14.

13. Ravi, P., Howe, J., Chattopadhyay, A., & Bhasin, S. (2021). Lattice-based key-sharing schemes: A survey. *ACM Computing Surveys (CSUR), 54*(1), 1–39.
14. Albrecht, M. R., Bernstein, D. J., Chou, T., Cid, C., Gilcher, J., Lange, T., Maram, V., von Maurich, I., Misoczki, R., Niederhagen, R., Persichetti, E., Peters, C., Sendrier, N., Szefer, J., Tjhai, C. J., Tomlinson, M., & Wang, W. (2022). Classic McEliece: Conservative code-based cryptography. Submission to the NIST Post-Quantum Cryptography Standardization Project. October 23, 2022.
15. Mohamed, M. S. E., & Petzoldt, A. (2017). RingRainbow–an efficient multivariate ring signature scheme. In *Progress in Cryptology-AFRICACRYPT 2017: 9th International Conference on Cryptology in Africa, Dakar, Senegal, May 24–26, 2017, Proceedings 9* (pp. 3–20). Springer International Publishing.
16. Sim, M., Eum, S., Song, G., Kwon, H., Jang, K., Kim, H., … & Seo, H. (2022). K-XMSS and K-SPHINCS+: Hash-based signatures with Korean cryptography algorithms. *Cryptology ePrint Archive*, Paper 2022/152.
17. Fouotsa, T. B., & Petit, C. (2021). SHealS and HealS: Isogeny-based PKEs from a key validation method for SIDH. In *Advances in Cryptology–ASIACRYPT 2021: 27th International Conference on the Theory and Application of Cryptology and Information Security,* Singapore, *December 6–10, 2021, Proceedings, Part IV 27* (pp. 279–307). Springer International Publishing.
18. Ashraf, Z., Sohail, A., & Yousaf, M. (2023). Robust and lightweight symmetric key exchange algorithm for next-generation IoE. *Internet of Things*, *22*, 100703.
19. Nahar, N., Andersson, K., Schelén, O., & Saguna, S. (2024). A survey on zero trust architecture: Applications and challenges of 6G networks. *IEEE Access*, 12, 94753–94764.
20. Bonkra, A., Dhiman, P., Goyal, S., Islam, S. M., Rana, A. K., & Sharma, N. (2023, June). An innovative new open computer vision framework via artificial intelligence with python. In *International Conference on Data Science and Network Engineering* (pp. 95–109). Springer Nature Singapore.
21. Tiwana, P. S., Kaur, L., Kaur, N., Wadhwa, H., Kumari, N., & Sharma, R. (2024). Mitigating DDoS attacks: A distributed blockchain-SDN secure IoT system enhanced by artificial neural networks. In S. Kaur, S. Kamboj, M. Kumar, A. Dagur, & D. K. Shukla (Eds.), *Computational Methods in Science and Technology* (pp. 467–472). London: CRC Press.
22. Khan, L. U., Saad, W., Han, Z., Hossain, E., & Hong, C. S. (2021). Federated learning for internet of things: Recent advances, taxonomy, and open challenges. *IEEE Communications Surveys & Tutorials*, *23*(3), 1759–1799.
23. Ofili, B. T., Erhabor, E. O., & Obasuyi, O. T. (2025). Enhancing federal cloud security with AI: Zero trust, threat intelligence, and CISA Compliance. World Journal of *Advanced Research and Review*, 25(2), 2377–2400.
24. Yao, H., & Wang, C. (2018, June). A novel blockchain-based authenticated key exchange protocol and its applications. In *2018 IEEE Third International Conference on Data Science in Cyberspace (DSC)* (pp. 609–614). IEEE.
25. Ziegler, V., Schneider, P., Viswanathan, H., Montag, M., Kanugovi, S., & Rezaki, A. (2021). Security and trust in the 6G era. *IEEE Access*, *9*, 142314–142327.
26. Mao, B., Liu, J., Wu, Y., & Kato, N. (2023). Security and privacy on 6G network edge: A survey. *IEEE Communications Surveys & Tutorials*, *25*(2), 1095–1127.
27. Deebak, B. D., & Hwang, S. O. (2023). Federated learning-based lightweight two-factor authentication framework with privacy preservation for mobile sink in the social IOMT. *Electronics*, *12*(5), 1250.
28. Stanley, M., Gui, Y., Unnikrishnan, D., Hall, S. R. G., & Fatadin, I. (2022, December). Recent progress in quantum key distribution network deployments and standards. In *Journal of Physics: Conference Series* (Vol. 2416, No. 1, p. 012001). IOP Publishing.

29. Bhardwaj, I., Kumar, A., & Bansal, M. (2017, September). A review on lightweight cryptography algorithms for data security and authentication in IoTs. In *2017 4th International Conference on Signal Processing, Computing and Control (ISPCC)* (pp. 504–509). IEEE.
30. Saha, S., Das, A. K., Wazid, M., Park, Y., Garg, S., & Alrashoud, M. (2024). Smart contract-based access control scheme for blockchain assisted 6G-enabled IoT-based big data driven healthcare cyber physical systems. *IEEE Transactions on Consumer Electronics*, 70(3), 256–268.
31. Zhu, X., & Badr, Y. (2018). Identity management systems for the internet of things: A survey towards blockchain solutions. *Sensors*, *18*(12), 4215.
32. Falade, P. V. (2023). Decoding the threat landscape: Chatgpt, fraudgpt, and wormgpt in social engineering attacks. *arXiv preprint arXiv:2310.05595.*
33. Junior, N. F., Silva, A. A., Guelfi, A. E., & Kofuji, S. T. (2021). Privacy-preserving cloud-connected IoT data using context-aware and end-to-end secure messages. *Procedia Computer Science*, *191*, 25–32.
34. Temirbekova, Z. E., Turken, G., & Barata, M. M. (2023). Using fully homomorphic encryption in IoT. In *DTESI (Workshops, Short Papers).*
35. Nag, A. (2025). The role of 6G technology in smart city development. In S. K. Singh, B. Virdee, R. Jadeja, S. Tanwar, & A. Khanna (Eds.), *Building Tomorrow's Smart Cities With 6G Infrastructure Technology* (pp. 1–38). Hershey, PA: IGI Global.
36. Kebande, V. R., Awaysheh, F. M., Ikuesan, R. A., Alawadi, S. A., & Alshehri, M. D. (2021). A blockchain-based multi-factor authentication model for a cloud-enabled internet of vehicles. *Sensors, 21*(18), 6018.
37. Hoffmann, M., Kunzmann, G., Dudda, T., Irmer, R., Jukan, A., Macher, G., ... & Voicu, A. M. (2023). A secure and resilient 6G architecture vision of the German flagship project 6G-ANNA. *IEEE Access*, *11*, 102643–102660.
38. Awad, A. I., Babu, A., Barka, E., & Shuaib, K. (2024). AI-powered biometrics for Internet of Things security: A review and future vision. *Journal of Information Security and Applications*, *82*, 103748.
39. Mazzocca, C., Acar, A., Uluagac, S., Montanari, R., Bellavista, P., & Conti, M. (2024). A survey on decentralized identifiers and verifiable credentials. *arXiv preprint arXiv:2402.02455.*
40. Yaacoub, E., & Alouini, M. S. (2020). A key 6G challenge and opportunity—Connecting the base of the pyramid: A survey on rural connectivity. *Proceedings of the IEEE*, *108*(4), 533–582.
41. Bourechak, A., Zedadra, O., Kouahla, M. N., Guerrieri, A., Seridi, H., & Fortino, G. (2023). At the confluence of artificial intelligence and edge computing in IoT-based applications: A review and new perspectives. *Sensors*, *23*(3), 1639.
42. Bernabe, J. B., Canovas, J. L., Hernandez-Ramos, J. L., Moreno, R. T., & Skarmeta, A. (2019). Privacy-preserving solutions for blockchain: Review and challenges. *IEEE Access*, *7*, 164908–164940.

2 Smart Healthcare
Leveraging IoT for Enhanced Medical Solutions

Anupam Bonkra, Damandeep Singh, Pummy Dhiman, and Gaganpreet Kaur

2.1 INTRODUCTION

The IoT stands for the connection of physical devices to the internet and is still emerging, but promises to revolutionize our environments [1]. According to some estimations, there will be more than 75 billion devices connected globally in 2025 that will significantly impact certain sectors, such as healthcare [2,3]. The IoT refers to interconnected web-connected devices, sensors, and apparatus that work online together, intercommunicating [4]. IoT technologies are being applied in the health sector [5] and are already permitting patients to have distinct treatment arrangements for administrations that were previously out of reach, patient monitoring that can take place from a distance, and much enhanced healthcare services delivery (Figure 2.1). IoT is revolutionizing not only the health and medical sector, but also industry as a whole, along with the transportation and agriculture sectors. Science has the capability to revolutionize civilization in the shortest time possible [6]. Thus, to enhance the favourable impact IoT might bring on society, it is imperative to keep abreast of the most recent innovations. IoT has been estimated to reduce the healthcare industry's expenses by relegating routine work, streamlining procedures, and reducing the rate of costly procedures. It makes it possible to gather and monitor data in real time, which enhances patient outcomes and reaction times. IoT also reduces

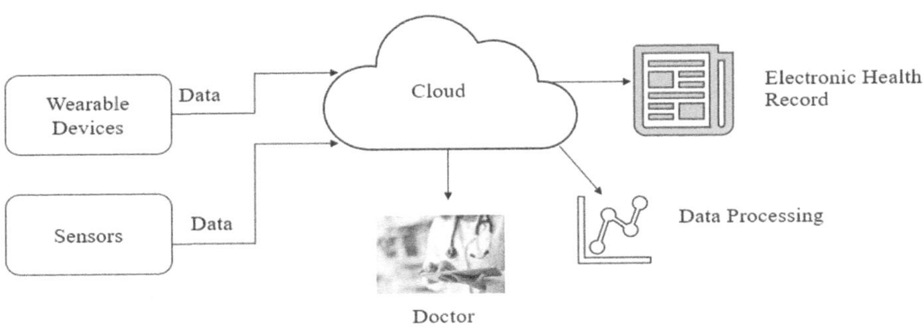

FIGURE 2.1 Backbone of smart healthcare.

DOI: 10.1201/9781003586456-2

the need for big administrative personnel, automates administrative activities, saves time, and reduces mistakes in regular tasks like prescription administration and vital sign monitoring, which simplifies healthcare operations. The identification and monitoring abilities of IoT mean that costly therapies, hospitalizations, and expensive medical issues are prevented earlier. IoT also helps to boost patient involvement in their healthcare [7]. Patients themselves can perform some vital signs using wearable gadgets and applications, and patients can use a patient portal for their records [8]. Teleconsultancy and advisory, remote working, and instant exchange and transfers of data are now enabled through teleconsultancy/telecare and telemonitoring.

Ways such as encrypted messaging solutions enhance contact with healthcare practitioners, improving the use of mobile applications for self-management. These technologies equally provide supremacy to people as a whole and engage them in a more active role in the provision of their care through optimally redefining the interaction between the healthcare providers and the patients. IoT in healthcare, however, comes with a set of challenges. Such concerns are the security and privacy of data, compatibility issues that affect integration, and the need to have standard policies and practices. Since the patients' sensitive data has to be safeguarded at the time of collection, as well as during transmission, patient privacy is paramount [9]. A huge number of vendors and technologies are used in IoT devices, and these might affect compatibility and may lead to rigid data segregation and weak communication. For existing healthcare systems, IoT devices need to operate through certain stipulated regulations and guidelines. The future of healthcare has been reshaped to a great extent through the use of IoT [10,11]. To unlock the full potential of the IoT in improving the delivery of health services and the healthcare systems, the researchers, policymakers, and caregivers need to remain informed and, at the same time, address emergent challenges. The purpose of this study is to further discuss IoT in healthcare and explore how the improvement of this technology should take place.

This study's primary contributions are listed below:

- It explores the smooth integration of cutting-edge technologies to provide accurate and customized medical care based on each patient's particular genetic, environmental, and lifestyle factors.
- It sheds light on how ML-based techniques are used with IoT and mobile health (m Health) solutions to provide tailored approaches for diagnosis purposes.
- It offers insights into the revolutionary potential of personalized medicine and promises creative solutions for customized medical treatment.

These contributions demonstrate the revolutionary potential of IoT and AI in smart healthcare and provide avenues for more efficient and secure customized medical solutions. The rest of the chapter is organized as follows: Section 2.2 provides an overview of the key components that make up smart healthcare. Section 2.3 explores healthcare solutions powered by IoT technologies. Section 2.4 discusses various applications of IoT in the healthcare domain. Finally, Section 2.5 concludes this study with a comprehensive discussion.

2.2 SMART HEALTHCARE

Smart health is defined as the digitized enhancement of the distribution of supporting technologies like IoT, big data, AI, and ML that foster the delivery of services in the health sector [12]. The following are key components of smart healthcare:

2.2.1 Sensor

In particular, sensors are vital parts of IoT devices in the medical field. Temperature, blood pressure, heart rate, and glucose levels are just a few of the health characteristics they measure and provide data on [13]. These sensors enable real-time health monitoring by being incorporated into medical equipment, wearable technology, and even human implants. As a result of this integration, patients can receive individualized treatment, faster interventions, and better results since vital signs can be continuously tracked without requiring frequent hospital visits. Table 2.1 lists common sensors used in smart healthcare.

The summary of the intended remote monitoring, locating systems, and personalized medicine research in AI literature is provided in Table 2.2. New generation substitutes for well-documented sensor technologies are under development through rigorous research. Some of them are VITALS, an online life sign sensor that utilizes clinical equipment to monitor four physiological indicators. It comprises a big data analysis system [26], a wireless connection module, and a microprocessor. From the lifelog, life signals are recorded every 30 minutes and passed to the analysis system. In a one-week test, the system was implemented and disclosed enhanced possibilities of care, especially to those with limited access to hospitals. In the proposed heart rate sensor, the MAX30102 sensor [27], which emits light in the range of 650–670 nm, is used. As a rather flat wave when worn at the wrist, it measures the heart rate signal and employs the detrend method to derive the correct data. Another study [28] aims at developing a disruptive blood pressure monitor that is based on the micro pyramid-assisted piezoelectric film (MPF) technology in a positive and non-contact manner. Each of the measurements taken with this MPF sensor is on par with those obtained using more recognized industrial products, the studies show.

TABLE 2.1
Common Sensors

Sensor Type	Measurement	Clinical Applications
Accelerometer	Body movement and orientation	Posture correction, physical activity analysis
EEG sensor	Brain's electrical activity	Neurological disorder monitoring (e.g., epilepsy)
EMG sensor	Muscle activity	Monitoring muscle health and rehabilitation
Wearable pressure sensor	Skin pressure levels	Bedsore prevention, prosthetics monitoring
pH sensor	Body fluids' acidity/ alkalinity	Monitoring acid-base balance in the blood or the digestive system

TABLE 2.2
Comparison of Research on Remote Monitoring, Locating Systems, and Personalized Medicine

Study	Year	Category	Communication	Advantages
[14]	2022	Remote Monitoring	Wi-Fi	By working with medical specialists, timely problem solving is made possible by ongoing monitoring of critical health indicators
[15]	2022	Remote Monitoring	Cloud	Alerts for crises and unusual occurrences that have been shown to be reliable and useful
[16]	2022	Remote Monitoring	Wi-Fi	Accurate performance evaluation and identification of the best body areas for temperature monitoring are ensured by an IoT connection with a medical promotion platform
[17]	2022	Remote Monitoring	Cloud	Promotes IoT integration in healthcare
[18]	2023	Remote Monitoring	Cloud	Effective data collection and mining are made possible by remote healthcare service delivery with professional assistance, improving the effectiveness of patient health surveillance as a whole
[19]	2022	Remote Monitoring	Cloud	Real-time data collecting and analysis, a variety of data input, and alert systems for prompt actions are all made possible by IoT cloud server integration
[20]	2023	Remote Monitoring	Cloud	Post-COVID co-morbidity monitoring
[21]	2023	Remote Monitoring	Cloud	With an emphasis on emergency medical services, it provides exceptional performance and high accuracy in vital sign monitoring while guaranteeing high dependability for IoT data transmissions
[22]	2023	Remote Monitoring	Cloud	Makes it possible to anticipate future states using data that has been saved, with an accuracy rate of 99.53%
[23]	2023	Remote Monitoring	Wi-Fi	Detection of influenza symptoms with a wide range of applications in different healthcare environments
[24]	2023	Locating Systems	Healthcare Resources	Uses a smart specimen transport box with cloud-based monitoring to improve tracking and management in place of barcode solutions
[25]	2023	Locating Systems	Patients	Monitors vital health indicators and, in response to anomalous readings, promptly makes medical judgements

This paper presents a novel graphical method for following blood pressure levels with a degree of persistency, which requires only one signal input and no additional sensing instruments or adjustments. Additionally, a pulse oximeter [29] to test body temperature, blood oxygen level, and heart rate has also been created. This gadget involves measuring temperatures in the body using light and sensors without touching the patient, the result being displayed in an LCD screen and a web browser.

2.2.2 Wi-Fi

Wireless internet connection is often utilized to establish internet connection on IoT gadgets; information transmission is also carried out over short distances. Since its introduction in 1997 [30], it has developed into one of the oldest technologies used in wireless communication, which provides a reliable and fast connection to billions of devices globally. Wireless technology is based on the IEEE 802.11 standard [31], which specifies the technical characteristics of WLANs. It involves enabling many devices to be connected and all communicate using Wi-Fi. In the last few years, Wi-Fi technology has advanced greatly, thanks to fresh standards like Wi-Fi 6 and Wi-Fi 6E. With these new standards, novel applications and services, which need fast connection and negligible delays, can be established; advances in speed, protection, and storage capacity are also possible. Medical professionals and various other economic sectors that require consistent and high-speed WWAN telecom services are just beginning to rely more on Wi-Fi as it evolves.

2.2.3 Bluetooth

Bluetooth is a wireless technology that connects devices, and the name Bluetooth comes from the Viking king Harald Bluetooth, who united Denmark and Norway in the 10th century. Bluetooth technology was invented in 1994 by the Swedish firm of Ericsson and is widely popular for connecting smart devices such as the IoT with Smartphones and other portable electronic gadgets. Frequency-hopping spread spectrum (FHSS) is employed in an effort to minimize interaction with other wireless devices operating in similar or the same frequency band [31].

2.2.4 Zigbee

Zigbee is a low-power wireless communication protocol that was mainly created for IoT applications that require low data rates and long battery lives, including remote medical monitoring systems. Its initial standard was published in 2004, having been developed in the early 2000s [32]. By using a mesh network design, Zigbee enables effective device-to-device communication. Low-power consumption, mesh networking capabilities, and cost-effective optimization for low data rate transfers are important characteristics. Because it can save energy by turning down radios when not sending or receiving data (a feature achieved through sleep modes), it's especially well-suited for IoT devices that run on batteries. In order to reduce the likelihood of data collisions in the network, Zigbee additionally uses a collision avoidance technology known as Carrier Sense Multiple Access with Collision Avoidance (CSMA-CA) [33].

2.2.5 Near-field Communication (NFC)

As seen in Figure 2.2, NFC [34] is a short-range wireless communication technique intended for data transfer in close proximity, usually within a few millimetres of each other. The foundation of this technology is Radio Frequency Identification (RFID), a wireless data transmission method between two devices that uses electromagnetic fields. Because NFC makes it possible to communicate data quickly and securely without requiring physical touch, it has become more and more common. In order for NFC to function, a communication link must be established between two active devices or between an active device (like a smartphone) and a passive device (such as an NFC tag or payment terminal). In just a few moments, this interaction takes place, enabling a smooth information flow with little assistance from the user. The NFC technology opens up a lot of uses.

NFC technology is utilized in contactless payment systems such as Apple Pay, Google Pay, and other mobile payment platforms. This technology enables users to safely complete transactions at payment terminals by only tapping their smartphones. NFC technology improves security and ease in access management by enabling admission into buildings or restricted areas using key cards or mobile apps. By placing devices near each other, NFC also makes data interchange between them easier, including transferring files, links, and multimedia material. Furthermore, NFC's broad application in consumer electronics, healthcare, and public transit can be attributed to its low energy consumption and intuitive interface. NFC can be utilized, for example, in patient monitoring, where medical records are kept on NFC-capable devices and promptly retrieved by medical staff. The technology is becoming a crucial component of contemporary smart settings due to its expanding ecosystem [34].

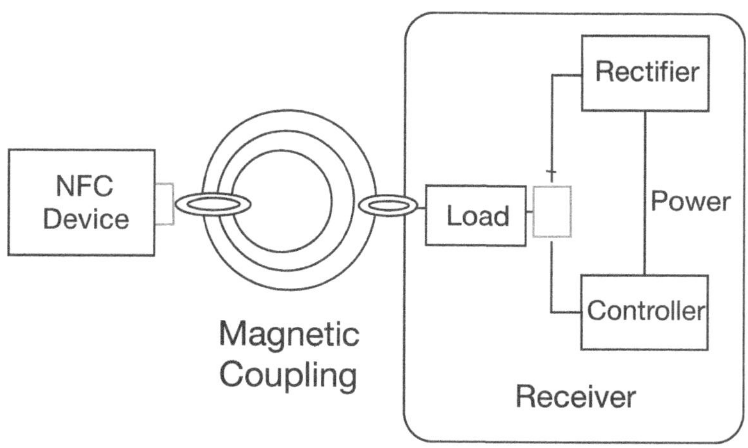

FIGURE 2.2 NFC functioning.

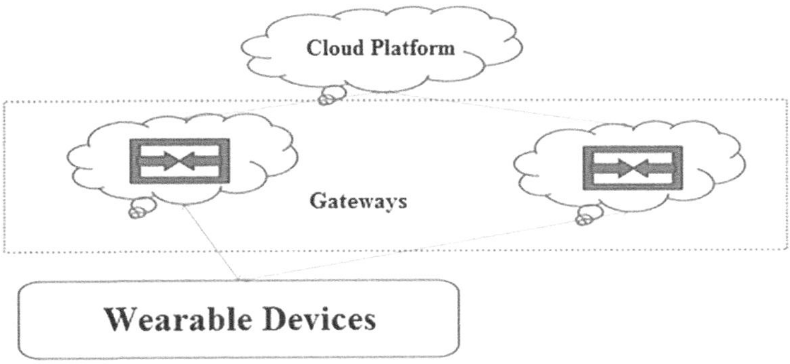

FIGURE 2.3 Communication between IoT and cloud.

2.2.6 Cloud

As seen in Figure 2.3, the cloud offers a potent platform for storing, managing, and analysing enormous volumes of healthcare data produced by IoT devices [35]. However, the cloud is not a communication technology in and of itself. Large-scale data handling and processing may now be done more effectively and affordably thanks to the cloud, which has completely changed how both consumers and enterprises access and use computer resources. This transformation supports IoT-driven healthcare innovations by enabling enterprises to store, process, and analyse healthcare data more effectively [36].

The earliest instances of utility computing were presented in the 1960s, which is when cloud computing initially emerged. But it wasn't until the late 1990s that cloud computing started to take on its modern form [37]. Cloud methodology plays a crucial role in the processing and storage of medical data with reference to the IoT environment in the health sector. Providers in the healthcare sector may curb expenses, improve patients' experiences, and, in the process, increase quality across this sector, stemming from innovative cloud technologies. For example, cloud computing makes it possible to store electronic health records (EHRs) securely, which makes it simpler to access and share patient data between different systems and organizations [37].

2.2.7 AI

There is an increased adoption of AI algorithms when used alongside big data from sensors in the IoT. Using the amount of data collected by these devices, AI algorithms can provide targeted health interventions as well as real-time data on patients' health. Another type of AI is machine learning models, and they can analyse IoT data concerning the disease and predict its future development as well as suggest the proper therapy for most patients. These models notice slight changes or deviations from the normal in a patient's status, which makes it easier for a patient's health status to be noticed if there is a problem before it becomes catastrophic. AI systems can monitor vital signs, level of activity, sleep patterns, and other important parameters of healthcare IoT through wearables and other devices. Using early warning signals

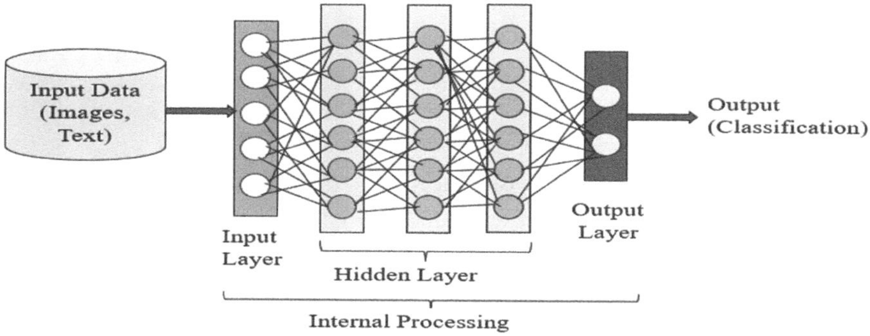

FIGURE 2.4 Deep learning technique for decision-making.

of health declines or relapses, chronic monitoring benefits doctors by letting them address complications as soon as possible. The use of AI can also be in developing individual care plans based on data from IoT devices [38]. These algorithms have the capacity to contemplate enormous volumes of data and offer exact and personalized chemotherapy regimens for patients, taking into consideration their characteristics, history, and data in the present time. Figure 2.4 presents the working of the AI-based technique in decision-making.

Both IoT and AI applied to the health area mean an innovative paradigm shift of focus on the patient. This synergy can be seen as a sign of possible development within the healthcare sector, where technology innovation and care pathways will be a key focus to extrapolate enhancement in the predicted accuracy and in the delivery of custom-made therapy.

2.2.8 Blockchain

Blockchain technology [39], in conjunction with IoT in the healthcare industry, has enormous promise for enhancing data security and expediting provider-to-provider data sharing. The decentralized and immutable characteristics of blockchain technology may be used to safeguard patient data against unauthorized access, alteration, and breaches. As seen in Figure 2.5, the distributed ledger system of blockchain makes sure that each transaction or data input is cryptographically protected and connected to earlier entries, creating an unchangeable chain. Along with preserving patient data integrity, this approach encourages responsibility and openness in data management.

Healthcare workers may safely access and share pertinent patient data via blockchain, bridging organizational and geographic borders. Care coordination is improved by this smooth data exchange, giving medical professionals access to a patient's complete medical history so they may make well-informed decisions. By allowing particular healthcare providers to access their data, patients in turn have greater control over their data. Blockchain's smart contracts further automate data-sharing agreements. These self-executing contracts promote confidence and accountability in healthcare data management by guaranteeing adherence to privacy standards and giving patients more transparency and control over the use of their data.

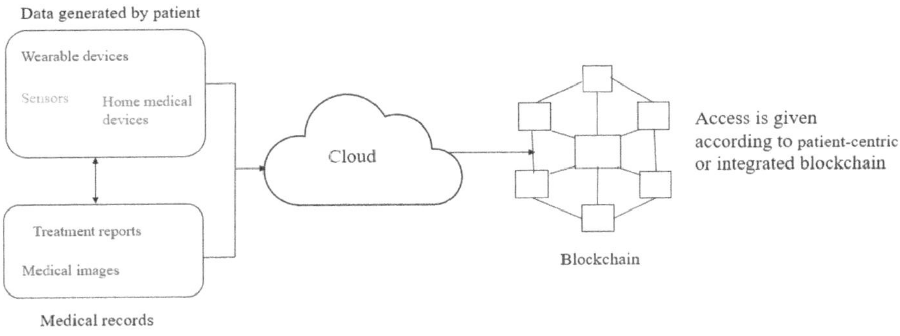

FIGURE 2.5 AI techniques in smart healthcare.

2.2.9 Edge Computing

Real-time data processing and performance improvements are critically required for the IoT in the healthcare system, as accomplished by edge computing. As such, the problem areas, including the delay of network latency and rapid response in the healthcare segment, are well-captured by edge computing. IoT devices communicate data for processing to cloud servers, which can cause latency in critical facilities, for instance. However, edge computing works on data at or very close to the networks' periphery, offering faster analytics. Net latency is a critical area where edge computing shines for the healthcare sector. Data is processed locally to allow near-real-time analysis because the time taken to send information to distant cloud servers is long. This is important in real-time monitoring, such as in emergency response, continuous health assessment and monitoring, as well as remote patient monitoring (RPM). Edge computing also facilitates resilience to network outages and offline operation [40]. Edge devices can keep working even when there is inadequate or no internet access, which enables the provision of healthcare services and other necessary functions. By guaranteeing constant care delivery under all circumstances, this capability improves the dependability of healthcare IoT systems.

2.2.10 5G

5G networks facilitate real-time communication in the field of healthcare IoT, enabling healthcare practitioners to continually monitor and engage with patients [41]. 5G's increased speed and low latency make data transfer almost instantaneous, allowing for prompt reactions to life-threatening medical emergencies. The ability of 5G networks to accommodate a large number of linked devices is one of their key features. This scalability is needed in healthcare facilities, where a large number of connected IoT devices are utilized. The effectiveness of smartphones enables efficient data aggregation and processing from the patients through 5G networks due to their high-capacity data transmissions. The dependability of 5G networks enhances their suitability, as many applications require continuous communication and monitoring. In restricted accessibility to healthcare, that is in distant areas and during emergencies, this dependability comes as something useful. Through the adoption of the new 5G technology, most healthcare providers may provide effective remote treatment so that a considerable distance between the patients and the physicians may not be witnessed. Patients' data can be rapidly accessed and assessed due to reduced data transfer latency; this results in more efficient diagnostics and treatment adjustments.

2.3 IOT-DRIVEN HEALTHCARE SOLUTIONS

Probably, thanks to IoT implementation in the devices applied in the healthcare sector, people can be monitored even outside the healthcare centre, the idea of keeping people healthy and safe is possible, and physicians can offer patients the utmost quality of beneficial healthcare services. IoT is indeed affecting the further development of the healthcare industry as it has changed the area of the devices and people relations concerning the appliances used to deliver healthcare solutions. Given the IoT, it has numerous use cases suitable for patients, their closest families, doctors, healthcare centres, and insurance service providers.

2.3.1 IoT Solutions for Indoor Healthcare

IoT devices are essential for tracking inventories, controlling medical equipment, and keeping an eye on patients in interior healthcare facilities like hospitals. In addition to monitoring temperature, humidity, and air quality, IoT sensors may also stop the spread of illnesses. Wearable technology, RFID tags [42], or smartphone apps make it easier to follow patients and medical personnel thanks to IoT technology. In the end, improved emergency response times, improved processes, and improved navigation are all results of this real-time monitoring, which improve patient care. For example, in this research, a deep Convolutional Neural Network – Long Short-Term Memory (CNN-LSTM) architecture is proposed for IoT healthcare systems' interior position estimation employing Passive Infrared (PIR) sensors. Further research emphasizes interior tracking and behavioural pattern recognition while examining audience behaviour at a resort. Indoor location technology also helps with medical equipment tracking and management. Healthcare professionals may efficiently monitor the location, usage, and maintenance requirements of medical devices by affixing RFID tags or using sensor networks. This guarantees their availability and lowers the price of replacing lost or misplaced equipment. In order to provide role-based access management and improved security, research suggests a non-intrusive indoor location framework for IoT healthcare that makes use of wireless sensor networks and blockchain-based smart contracts. Systems for indoor localization powered by the IoT can support environmental monitoring in healthcare facilities. Real-time monitoring of variables like temperature, humidity, air quality, and occupancy is possible with the deployment of sensor networks. This enables automatic modifications that promote energy economy and lower the risk of infections linked to healthcare while guaranteeing a cosy and secure atmosphere for patients and personnel. In one study [43], a Multiple-Input Multiple-Output Visible Light Communication (MIMO VLC) system that gathers data and uses a variety of modulation techniques for dependable communication is examined as part of an indoor IoT system. This study provides a useful solution for monitoring a variety of data types in healthcare applications.

2.3.2 IoT Solutions for Outdoor Healthcare

Individual well-being is greatly improved by the integration of smart and IoT-enabled equipment in outdoor healthcare contexts, in addition to personal localization and RPM. A growing number of public venues, including parks, leisure centres, and transit hubs, are outfitted with cutting-edge healthcare facilities that utilize IoT technology to provide seamless access to healthcare [44]. The public's varied healthcare demands are effectively met by the design of these smart facilities.

Using IoT-enabled kiosks is a noteworthy aspect of smart outdoor healthcare environments. These interactive kiosks serve as information centres, offering up-to-date information about local pharmacies, emergency contact numbers, and healthcare providers. People may find the closest medical facilities, obtain individualized health information, and get emergency help right now. Future fitness stations with IoT-connected workout equipment are expected to be seen in smart parks and recreational spaces. These intelligent exercise monitors might measure important health parameters, keep an eye on users' activity levels, and offer tailored workout recommendations. Through smartphone applications, users may gain access to these facilities, providing a customized workout experience based on personal preferences and health goals. Hubs for transportation gain from IoT integration as well. These days, IoT sensors are used in automated external defibrillators (AEDs) [45] to continually check on their operation. This life-saving equipment is constantly prepared for use since automated notifications are issued to healthcare personnel in the event of any problems or failures.

2.3.3 Real-Time RPM

Patient health may now be tracked outside of traditional healthcare settings thanks to IoT technologies like wearables and remote monitoring systems. By enabling early identification of changes in a patient's health state and prompt actions by healthcare professionals, RPM enhances patient outcomes [46]. RPM uses technology to collect pertinent data, enable virtual consultations, and continually monitor patients' health. Some of the examples of this technology are wearables, sensors, and communication. This means that frequent face-to-face follow-ups to check on patients' vital signs, adherence to prescribed medication, signs/symptoms, and any other feature of health as outlined herein are not required using RPM. Having engaged RPM, it is one of the significant strengths of the method that allows for healthcare services to be provided to individuals in underdeveloped or remote regions. Patients who live in rural areas or extended villages often lack access to specialized care. RPM comes in handy to deal with this challenge as a way of offering monitoring and consultations from a distance, enabling healthcare professionals to offer timely support and corrective measures to be provided. In addition, RPM is useful, especially for patients with limited mobility, who have long-term medical illnesses or injuries that make it challenging to make visits. "These are patients who can get their medical needs met from the comfort of their homes with the help of remote monitoring equipment." This allows healthcare practitioners to reassess their state without needing to go to the hospital often, change doses, and offer support; RPM is a bridge to healthcare for paralysed patients and those with other mobility issues. They can monitor those important signs, like blood pressure, glucose level, and heart rate, by using wearables such as watches or sensors. The collected data is accessed safely to allow the healthcare providers to analyse it and take the essential medical steps. Health care providers are in a position to reduce problems and readmission by ensuring they use RPM to report any health complications or abnormalities accordingly. Besides enhancing patient results, it also enhances the general well-being of those who have restricted mobility in care settings.

2.3.4 Connected Urban Living: Smart Cities

Smart traffic lights, air quality monitors, noise sensors, and other gadgets that encourage healthy living are made possible by IoT technology. By lowering the danger of respiratory ailments, noise pollution, and accidents, these technologies contribute to the development of a safer, more liveable urban environment. The ability of RPM to raise patient happiness, save costs, and improve healthcare outcomes has been acknowledged by governments. As a result, they have put in place laws and guidelines that give patient data security and privacy a priority inside RPM systems [47]. Governments guarantee the encryption and authentication of patient data by implementing data protection laws and formulating standards for safe data transfer. Governments finance healthcare organizations in addition to creating policies to aid with the incorporation of RPM into healthcare systems. Infrastructure investments, such as those in telephone networks and internet connections, guarantee that RPM services are available even in isolated or underdeveloped locations. It is also recommended that healthcare providers, technology businesses, and academic institutes work together to standardize RPM procedures, foster innovation, and exchange best practices. These group initiatives show a strong dedication to increasing RPM use and enhancing the provision of healthcare. Through wearable health technology and telemedicine, smart cities provide individuals with individualized healthcare experiences that include real-time health monitoring and remote access to medical specialists. These technologies provide accessibility to healthcare and enable people to play an active role in managing their well-being. Smart cities use data-driven insights to improve population health on a larger scale. Smart cities encourage proactive health management through the detection and mitigation of environmental issues such as air pollution and possible disease outbreaks. Green areas and effective transit networks are included in smart urban design to promote active lives. A healthy urban environment is further enhanced by public health monitoring tools, including real-time air quality warnings and early warning systems for disease epidemics. In the end, smart cities support a healthcare ecosystem that improves general well-being by making healthcare more accessible, proactive, and responsive for the benefit of both people and the public.

2.4 IOT APPLICATION IN HEALTHCARE

IoT offers real-time health monitoring and data collection through linked devices like wearables and sensors. These gadgets send data to the cloud for analysis, facilitating better patient care, personalized medicine, and remote monitoring.

2.4.1 Real-Time Remote Monitoring

By developing a setting in which the portable physiological monitoring gadgets can seamlessly interconnect with expert systems, the application of IoT in RHM is revolutionizing patient care and medical history record-keeping systems, among other aspects. Based on the research done in recent years, cloud computing based systems with real-time data transfer and detection of anomaly is only way to healthy monitoring due to the use of IoT. For instance, one system uses Wi-Fi and infrared to determine the body temperature of the patients, while another guarantees reliability of received real-time patient data alongside patient vital signs. IoT systems leverage mobile apps and wearables to keep track of patients' biometrics while implementing secure login procedures and reliable channels of conveying data. Moreover, Machine Learning System (MLS) helps to facilitate improved elderly monitoring with the use of Information Technology, leading to improved data analysis and health services delivery. Even attending a hospital has become a rare occasion due to the usage of IoT-based electrocardiogram (ECG) apparatus for remote analysis and quick notification. One of the areas of innovation in the post-COVID world is the use of ML algorithms in combination with IoT gadgets to deliver exigent medical therapies. In regard to patient tracking, systems with connectivity to the cloud and a private backend also have high accuracy and reliability. IoT has also been employed to track COVID-19 and influenza signs for faster health care and to track the movement of patients during high-risk conditions. These advancements in technology explain how IoT may offer precise, timely, and individual-centred, or patient-oriented solutions for optimizations in health to virtually increase effectiveness and performance.

2.4.2 Smart Locating Solutions

Some of the applications of real-time locating systems (RTLS) include the ability to identify a patient's whereabouts, which is a key role for the medical personnel, particularly in emergency situations. It enables early treatments to be done and minimizes the chances of the patient getting an injury. Besides patient tracking, RTLS can give accurate and up-to-date information concerning the availability and location of critical medical inventory, equipment, and even drugs. This has the advantage of enhancing the management of total resources since search times are reduced while some inventories run out. RTLS research area has now evolved into the best technologies, such as RFID and Narrowband Internet of Things (NB-IoT), for tracking medical resources, particularly on specimen transport within hospitals, deep neural networks, and spatial spectrum fitting for precise source location within unmanned aerial vehicle networks. Moreover, it is recommended that wearable RTLS devices track people's motion, such as that of Alzheimer's disease patients, and, in case of deviation, increase safety through alarms for carers. Another strategy adopts the mode of unobtrusive and sensor-independent interfaces and ontology components suitable for wellness monitoring and environmental sensing applications, while connected health care devices and appliances have proven useful in real-time waste level monitoring in communal bins, wellness parameter monitoring, and timely health management decisions.

2.4.3 Smart Solutions for Personalized Medicine

In order to provide more accurate and efficient care, personalized medicine has become a ground-breaking method in the healthcare industry. It centres on the customization of the medical procedures depending on the individual's genes, environment, and behaviour. Recent literature and innovations in this field explain how certain technologies could enhance the delivery of personalized care. For instance, there is research evidence that supports the idea that the use of IoT and patient care engagement are related. These studies have also presented a model that addresses patient self-management alongside patient privacy and security in healthcare organizations. Using the Latent Dirichlet Allocation (LDA) model in diabetes management represents a significant leap in the improvement of treatment decision-making in the healthcare system because it provides predictions that are specific to the future. In a different research, blood glucose levels are examined in relation to physical activity using IoT and mHealth technologies, allowing physicians to provide customized food recommendations for patients with paediatric diabetes. Similar to this, machine learning models have been used to develop a customized dental caries prediction system that makes it easier for patients and dentists to estimate risk by enabling the early detection of possible dental problems. The utilization of 5G technology in real-time diabetes patient monitoring pushes the boundaries of individualized treatment by offering early warnings derived from ongoing health data. Furthermore, a method for individualized illness monitoring that makes use of exhaled volatile organic compounds (VOCs) has been created. Some of the benefits associated with this system include early diagnosis, reduced expense, and quality treatment. Finally, the strategy of adapting meta-learning for a customized health monitoring approach that has been presented here performs better than traditional methods in detecting vitals and human movements. Collectively, the derived findings explain the achievable revolution in treating diseases and disorders through customized therapeutic approaches made possible by advanced technologies of personalized medicine.

2.5 DISCUSSION

Industry 4.0 approaches have therefore merged with personalized solutions in delivering healthcare through higher technologies such as IoT, AI, and machine learning. As shown throughout this investigation, these advances make it feasible to provide individualized, precise vigilance and treatment. Even though machine learning models have increased the chances of forecasting and handling of particular health disorders, IoT-based frameworks are effective in increasing patient engagement and guaranteeing timely treatment of different chronic diseases. Before having symptoms and illnesses, symptom identification and disease treatment are also ensured by inhaled VOC analysis and the use of specialist technologies like real-time monitoring through 5G. Moreover, by using meta-learning strategies for analysing different antecedent experiences, there have been improvements in detecting parameters of health and vital signs that will guarantee an even more reliable and personalized system of healthcare. It is now clear that more has to be done in order to harness the benefits of these solutions, and only increased collaboration between policymakers, technologists, and healthcare professionals will suffice. By using a more solid IoT and AI architecture in the development of additional personalized medicine, the future could completely redefine the existing rules of care through tailored and patient-centred medical approaches. Customized medicine has the ability to drastically change the healthcare sector because of advancements in IoT, AI, and machine learning. With the advent of 5G and edge computing, real-time monitoring and prompt decision-making will be more effective, enabling continuous patient surveillance and preventative interventions. Blockchain technology with the IoT may make data transfer simple, secure, and private while preserving data confidentiality. Moreover, the development of precision medications and individualized preventative healthcare regimens may be aided by the integration of AI with genetic and environmental data. These advancements will transform personalized medicine and enhance patient-focused, precise, and anticipatory healthcare.

REFERENCES

1. Tiwana, P. S., Kaur, L., Kaur, N., Wadhwa, H., Kumari, N., & Sharma, R. (2024). Mitigating DDoS attacks: A distributed blockchain-SDN secure IoT system enhanced by artificial neural networks. In Sukhpreet Kaur, Sushil Kamboj, Manish Kumar, Arvind Dagur, & Dhirendra Kumar Shukla (Eds.), *Computational Methods in Science and Technology* (pp. 467–472). London: CRC Press.
2. Dang, V. A., Vu Khanh, Q., Nguyen, V. H., Nguyen, T., & Nguyen, D. C. (2023). Intelligent healthcare: Integration of emerging technologies and Internet of Things for humanity. *Sensors*, 23(9), 4200.
3. Shehada, D., Gawanmeh, A., Yeun, C. Y., & Zemerly, M. J. (2022). Fog-based distributed trust and reputation management system for internet of things. *Journal of King Saud University-Computer and Information Sciences*, 34(10), 8637–8646.
4. Bonkra, A., & Dhiman, P. (2021, December). IoT security challenges in cloud environment. In *2021 2nd International Conference on Computational Methods in Science & Technology (ICCMST)* (pp. 30–34). IEEE.
5. Alshammari, H. H. (2023). The internet of things healthcare monitoring system based on MQTT protocol. *Alexandria Engineering Journal*, 69, 275–287.
6. Zeng, H., Dhiman, G., Sharma, A., Sharma, A., & Tselykh, A. (2023). An IoT and blockchain-based approach for the smart water management system in agriculture. *Expert Systems*, 40(4), e12892.
7. Krishnamoorthy, S., Dua, A., & Gupta, S. (2023). Role of emerging technologies in future IoT-driven Healthcare 4.0 technologies: A survey, current challenges and future directions. *Journal of Ambient Intelligence and Humanized Computing*, 14(1), 361–407.
8. Bhatt, V., & Chakraborty, S. (2023). Improving service engagement in healthcare through internet of things based healthcare systems. *Journal of Science and Technology Policy Management*, 14(1), 53–73.

9. Alazzam, M. B., Alassery, F., & Almulihi, A. (2022). Retracted Federated deep learning approaches for the privacy and security of IoT systems. *Wireless Communications and Mobile Computing*, 2022(1), 1522179.
10. Albouq, S. S., Abi Sen, A. A., Almashf, N., Yamin, M., Alshanqiti, A., & Bahbouh, N. M. (2022). A survey of interoperability challenges and solutions for dealing with them in IoT environment. *IEEE Access*, 10, 36416–36428.
11. Kaur, A., & Tiwana, P. S. (2023, December). Analysis of IoT devices data using bayesian learning on fog com-puting. In *Artificial Intelligence, Blockchain, Computing and Security Volume 2: Proceedings of the International Conference on Artificial Intelligence, Blockchain, Computing and Security (ICABCS 2023)*, Gr. Noida, UP, India, 24–25 February 2023 (p. 194). CRC Press.
12. Nasr, M., Islam, M. M., Shehata, S., Karray, F., & Quintana, Y. (2021). Smart healthcare in the age of AI: Recent advances, challenges, and future prospects. *IEEE Access*, 9, 145248–145270.
13. Haghi, M., Thurow, K., & Stoll, R. (2017). Wearable devices in medical internet of things: Scientific research and commercially available devices. *Healthcare Informatics Research*, 23(1), 4–15.
14. Valsalan, P., ul Hasan, N., Baig, I., & Zghaibeh, M. (2022). Remote healthcare monitoring using expert system. *International Journal of Advanced Computer Science and Applications*, 13(3), 593–599.
15. Sahu, M. L., Atulkar, M., Ahirwal, M. K., & Ahamad, A. (2022). Cloud-based remote patient monitoring system with abnormality detection and alert notification. *Mobile Networks and Applications*, 27(5), 1894–1909.
16. Boonsong, W., & Senajit, N. (2022). Remote patient body temperature monitoring based-IEEE802. 11a Internet of Things (IoT). *Przegląd Elektrotechniczny*, 98(6), 95–98.
17. Ahmed, M. I., & Kannan, G. (2022). Secure and lightweight privacy preserving Internet of things integration for remote patient monitoring. *Journal of King Saud University-Computer and Information Sciences*, 34(9), 6895–6908.
18. Deepa, S., Sridhar, K. P., Baskar, S., Mythili, K. B., Reethika, A., & Hariharan, P. R. (2023). IoT-enabled smart healthcare data and health monitoring based machine learning algorithms. *Journal of Intelligent & Fuzzy Systems*, 44(2), 2927–2941.
19. Bushnag, A. (2022). A wireless ECG monitoring and analysis system using the IoT cloud. *Intelligent Automation & Soft Computing*, 33(1), 51–70.
20. Rahman, S., Parveen, S., Sofi, S. A., & Zahoor, S. (2023). Post-covid remote patient monitoring using medical internet of things and machine learning analytics. *Scalable Computing: Practice and Experience*, 24(1), 1–16.
21. Ali, F. I., Ali, T. E., & Al-Dahan, Z. T. (2023). Private backend server software-based telehealthcare tracking and monitoring system. *International Journal of Online and Biomedical Engineering*, 19(1), 119–134.
22. Balasubramanian, K., Prabu, A. V., Shaik, M. F., Naik, R. A., & Suguna, S. K. (2023). A hybrid deep learning for patient activity recognition (PAR): Real time body wearable sensor network from healthcare monitoring system (HMS). *Journal of Intelligent & Fuzzy Systems*, 44(1), 195–211.
23. Abdulhussein, T. A., Al-Falahi, H. A., Smait, D. A., Alani, S., Mahmood, S. N., & Mustafa, M. S. (2023). Early coronavirus disease detection using internet of things smart system. *International Journal of Electrical and Computer Engineering (IJECE)*, 13(1), 1161–1168.
24. Le, N. T., Thwe Chit, M. M., Truong, T. L., Siritantikorn, A., Kongruttanachok, N., Asdornwised, W., … & Benjapolakul, W. (2023). Deployment of smart specimen transport system using RFID and NB-IoT technologies for hospital laboratory. *Sensors*, 23(1), 546.

25. García-Requejo, A., Pérez-Rubio, M. C., Villadangos, J. M., & Hernández, Á. (2023). Activity monitoring and location sensory system for people with mild cognitive impairments. *IEEE Sensors Journal*, 23(5), 5448–5458.
26. Banu, E. A., & Rajamani, V. (2023). Design of online vitals monitor by integrating big data and IoT. *Computer Systems Science & Engineering*, 44(3), 2469–2487.
27. Gunawan, R., Andang, A., & Ridwan, M. (2023). Performance comparison for hearth rate signal detection for different location in fingertip and wrist using sensor MAX30102. *Journal of Biomimetics, Biomaterials and Biomedical Engineering*, 59, 131–143.
28. Kim, Y., Lee, J., Hong, H., Park, S., & Ryu, W. (2023). Self-powered wearable micropyramid piezoelectric film sensor for real-time monitoring of blood pressure. *Advanced Engineering Materials*, 25(2), 2200873.
29. Arifah, A. G., Costrada, A. N., & Harmadi, H. (2023). Inovasi Pulse Oximeter dengan Sumber Cahaya LED Merah dan Inframerah yang Dilengkapi Suhu Tubuh Menggunakan Sensor MLX90614 Berbasis IoT. *Jurnal Fisika Unand*, 12(2), 200–206.
30. Dongus, S., Jalilian, H., Schürmann, D., & Röösli, M. (2022). Health effects of WiFi radiation: A review based on systematic quality evaluation. *Critical Reviews in Environmental Science and Technology*, 52(19), 3547–3566.
31. Ye, J., Gharavi, H., & Hu, B. (2023). Fast beam discovery and adaptive transmission under frequency selective attenuations in Sub-Terahertz bands. *IEEE Transactions on Signal Processing*, 71, 727–740.
32. Cheng, X., & Fan, Y. (2022). Research and design of intelligent speech equipment in smart English language lab based on Internet of Things technology. *Procedia Computer Science*, 198, 505–511.
33. Tanash, R., AlQudah, M., & Al-Agtash, S. (2023). Enhancing energy efficiency of IEEE 802.15. 4-based industrial wireless sensor networks. *Journal of Industrial Information Integration*, 33, 100460.
34. Yu, W. W., & Fang, C. Y. (2023). The role of near-field communication mobile payments in sustainable restaurant operations: A restaurateur's perspective. *Sustainability*, 15(16), 12471.
35. Pericherla, S. S. (2023). Cloud computing threats, vulnerabilities and countermeasures: A state-of-the-art. *The ISC Int'l Journal of Information Security*, 15(1), 1–58.
36. Khamis, A. A., Idris, A., Abdellatif, A., Mohd Rom, N. A., Khamis, T., Ab Karim, M. S., … & Abd Rashid, R. B. (2023). Development and performance evaluation of an IoT-integrated breath analyzer. *International Journal of Environmental Research and Public Health*, 20(2), 1319.
37. Gupta, A., Mazumdar, B. D., Mishra, M., Shinde, P. P., Srivastava, S., & Deepak, A. (2023). Role of cloud computing in management and education. *Materials Today: Proceedings*, 80, 3726–3729.
38. Mozumder, M. A. I., Armand, T. P. T., Imtiyaj Uddin, S. M., Athar, A., Sumon, R. I., Hussain, A., & Kim, H. C. (2023). Metaverse for digital anti-aging healthcare: An overview of potential use cases based on artificial intelligence, blockchain, IoT technologies, its challenges, and future directions. *Applied Sciences*, 13(8), 5127.
39. Khan, S., Kaushik, M. K., Kumar, R., & Khan, W. (2023). Investigating the barriers of blockchain technology integrated food supply chain: A BWM approach. *Benchmarking: An International Journal*, 30(3), 713–735.
40. Bourechak, A., Zedadra, O., Kouahla, M. N., Guerrieri, A., Seridi, H., & Fortino, G. (2023). At the confluence of artificial intelligence and edge computing in IoT-based applications: A review and new perspectives. *Sensors*, 23(3), 1639.
41. Suleiman, T. A., & Adinoyi, A. (2023). Telemedicine and smart healthcare—The role of artificial intelligence, 5G, cloud services, and other enabling technologies. *International Journal of Communications, Network and System Sciences*, 16(3), 31–51.

42. Wang, W., Asci, C., Zeng, W., & Sonkusale, S. (2023). Zero-power screen printed flexible RFID sensors for Smart Home. *Journal of Ambient Intelligence and Humanized Computing*, 14(4), 3995–4004.
43. Sejan, M. A. S., & Chung, W. Y. (2022). Performance analysis of a long-range MIMO VLC system for indoor IoT. *IEEE Internet of Things Journal*, 10(8), 6999–7010.
44. Nabi, M. N. U., Zohora, F. T., & Misbauddin, S. M. (2023). Social media links with social capital to trust in healthcare facilities: Empirical evidence from Bangladesh. *Library Hi Tech*, 41(1), 210–228.
45. Berglund, E., Hollenberg, J., Jonsson, M., Svensson, L., Claesson, A., Nord, A., … & Ringh, M. (2023). Effect of smartphone dispatch of volunteer responders on automated external defibrillators and out-of-hospital cardiac arrests: The SAMBA randomized clinical trial. *JAMA Cardiology*, 8(1), 81–88.
46. Shaik, T., Tao, X., Higgins, N., Li, L., Gururajan, R., Zhou, X., & Acharya, U. R. (2023). Remote patient monitoring using artificial intelligence: Current state, applications, and challenges. *Wiley Interdisciplinary Reviews: Data Mining and Knowledge Discovery*, 13(2), e1485.
47. Xia, L., Semirumi, D. T., & Rezaei, R. (2023). A thorough examination of smart city applications: Exploring challenges and solutions throughout the life cycle with emphasis on safeguarding citizen privacy. *Sustainable Cities and Society*, 98, 104771.

3 Innovating Securely

6G and IoT for Immersive Futures

Neha Garg, Priyanka Datta, Amitesh Aggarwal, Amanpreet Kaur, and Yonis Gulzar

3.1 INTRODUCTION

The technology that has brought revolution in the digital world is immersive. Immersive technology is further classified into AR, VR, and Mixed Reality (MR) [1], as shown in Figure 3.1. Virtual elements are overlaid onto real objects in AR, enhancing the real-world environment through auditory and visual effects [2]. For instance, AR games like Pokemon Go assimilate virtual elements with physical spaces. On the other hand, in VR, users are completely involved in a virtual environment away from the world using devices like Cave Automatic Virtual Environment (CAVE) or VR headsets like Oculus Quest, HTC Vive, and so on. Users are unable to interact with the virtual object while using AR applications, but when using MR applications, users can visualize and interact with the virtual object while it exists in the real world [3]. These technologies are becoming increasingly relevant in industries such as education, entertainment, and healthcare because they can boost user involvement and engagement.

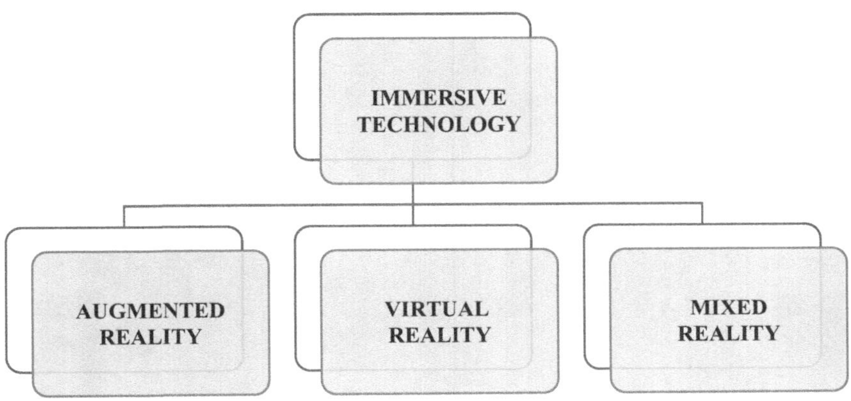

FIGURE 3.1 Types of immersive technology.

DOI: 10.1201/9781003586456-3

3.1.1 The Role of 6G in Immersive Technology

The evolution to 6G from 5 Generation (5G) opens a transformative phase for immersive technologies [4]. Though 5G provides greater reliability, assistance for connected devices, and faster data speeds, 6G has the potential to revolutionize connectivity by allowing peak rates of data transfer exceeding 1 Terabit per Second (Tbps), permitting ultra-low latency and unparalleled support for large-scale IoT ecosystems. These improvements allow immersive applications to work seamlessly in real time under conditions of high network demand.

Example:

Figure 3.2. demonstrates a futuristic remote surgery scenario using 6G technology. The main features included are as follows:

- **Robotic Assistance:** IoT devices help to connect robot arms and help in surgery, controlled remotely by surgeons.
- **Surgeon with AR Glasses:** A surgeon remotely operates using advanced AR glasses, which overlay important real-time data on their view.
- **Real-Time Data Transmission:** The 6G technology confirms ultra-low latency, helping the surgeon's movements and also providing instant and accurate commands to be executed.
- **AI and Edge Computing:** This feature analyses and processes data locally to improve responsiveness and accuracy in their actions, and helps to make decisions instantly [5].

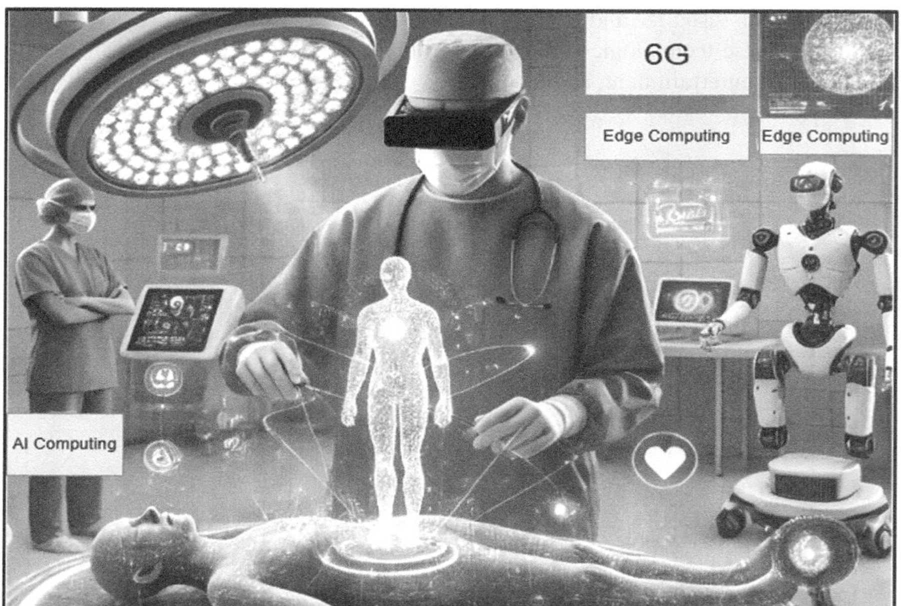

FIGURE 3.2 Remote surgery scenario using 6G.

- **Modern and Clean Environment:** The clinical setup highlights innovative medical technology.
- **Holographic Interface:** The holographic human body projection helps the surgeon with a visual response on the status of the patient and accuracy in their actions.

Figure 3.2 describes the importance of 6G and AR in revolutionizing remote surgical procedures and telemedicine.

3.1.2 IoT's Role in Immersive Technologies

IoT devices act as an important enabler of immersive technology by connecting virtual objects with the real world. IoT makes it possible to gather and interconnect data in real time by a network of connected devices and sensors, allowing immersive technology to engage with its environment in real time.

Example:

Experts in industries like healthcare, manufacturing, and logistics can work in an AR-enabled environment due to IoT-enabled smart or AR glasses that can overlay virtual objects onto physical objects in real time. In a similar context, training simulators can become more effective and realistic by simulating real-world scenarios in VR environments improved with IoT data. Delivering context-aware and hyper-personalized experiences needs the collaboration of immersive technologies and IoT devices.

3.1.3 The Importance of Securing 6G and IoT for Immersive Experiences

Nevertheless, despite their potential, there are severe security issues with the amalgamation of 6G and IoT into immersive technologies. Various points of vulnerability are developed by the interconnectedness made possible by IoT devices and the huge scale and complexity of 6G networks. Threats like device illegal access, device spoofing, and data breaches can endanger the integrity of immersive applications and jeopardize user privacy and safety.

Example:

A smart home AR interface, powered by 6G technology and IoT devices, provides homeowners with an innovative way to control and monitor their interconnected smart devices, such as thermostats and cameras. With AR, users can flawlessly interact with their smart devices in real time. However, this advanced functionality depends heavily on the security of the 6G network. If the network is not safe, attackers can intercept AR commands or manipulate data generated from IoT devices, leading to unauthorized data breaches or access. To solve this vulnerability, assimilating blockchain-based validation into the smart devices can guarantee that only authenticated commands are implemented. These techniques enhance trust and deliver a robust layer of security, providing smart homes with reliable and safer.

The goal of this chapter is to provide an introductory understanding of security issues related to 6G IoT convergence and the innovative solutions that resolve them.

The remainder of the chapter is organized as follows: Section 3.2 discusses the main security challenges in 6G IoT convergence, focusing on key threats and vulnerabilities in this domain. Section 3.3 illustrates the role of AI in enhancing security, concentrating on its applications in mitigation, predictive analysis, and threat detection. Section 3.4 explores the integration of blockchain and quantum cryptography as fundamental approaches to strengthen 6G IoT security. Section 3.5 discusses case studies focusing on the application of security frameworks in 6G IoT immersive technologies. Section 3.6 presents the experimental setup and results, focusing on the practical implementation and evaluating the performance of the proposed solutions. Section 3.7 discusses future research directions and emerging trends in 6G IoT security, offering insights into the potential evolution of the field. Lastly, Section 3.8 concludes with a brief summary of the findings and their implications for advancing secure 6G IoT ecosystems.

3.2 SECURITY CHALLENGES IN 6G IOT CONVERGENCE

The integration of 6G technology with the IoT presents revolutionary opportunities for connectivity, speed, and efficiency. However, this convergence introduces critical security challenges that demand attention. Below is a detailed exploration of these challenges:

3.2.1 Larger Attack Surface

With 6G, the range of connected devices within the IoT will also increase and include not only smart homes, healthcare, industrial automation, and urban infrastructure. This is because, with each new device that is connected to the network, the risk of attack also increases due to the enlarged attack surface. Each device is a possible point of entry for threats, hence enlarging the risks [6]. For example:

- **Challenges in Management:** This includes the large number of devices, which makes it difficult to track and manage security.
- **Device Heterogeneity:** Most of the IoT devices are developed by different vendors and have different security measures in place, thus creating vulnerabilities in the system.
- **Shadow IoT:** Default or unauthorized devices are devices that can be used by attackers as a point of entry.

This can be achieved by ensuring that there is a single security protocol and implementing regular monitoring to detect and address threats

3.2.2 Data Privacy Concerns

6G IoT systems can deal with sensitive information like personal health records, financial transactions, and real-time location information. The next generation 6G networks and the services that come with it, such as AR/VR and the tactile internet, amplify the privacy risks because of [7]:

- **Data Collection Intensity:** Many smart devices and systems incorporate sophisticated sensors and real-time processing to gather large quantities of data [8].
- **Cross-Border Data Transfers:** This is because the use of IoT networks is cross-border, which creates issues with data privacy laws.
- **Potential for Misuse:** If proper measures are not taken, sensitive information can be accessed, altered, or used in the wrong manner without the victim's knowledge.

To mitigate the aforementioned risks, there is a need to put in place strong privacy-preserving techniques like differential secure and privacy multi-party computation. Also, it is important to follow the set international privacy policies, such as General Data Protection Regulation (GDPR) or Health Insurance Portability and Accountability Act (HIPAA).

3.2.3 Network Vulnerabilities

The main pillar of 6G IoT systems is ultra-reliable low-latency communication (URLLC), and thus, network vulnerabilities are crucial. Some of the major issues of concern include:

- **Distributed Denial of Service (DDoS) Attacks:** Some of the threats that can be levelled against IoT networks include DDoS, where the attacks flood the networks and therefore cause havoc.
- **Spoofing and Eavesdropping:** It is a common tactic where the intruder poses as another device or snoops on the data being exchanged to get into the network.
- **Routing Attacks:** Alteration of the data routing can divert the traffic to any desired location where the attacker intends to cause harm or congestion.

Some of the mitigation measures that can be put in place include the use of efficient firewalls, AI in threat identification, and network slicing in the management of the network to avoid interference, especially on the key nodes [9].

3.2.4 Encryption Needs

Encryption is crucial for the security of communication in 6G IoT systems. This is because there will be high data throughput and a large amount of data to be transmitted, and, therefore, encryption has to be both secure and fast [10]. Some of the important factors that need to be taken into account include

- **End-to-End Encryption (E2EE):** Ensures data integrity and confidentiality throughout its journey, from one device to another.
- **Post-Quantum Cryptography:** This is done with a view to the future where quantum computing will be dominant and may render current forms of encryption ineffective.

- **Lightweight Encryption:** This is because most IoT devices have constraints in processing power, and hence encryption algorithms used must be secure yet efficient [11].

Therefore, it is important to ensure that encryption standards are consistent throughout the 6G IoT network to guarantee the security of information exchanged.

3.3 AI FOR ENHANCED SECURITY

Due to its capability of processing huge data, learning new threats, and predicting malicious activities, AI has been integrated into security frameworks as a vital component. Here are the specific ways through which AI enhances security:

3.3.1 AI-Driven Threat Detection

AI is particularly efficient in detecting anomalies and thus preventing security incidents with the help of data processing and analysis of large datasets in real time.

- **Anomaly Detection:** In this case, AI technologies incorporate machine learning (ML) to set baselines for the normal activity of networks or devices. Such deviations are potential threats and are, therefore, highlighted.
- **Early Breach Identification:** AI is capable of identifying some of the subtle cues that are associated with cyberattacks, for instance, unusual login points or high levels of traffic that are not normal, thus enabling quick action.

This proactive approach also minimizes the time it takes to identify and mitigate threats, which would otherwise be taken through the traditional approach.

3.3.2 Predictive Analytics in Security

AI applies predictive analytics to predict future security threats by analysing past events and current events data.

- **Threat Forecasting:** As such, AI is capable of recognizing the cyberattack behaviours that create similarities with past incidents and, therefore, can foresee threats that are only in the making [12].
- **Proactive Defence:** Predictive analytics allows organizations to implement countermeasures well in advance, thus enhancing their protection mechanisms.

This capability is especially critical for combating advanced persistent threats (APTs) and zero-day vulnerabilities

3.3.3 Adaptability of AI in Dynamic Environments

Because AI systems are dynamic and progressive, they are very useful in the ever-changing and complicated security environment.

- **Self-Improvement:** AI revises its algorithms according to the new dataset it encounters and, therefore, enhances its probability of identifying hitherto unseen threats.
- **Context Awareness:** It can therefore modify its replies according to variations in the network environment, user activities, or even the nature of the threats [13].
- **Scalability:** AI's flexibility makes it possible to accommodate large data sets and multiple threat profiles in real time.

This is the reason why AI is crucial to present-day cybersecurity solutions that have to deal with constantly appearing threats.

3.4 BLOCKCHAIN AND QUANTUM CRYPTOGRAPHY IN 6G IOT

In this chapter, the challenges of 6G IoT systems are discussed. Integration of blockchain technologies and new quantum cryptography methods has been explored as a potential solution for securing information and ensuring the security and trustworthiness of data in a connected and high-bandwidth environment.

3.4.1 Blockchain for IoT Security

Because the data stored in the blockchain cannot be altered, the blockchain provides an excellent solution for increasing the security of the IoT [14].

- **Decentralized Structure**: Blockchain gets rid of the central point of failure by having multiple copies of data stored across different nodes in a network to avoid data corruption or fraud.
- **Data Authenticity:** Every transaction or data in the blockchain is encrypted and thus is not able to be altered, making it authentic.
- **Smart Contracts:** The use of blockchain provides the added advantage of device authentication and ensures that only authorized parties transact through automated agreements.

With the use of blockchain, 6G IoT systems can be assured of reliable, clear, and undoubted operations.

3.4.2 Quantum Cryptography for Future Security

Quantum cryptography presents the latest forms of encryption that are difficult to break even by quantum computers, which is a future risk to current cryptographic systems [15].

- **Quantum Key Distribution (QKD):** QKD guarantees the secure exchange of keys as a result of the use of quantum mechanics and the principles based on it, where any attempt made to intercept the key would affect its state and raise an alarm to the users about possible intrusion.

- **Quantum-Safe Algorithms:** These are types of algorithms that are developed in such a way that they cannot be hacked by using the quantum computers that might be in use in the future.
- **Future-Proofing Security:** By implementing quantum cryptography with the 6G IoT systems, it is possible to safeguard the data from upcoming threats.

These technologies are crucial changes in creating trust and the ability to adapt to threats that are being seen today in cyberspace.

3.5 CASE STUDIES: 6G IOT SECURITY IN IMMERSIVE TECHNOLOGIES

3.5.1 Healthcare Application

AR has introduced significant transformations to the healthcare sector, especially in the areas of remote surgeries and diagnostics. By superimposing digital graphics onto actual environments, AR enables surgeons to access real-time, immersive three-dimensional medical data, enhancing the accuracy and efficiency of procedures and operations. Devices such as AR-enabled smart glasses and headsets, like Microsoft HoloLens, allow surgeons to see the anatomy of the patient, for example, organs, tissues, and blood arteries, as if they were right in front of them [16].

Additionally, AR also connects geographical distances and facilitates remote surgeries. In this instance, a surgeon located in one nation can help or perform procedures or operations using robotic arms and AR technology while applying 3D overlays and live video feeds. For example, during crucial surgeries, AR can overlay Magnetic Resonance Imaging (MRI) or Computed Tomography (CT) scan data onto a patient's body, providing the surgeon with extremely accurate guidance.

As shown in Figure 3.3, there are six main real-life applications of AR in healthcare: surgery visualization and assistance, patient care and rehabilitation, vein visualization, medical imaging, symptom detection, and patient self-care.

To complete such necessary activities, 6G networks are essential. In contrast to 5G, 6G delivers ultra-high bandwidth (up to 1 Tbps) and ultra-low latency (less than 1 ms), guaranteeing uninterrupted real-time communication between AR systems and distant doctors without any delay. For instance, during surgery performed remotely, even a microsecond of latency can jeopardize patient safety. With 6G, AR tools can easily deliver complex medical images and real-time biometric information collected by IoT devices such as connected surgical instruments, patient monitors, and wearable health trackers [17].

IoT improved AR's functionality by continually gathering patient vitals like heart rate, blood pressure, and oxygen levels, which are sent to the surgeon in real time. This assures that even during high-risk surgeries and remote procedures, surgeons are kept completely informed, enabling swift intervention in the case that patient conditions worsen.

As AR-driven surgeries and diagnostics depend significantly on sensitive medical data, its security must be guaranteed. Any breach of live operational data could have disastrous outcomes, including patient injury and data theft. To counter these

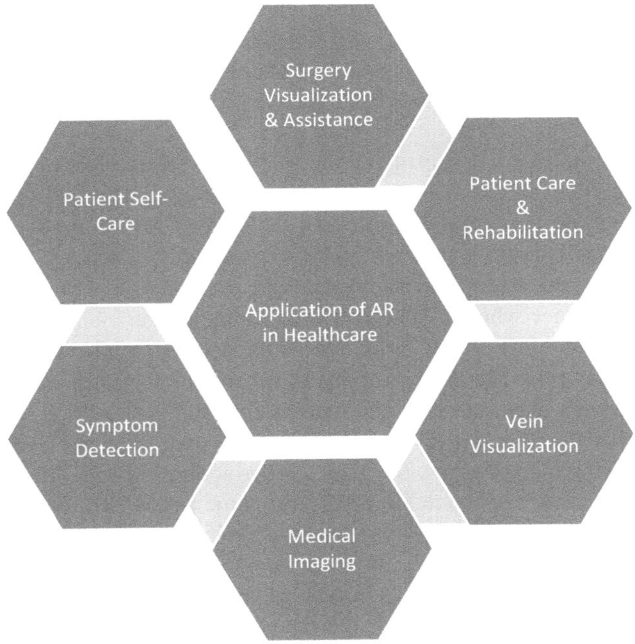

FIGURE 3.3 Real-life applications of AR in healthcare.

hazards, several security measures are necessary, such as E2EE, multi-factor authentication (MFA), blockchain technology, and AI-driven monitoring systems.

For instance, remote surgery helped by AR can use 6G-enabled encryption techniques safely to send images from IoT equipment (like robotic surgical tools) to surgeons' AR displays. Additionally, AI algorithms monitor data flow to spot anomalies or cyberattacks.

In conclusion, the convergence of AR, IoT, and 6G will revolutionize healthcare delivery, especially in the areas of remote surgeries and diagnostics. However, giving priority to data security and authentication is indispensable to guarantee patient safety and preserve confidence in these innovative technologies.

3.5.2 Entertainment and Gaming

The entertainment and gaming sectors have seen enormous innovation with the arrival of VR and IoT technologies. VR-based gaming platforms offer completely immersive digital experiences, allowing players to engage with incredibly lifelike settings. IoT devices like haptic suits, motion controllers, and VR headsets capture user movements, gestures, and biometrics, allowing players to feel connected to virtual worlds in real time.

While 5G networks offer greater data speeds, 6G networks are set to completely transform VR gaming by providing speeds up to 1 Tbps and latency under 1 ms.

This makes it possible for VR environments to react to human actions instantaneously and produce photorealistic images. Multiplayer VR games, which frequently suffer from latency because of their large data requirements, will be extremely beneficial from 6G's ability to process enormous volumes of data in real time [18].

IoT devices improve VR gaming by facilitating a variety of physical interactions. Wearable equipment such as haptic gloves replicates physical feelings, motion sensors monitor body movement, and eye-tracking technology allows realistic interactions with VR items. IoT devices send user data like gestures, reactions, and movements over the 6G network to synchronize games in real time. This results in a highly immersive experience where virtual and real worlds merge.

For instance, in a VR racing game, IoT-enabled steering wheels and haptic feedback devices may replicate the vibrations of automobile engines, while 6G guarantees that all players play in unison and without interruption.

A game like Fortnite with VR capabilities uses IoT devices to monitor player movements. Using 6G networks, the game can provide a lag-free, lifelike experience for thousands of players at once. To protect in-game assets like digital skins, blockchain technology guarantees secure ownership verification and safe transactions.

In conclusion, the convergence of 6G networks and IoT devices is becoming an incredibly realistic, real-time experience. However, strong security measures must be put in place to preserve user privacy, guarantee fair gameplay, and secure priceless digital assets.

In Figure 3.4, it is shown that 6G serves as the basis for networking and data transfer. IoT devices gather user input, including movements, gestures, and feedback, and send this data to the VR gaming system. It processes input from IoT devices such as movement and gestures and creates a realistic virtual world. The collaboration of 6G networks and IoT devices provides smooth, real-time communication.

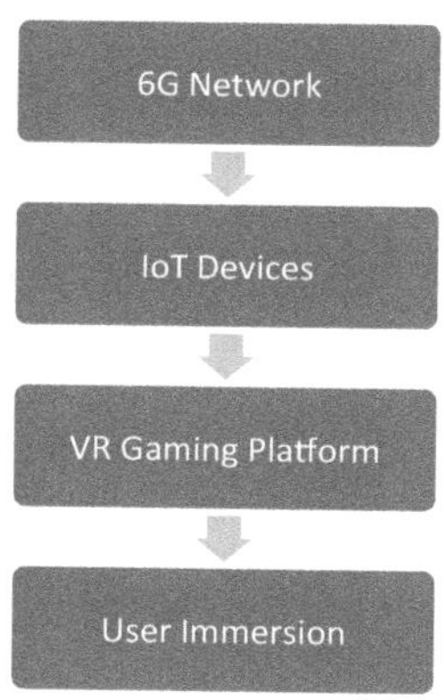

FIGURE 3.4 Role of 6G and IoT in enabling VR-based gaming experiences.

3.6 EXPERIMENTATION

Objective: To evaluate the effectiveness of AI-based anomaly detection systems in securing IoT devices within a 6G-enabled immersive environment, mainly targeting VR applications. The experiment compares AI-improved security with a non-AI setting to demonstrate how AI can avoid malicious activities like DDoS attacks.

3.6.1 Experiment Design

Setup:

i. **Environment:** A simulated 6G-based IoT network involving multiple IoT devices (like cameras, sensors, and VR headsets) interacting within a VR application.
ii. **Security Configurations: Two Setups Are Prepared:**
   a. **AI-Secured Setup:** Uses an AI-based anomaly detection system capable of identifying unusual traffic patterns and responding in real time.
   b. **Non-Secured Setup:** No AI-based security mechanism, depending on basic firewalls and encryption.

Procedure:

i. **Phase 1: Normal Traffic Monitoring:**
   a. Generate standard data exchange among IoT devices in both setups, simulating regular user interaction in the VR environment.
   b. Monitor and log baseline metrics such as data throughput, network activity, and response times for 10 minutes.
ii. **Phase 2: Malicious Attack Simulation:**
   a. Introduce simulated DDoS attacks into both setups by flooding the network with malicious traffic targeting the IoT devices.
   b. Measures how both setups respond to this abnormal activity in terms of system downtime, network latency, and detection time.

Metrics:

i. **Detection Time:** The time taken by the AI system to detect the DDoS attack in the secured setup compared to the reaction time of the non-secured setup.
ii. **Network Latency:** Changes in the system's response time during the attack, particularly focusing on the performance degradation in the VR environment.
iii. **System Downtime:** The time for which the system is disrupted by the DDoS attack before normal functionality is restored.
iv. **False Positives:** Instances where the AI misclassifies legitimate traffic as malicious.

TABLE 3.1
Output of the Experiment

Metric	AI-Secured Setup	Non-Secured Setup
Detection time	~1–2 seconds	No detection (manual intervention)
Network latency increase	10%	50%
System downtime	5 seconds (recovery)	1–2 minutes
False positives	2%	Not applicable

Output:

Results:

i. **Detection Time**: The AI-driven setup detects the DDoS attack within 1–2 seconds, significantly faster than the non-secured setup, which fails to automatically detect the attack without manual intervention, as shown in Table 3.1.

ii. **Network Latency**: In the AI-secured setup, network latency increases by about 10% during the attack but quickly returns to normal after the anomaly is mitigated. In contrast, the non-secured setup experiences a much higher increase in latency (~50%), causing notable disruptions in the VR application's user experience.

iii. **System Downtime**: The AI-secured system can mitigate the DDoS attack within 5 seconds, whereas the non-secured setup requires up to 2 minutes of manual intervention, causing prolonged service disruptions.

iv. **False Positives**: The AI setup produces a 2% false positive rate, meaning a small fraction of legitimate traffic was flagged as malicious. This can be fine-tuned by improving the AI algorithms.

Conclusion:

The experiment demonstrates that AI-enhanced security provides real-time detection and response, mitigating malicious attacks far more efficiently than traditional security approaches in a 6G-IoT immersive environment. The results suggest that AI-driven systems offer faster response times, lower system downtime, and better overall protection for sensitive applications like VR.

The findings emphasize the critical role AI plays in securing future 6G-IoT applications, particularly where immersive technologies are involved, and highlight the need for continuous refinement to minimize false positives while maintaining robust security.

3.7 FUTURE RESEARCH AND EMERGING TRENDS

Zero Trust Architecture (ZTA) is a cybersecurity paradigm that was built to replace weak old security techniques, which were assumed to be safe but were actually not fully safe. In ZTA, no one can be trusted, not even a person or a device from that network. If any person or any device needs access to the network, first, they have to get checked and verified, and every time they need access, they have to follow this process of checking and verification. All these plans can lead to minimizing the risk of attacks by hackers on the system.

ZTA gives the user and the devices limited access to the system, which means that it only gives access to the location where they have to work, and at the same time, they continuously keep an eye on their activity and also recheck their identity. By using ZTA, organizations can increase the security of the data, which is very important and rarely used, by regularly checking. It is very important for the new and upcoming technologies because users and devices have to communicate.

MFA is very important to make new and upcoming platforms and networks more secure. In MFA, users have to give more proof to identify and log in because logging in using just a username and password can be risky, so to make it more secure, we have to give more proof. Usually, MFA uses a password, a phone, a security token, and lastly a face or fingerprint. Because of this, if one of the login security methods gets hacked, to get access, hackers need to hack other login security methods.

In the setting of immersive platforms such as AR, VR, and IoT-enabled surroundings, MFA assists in protecting private user data, proprietary information, and device access from cyberattacks. These platforms frequently manage substantial amounts of private and confidential data, making them appealing targets for attackers. By using MFA, organizations may improve user identity verification, drastically lowering the risk of credential-based incidents, and offering secure user experiences. For instance, a corporate AR application that facilitates remote collaboration may require them to authenticate using a smartphone app or biometric verification before accessing MFA.

Edge and fog computing are innovative technologies that bring computing and security features within the reach of IoT devices, optimizing both performance and safety. Edge computing analyses information at or close to the point of data generation, like IoT sensors or smart devices, lowering latency and speeding up reaction times. Fog computing works by dividing the whole work into two parts: The first one is the cloud, which is the central server, and the second one is edge devices like sensors. This technology is very important in many fields, mainly in IoT, because of this, many devices can create and share data very easily across the world. It processes the data near the place where it was created, which reduces the risk of hacking and spying, as they don't have to send it far away. Security processes like firstly identifying and then giving limited control access, and also keeping an eye on unusual activity, keep the data more secure.

Edge and fog computing allow the system to handle all the tasks and devices easily by making the IoT system more flexible and efficient. There is a feature in industrial settings in which edge devices can easily and quickly detect cyberattacks and problems and quickly reply if any issue is found, without waiting for a central contract. All these steps can lead to a decrease in the chance of getting hacked and ensure that all tasks perform smoothly, in conditions where there is no network. To strengthen the protection of IoT devices, some security measures can be used, such as MFA and ZTA. These technologies collectively can make a system very secure.

Decentralized identity management is becoming important and growing quickly in this cybersecurity field. These systems allow people and devices to manage their credentials on their own without relying on central organizations. These technologies, like blockchain, are used to make data unchangeable, transparent, and secure. These are mainly used in the IoT platforms because many devices and users need to connect securely.

AI is changing the default setting of the company to defend its systems and networks from security breaches. AI uses many techniques, such as ML and real-time data, to spot patterns, detect all unusual activities, and find all the threats by predicting them before they occur. With the help of this, organizations can quickly protect their data from hackers with very little harm.

3.8 CONCLUSION

The combination of IoT and 6G is opening many immersive technologies, such as AR, VR, and XR. 6G network provides super-fast speeds, which decreases the delay in uploading and downloading the data and increases the ability to connect many devices, which leads to non-stop interactions in many fields like healthcare and entertainment. However, there is a drawback that these advanced technologies bring new risks, such as giving more way for hackers to hack the system, protection of personal data, and unreliable networks that can weaken security and trust in the system.

This chapter discusses how new technologies like AI, blockchain, and quantum cryptography can help in solving many modern security challenges. AI needs to find any unusual behaviour, predict all the potential threats, and automatically change security situations to stop attacks like DDoS. Blockchain takes care that data is accurate and trustworthy, while quantum cryptography uses advanced encryption that cannot be broken by new techniques.

The experimental results in this chapter further validate AI's ability to protect 6G-IoT systems. By matching AI-driven security configurations to conventional methods, the study shows notable improvements in discovery time, decreased system outages, and improved resilience against fraudulent activities in immersive environments. Facing minor issues, like false positives, the results confirm AI's ability to provide proactive, efficient, and dependable security solutions.

Looking ahead, future developments such as ZTA, MFA, and edge and fog computing have the potential to significantly transform paradigms further. ZTA makes sure that every organization within the network is thoroughly verified, while edge and fog computing bring intellectual and security features to more IoT devices, improving both efficiency and safety.

ACKNOWLEDGEMENT

The Deanship of Scientific Research and the Vice Presidency granted support for this Graduate Studies and Scientific Research, King Faisal University, Saudi Arabia, under the project (KFU250175).

REFERENCES

1. A. M. Al-Ansi, M, Jaboob, A. Garad, and A. Al-Ansi, "Analyzing augmented reality (AR) and virtual reality (VR) recent development in education," *Social Sciences and Humanities Open*, vol. 8, no. 1, 2023, doi: 10.1016/j.ssaho.2023.100532.
2. K. G. Nalbant, "The importance of artificial intelligence in education: A short review," *Journal of Review in Science and Engineering*, vol. 15, p. 2106302112361, 2021.

3. A. Ambika, H. Shin, and V. Jain, "Immersive technologies and consumer behavior: A systematic review of two decades of research," *Australian Journal of Management*, 2023, doi: 10.1177/03128962231181429.
4. R. E. Petruse, V. Grecu, M. Gakić, J. M. Gutierrez, and D. Mara, "Exploring the efficacy of mixed reality versus traditional methods in higher education: A comparative study," *Applied Sciences*, vol. 14, no. 3, 2024, doi: 10.3390/app14031050.
5. H. Sabireen and V. Neelanarayanan, "A review on fog computing: Architecture, fog with IoT, algorithms and research challenges," *ICT Express*, vol. 7, no. 2, pp. 162–176, 2021, doi: 10.1016/j.icte.2021.05.004.
6. V. L. Nguyen, P. C. Lin, B. C. Cheng, R. H. Hwang, and Y. D. Lin, "Security and privacy for 6G: A survey on prospective technologies and challenges," *IEEE Communications Surveys and Tutorials*, vol. 23, no. 4, pp. 2384–2428, 2021, doi: 10.1109/COMST.2021.3108618.
7. A. Pavithra, J. Kowsalya, S. K. Priya, G. Jayasree, and T. K. Nandhini, "An emerging immersive technology-a survey," *International Journal of Innovative Research and Growth*, vol. 6, no. 8, pp. 119–130, 2020.
8. L. Mohjazi, B. Selim, M. Tatipamula, and M. A. Imran, "The journey toward 6G: A digital and societal revolution in the making," *IEEE Internet of Things Magazine*, vol. 7, no. 2, pp. 119–128, 2024, doi: 10.1109/IOTM.001.2300119.
9. J. Karande and S. Joshi, "Real-time detection of cyber attacks on the IoT devices," *2020 11th International Conference on Computing, Communication and Networking Technologies, ICCCNT 2020*, 2020, doi: 10.1109/ICCCNT49239.2020.9225487.
10. I. Ara and B. Kelley, "Secure and ultra-reliable 6G IoT algorithms with AI-enhanced shared key physical layer security," *2024 IEEE 15th Annual Ubiquitous Computing, Electronics & Mobile Communication Conference (UEMCON)*, pp. 148–157, 2024.
11. P. Deepanramkumar and A. Helensharmila, "AI-enhanced quantum-secured IoT communication framework for 6G cognitive radio networks," *IEEE Access*, vol. 12, August, 2024, doi: 10.1109/ACCESS.2024.3471711.
12. R. Kaur, D. Gabrijelčič, and T. Klobučar, "Artificial intelligence for cybersecurity: Literature review and future research directions," *Information Fusion*, vol. 97, March, 2023, doi: 10.1016/j.inffus.2023.101804.
13. L. Devane, "Adaptive AI systems in autonomous environments: Real-time decision making adaptive AI systems in autonomous environments: Real-time decision making and resource allocation through cloud-based reinforcement learning," *ResearchGate Preprint*, October, 2024, doi: 10.13140/RG.2.2.21638.18241.
14. R. Kaur and A. Ali, "Implementation of blockchain in IoT," *Lecture Notes in Electrical Engineering*, vol. 841, pp. 149–161, 2022, doi: 10.1007/978-981-16-8774-7_13.
15. M. An et al., "Blockchain technology research and application: A literature review and future trends," *Journal of Data Science and Intelligent Systems*, pp. 1–13, 2023, doi: 10.47852/bonviewjdsis32021403.
16. D. C. Nguyen et al., "6G Internet of Things: A comprehensive survey," *IEEE Internet of Things Journal*, vol. 9, no. 1, pp. 359–383, 2022, doi: 10.1109/JIOT.2021.3103320.
17. Z. Qadir, K. N. Le, N. Saeed, and H. S. Munawar, "Towards 6G Internet of Things: Recent advances, use cases, and open challenges," *ICT Express*, vol. 9, no. 3, pp. 296–312, 2023, doi: 10.1016/j.icte.2022.06.006.
18. X. Shen et al., "Toward immersive communications in 6G," *Frontiers in Computer Science*, vol. 4, 2023, doi: 10.3389/fcomp.2022.1068478.

4 Machine and Deep Learning-Based Anomaly Detection in 6G Smart Cities

Ketan Sarvakar, Dolly R. Prajapati, and Francisca Rosique Contreras

4.1 INTRODUCTION

The integration of 6G networks with smart city (SC) construction holds the potential to revolutionize urban living by providing unparalleled speeds, low latency, and wide-ranging device connectivity, all through data-driven management. In general, these developments can improve living conditions. Improving public services and allocating resources appropriately, but as IoT devices and sensor networks expand, it has become increasingly difficult to identify and control anomalies that could harm critical infrastructure and services. To ensure the efficiency, safety, and reliability of operations in 6G-enabled SCs, this chapter highlights the important role that DL and ML techniques play in testing and focuses on their applications in anomaly detection.

4.1.1 Overview of 6G and SCs

Featuring unrivaled data rates, ultra-low latency, and seamless connectivity, 6G networks are poised to transform SCs and enable real-time data processing, sophisticated IoT features, and artificial intelligence (AI)-driven apps. Service improvements include urban infrastructure management and sustainability, which benefits public safety, energy, healthcare, and transportation sectors. Ultimately, 6G will make cities more connected, more efficient, more productive, and more sustainable, with large-scale IoT deployments and automation. These will promote resilient urban environments and innovation through technologies like remote healthcare, autonomous vehicles, and real-time environmental monitoring.

Figure 4.1 illustrates the rapid evolution of wireless communication. From 1G to 5G, there were improvements in speed and services such as texting and mobile internet. However, the current generation seeks to respond almost instantly to emerging technology. The upcoming 6G network promises even more capabilities.

DOI: 10.1201/9781003586456-4

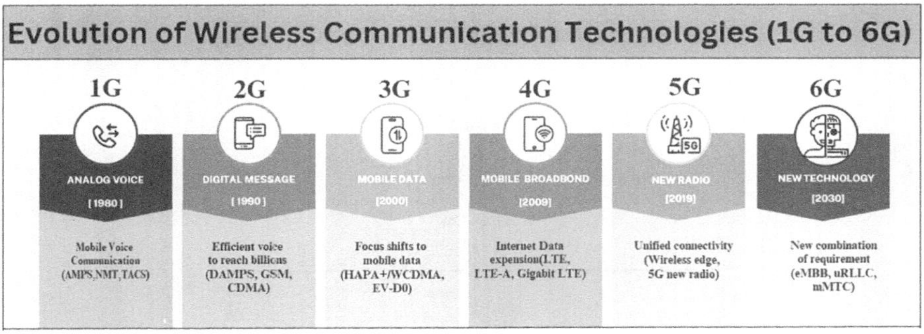

FIGURE 4.1 Evolution of wireless communication technologies.

Figure 4.2 shows how an SC incorporates technology into public services, infrastructure, transportation, and other urban areas. Real-time monitoring and efficiency are improved through the use of sensors and data processing.

4.1.2 Importance of Anomaly Detection

The size, complexity, and importance of urban infrastructure and services make anomaly detection important in the context of 6G-enabled SCs. SCs can take advantage of the unmatched speed and ultra-low latency of 6G networks, and widespread connectivity can improve operational efficiency and create safety and sustainability. But the development of interconnected IoT devices and sensor networks has also made them more prone to malfunctions. These problems are increasingly being addressed with ML and DL algorithms that can find them automatically. Anomalies in the real-time database may indicate abnormal environmental conditions, system malfunctions, or cyberattacks. ML- and DL-based anomaly detection systems help cities reduce risks, prevent potential disruptions, and provide consistent service by enabling rapid detection and response.

ML and DL models enable predictive analytics, helping cities identify future trends and abnormalities based on past data. This proactive approach helps optimize infrastructure management, resource allocation, and decision-making. Anomaly detection can also be used in the transportation sector to optimize traffic flow, predict equipment damage in mass transit systems, or identify congestion patterns. The healthcare industry can monitor patient data for early signs of declining health. This guarantees faster processing. Powerful anomaly detection uses cutting-edge AI technology to protect critical infrastructure, increase operational efficiency, and improve the general quality of life of the city. Essentially supporting the resiliency and sustainability of 6G SCs, integrating ML- and DL-based anomaly detection systems, SC the difficult barriers will be necessary for revolutionary social benefits to be eliminated and for capabilities to fully use 6G technology.

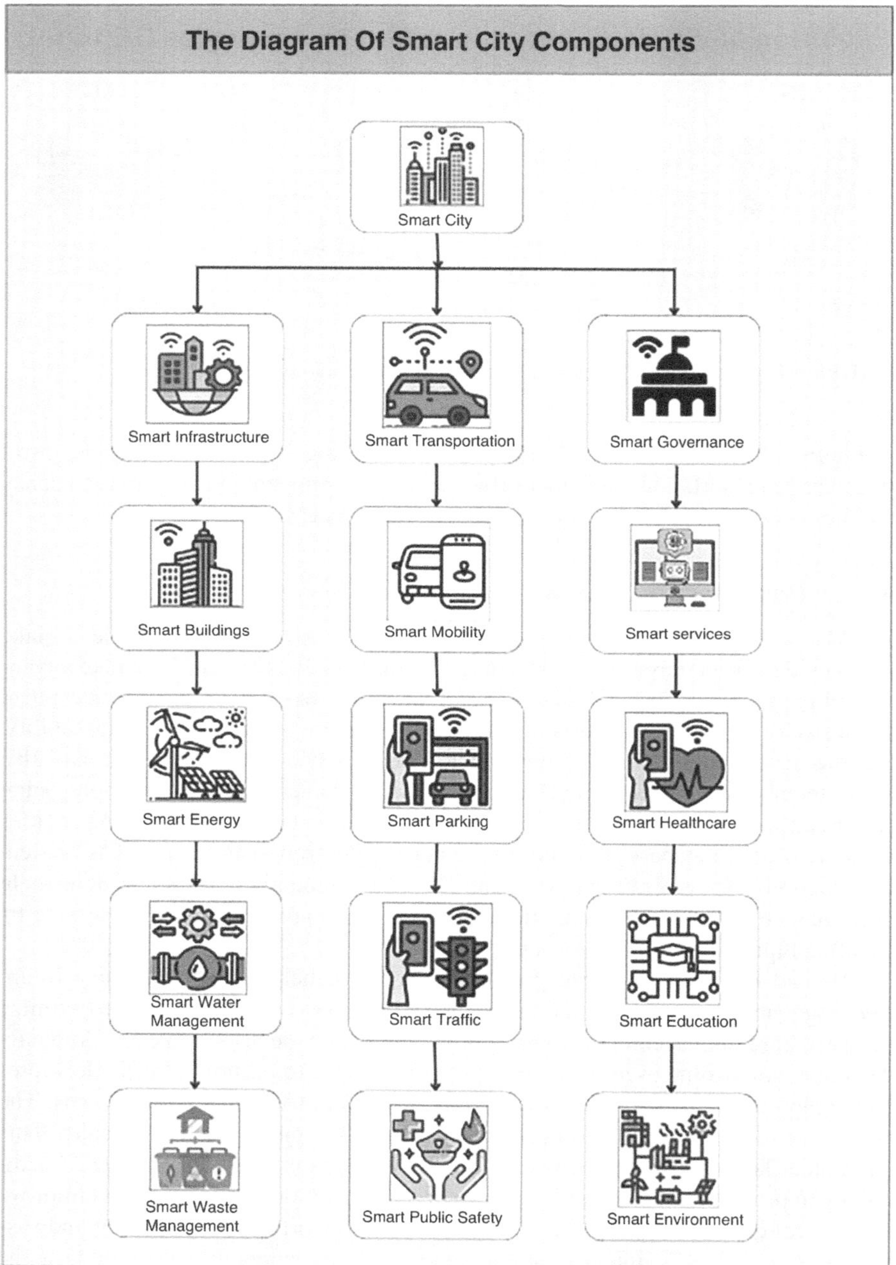

FIGURE 4.2 Diagram of a 6G-enabled SC.

4.1.3 Role of ML and DL in Anomaly Detection

Anomaly detection in 6G SCs relies heavily on ML and DL, which use complex algorithms to examine large and complex databases and identify anomalies in the behavior pattern during the search. Unsupervised ML methods are used for the anomaly detection task. Examples of these algorithms include decision trees, nearest neighbors (k-NN), and support vector machines (SVM), while those of unsupervised learning algorithms include methods based on density and organization. Group anomalies are identified without any prior training data. This makes it useful for exploring. Supervised learning algorithms identify and label anomalies based on patterns learned from learning data. DL techniques include Convolutional Neural Networks (CNNs), Recurrent Neural Networks (RNNs), and autoencoders. Anomalies can be detected by extracting complex features and creating hierarchical representations from large data sets.

CNN specializes in examining geographic data. This includes visually distinguishing it from systems aimed at SCs. However, based on models such as long short-term memory (LSTM) networks and RNNs, data can be analyzed sequentially to detect anomalies in serial data, such as time-series data from IoT sensors and vice versa. Together, they facilitate the identification of abnormalities because they can imitate specific patterns and then look for differences that might indicate discrepancies. Real-time monitoring and proactive response to irregularities in various sectors, including public safety, energy management, transportation, and healthcare, are enabled by ML- and DL-based anomaly detection systems in 6G SCs. In the era of 6G technology, ML and DL algorithms make city environments safer, more powerful, and adaptable, as they continually learn and adapt from new data formats. This increases the flexibility and sustainability of SC infrastructure.

Traditional and ML/DL anomaly detection techniques are different in Figure 4.3: Traditional methods only process data and use models. In contrast, ML/DL methods involve more complex processes such as feature extraction and model training. This shows that ML/DL methods are more complex.

4.2 LITERATURE SURVEY

The design of Misbehavior Detection Systems (MDS) is heavily reliant on ML, which is crucial for the security of 5G-V2X networks [1]. This study analyzes and categorizes different ML-based MDSs for 5G-V2X and performs a security and ML analysis on them [1]. It also addresses open research and standards problems that are essential for creating successful ML-based MDSs, as well as lessons learned.

Urban life and work are set to be revolutionized by 6G technologies and SCs; however, integrating these technologies presents several issues, including infrastructure investment, data protection, interoperability [2], and the digital divide. Other challenges include seamless connection between devices, security management of large amounts of sensitive data, and past adoption and access barriers. These issues are central to SCs. Similar obstacles include spectrum availability, high costs, security concerns, and regulatory barriers. Addressing the 6G problem will require significant infrastructure upgrades and international cooperation [2]. Despite these obstacles,

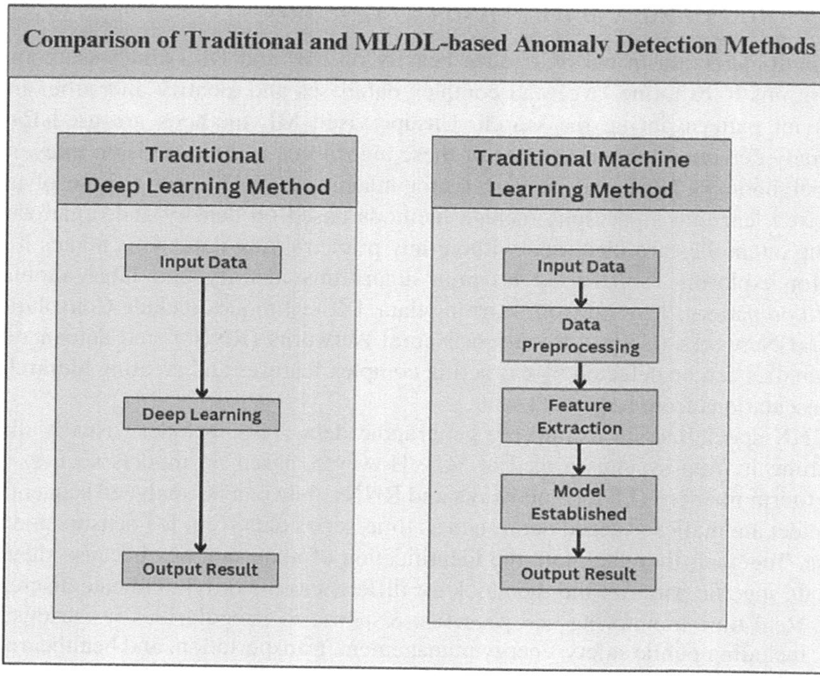

FIGURE 4.3 Traditional and ML/DL-based anomaly.

6G is expected to improve SC through faster data transfer, reduced latency, and increased connectivity, benefiting public safety, health care, power management, and transportation.

Building a 6G network should use new technologies to improve speed, security, and reliability. Traditional model-driven methods are limited by high computational complexity and inaccuracy [3]. This is especially true in difficult situations. For this reason, ML algorithms are important for optimization analysis [3]. These details show that the problem lies in the workload, specifically how to secure communication spectrum and energy efficiency through reinforcement learning (RL), deep reinforcement learning (DRL), ML, and DL algorithms. However, further studies are needed to effectively use these devices for 6G improvements, such as increasing network coverage and reducing the integration latency of ground technology. This study aims to stimulate continued discussion and creativity in this important area.

Through the introduction of new applications, the improvement of technological experiences, and the promotion of economic growth, 6G networks will improve global communication services [4]. This overview critically evaluates current 6G advancements, solutions, and problems, going over its goals of widespread coverage, a wide range of spectral bands, appealing applications, ubiquitous digital intelligence, and robust security. The survey explores key performance indicators (KPIs) for different usage scenarios [4], focusing on technological advancements and business performance. It reviews the progress of 6G trials and identifies key research challenges. The survey highlights the need for continued research to unlock 6G's full potential and promote continuous innovation in this sector.

AI is expected to be a fundamental aspect of 6G capabilities, rather than an optional component. This research reviews the risks and protections of federated learning (FL) in 6G-enabled IoT [5]. Drawing attention to the fact that the established edge learning models, training at the edge of the network via FL is important. However, they can be a security risk if they are not properly protected. This chapter evaluates existing security strategies and divides them into eight threat categories, adversaries, and covert attacks [5]. This chapter concludes by summarizing current research problems in 6G-enabled IoT security and the need for further study of new threats and prevention techniques.

Concerns about privacy and communication costs in present and future mobile networks must be addressed using FL [6]. By contrasting different ML architectures and looking at FL principles, methods, and frameworks, this research investigates the function of FL in wireless communication networks [6]. KPIs and communication across protocol stacks have been improved with recent advancements in Florida. It highlights the huge potential and impact of FL on the future development of mobile networks, as well as the difficulties in exploring the adoption of FL in 5G, 6G, etc.

Five major study areas for next-generation wireless technologies are examined in this article: DL applications in 6G networks, architecture, spectrum, networking, and IoT [7]. It draws attention to the transition in multi-component optimization from maximizing bandwidth efficiency to balancing other goals, such as delay, power efficiency, and complexity. The study emphasizes the need for Pareto optimization to deal with trade-offs between different indicators, including latency, processing volume, and bit error rate [7]. It also covers how complex ML methods, such as RL and FL, can be used and how they help achieve these complex optimizations. To answer open research questions and advance multi-component system optimization in next-generation wireless technology, this chapter concludes with a call for cooperation.

Enhancing interpretability in DL systems, especially for 6G communications, is a critical challenge. Meng et al. introduce Metis, a framework that improves interpretability by using decision trees and hypergraphs to transform DL policies into rule-based controllers and identify crucial components [8]. METIS has shown its effectiveness in facilitating the understanding of complex DL models with minimal loss of performance, in addition to causal ML [8], which focuses more on causality. This correlation alone offers a promising approach to solving the interpretability problem in the development of wireless communication systems.

This study develops a novel approach utilizing hybrid techniques from Rough Set (RS) and SVM algorithms to overcome the inadequacies of current Intrusion Detection Systems (IDSs) in wireless scenarios [9]. The strategy uses a Correlation-based Feature Selection (CFS) technique to optimize feature selection and manage high-dimensional, imbalanced traffic with lower detection rates. The approach outperformed existing classification algorithms, achieving an excellent accuracy of 99.9% with a low false alert rate of 0.00102 when tested on the CICDDOS2019 dataset [9]. In the IDS market, this creative approach provides a considerable competitive advantage.

This is because next-generation networks are complex and dynamic. Establishing reliable and efficient communication using traditional methods is therefore challenging [10]. This highlights the importance of intelligent routing. 6G networks with ML support are capable of overcoming these challenges [10]. To overcome these obstacles, this study covers several ML-based network routing systems, explores key 6G parameters, and discusses ML techniques to design reliable transmission lines. It explains the difficulties of research in this area and also describes possible future directions.

Improving people's quality of life through the use of cutting-edge technology has always been the goal of cutting-edge web development. 5G will enable intelligent IoT systems and improved services [11]. But the limitations of 5G push the exploration beyond 5G (B5G) and 6G protocols as user devices and bandwidth requirements increase. This study examines the integration of AI/ML in communication systems to enable the future IoT, recognizing its important role in network automation [11]. How are these systems evolving? The importance of AI/ML is related to specific algorithms and current research. This chapter highlights the possibilities and unresolved issues in AI/ML-based IoT communications.

Critical infrastructures will need to be connected by 6G communication networks; hence, these networks must be dependable in terms of availability, safety, and security [12]. The significance of reliability in 6G networks is emphasized in this essay, which also examines current issues, many of which arise because 6G is distributed, and makes recommendations for possible fixes [12]. To improve the reliability of 6G networks, research will focus on edge computing, FL mobile software, and web services. This chapter also points to areas where further research is needed and recommends ways to strengthen the security and latency of 6G networks.

To handle imbalanced, high-dimensional traffic, this study develops a unique ML technique to solve the inefficiency of current IDSs in wireless networks [13]. The technique demonstrated low false alarm rates and great accuracy across numerous datasets using CICDDOS2019 datasets. This approach is better than current algorithms to provide a competitive advantage for IDS, but additional work is needed to increase its accuracy and effectiveness and improve the system's resilience to potential threats [13]. The study also highlights the importance of robust ML models for 6G network security.

This study presents a comprehensive analysis of 6G-enabled IoT, highlighting open challenges and possible directions. A total of 36 key articles from 190 publications (2018–2021) were reviewed using a systematic literature review (SLR), revealing the increasing importance of IoT [14]. Research area studies highlight key 6G technologies for the Internet, including increased bandwidth, complex massive MIMO, and evolving network topologies [14], and also highlight the emergence of new 6G service classes that aim to improve traditional 5G services. The increasing amount of research in this area has made it possible to evaluate each contribution, although a deeper investigation is warranted.

A comparative study of two periods of data collection emphasizes the importance of adaptive techniques to free up resources and enhance network security [15]. Fast threat detection and effective incident response are guaranteed by the system's dynamic detection capabilities [15]. This study highlights the value of using pre-trained models to improve network behavior predictions and recommends continuing education work in the future. The results highlight the need for anomaly detection to balance temporal details and resource consumption. Further studies are expected to improve scalability and performance.

This chapter analyzes 255 published papers on topics such as security, privacy, robustness, wireless networks, etc., to observe the impact of DL on 6G technology in detail. While device-to-machine communication remains personal [16], 6G provides deep intelligence. It also emphasizes the important impact of learning [16]. From the conclusions of this report, DL and AI approaches will become increasingly important in meeting the many 6G needs of people, businesses, and industries.

This research article provides a detailed analysis of 6G technology and how it integrates with other cutting-edge fields such as AI, blockchain, VLC, THz communication, IoT, ML, and security issues, and emphasizes that they can help [17]. A taxonomy of 6G's fundamental technologies and applications is presented, along with an examination of the advantages and disadvantages of these technologies, as well as a look at cutting-edge methods like green energy harvesting [17]. In their conclusion, the authors summarize solutions to 6G issues and indicate that they plan to investigate RL and Q-learning in more detail.

Focusing on AI, this chapter explores the creation of a flexible and self-learning 6G network that facilitates the deployment of dynamic SCs [18], focusing on the technological needs that determine the growth of each generation [18]. The creation of wireless networks in this chronological order also provides an overview. The authors provide a taxonomy of applications related to AI-based 6G network SCs, focusing on self-learning models and SLA and QoS requirements for these applications to create intelligent, interconnected 6G-enabled SC digital ecosystems. Future research directions are covered in the conclusion of this article.

This chapter provides a detailed analysis of ML applications in 6G wireless networks, including state-of-the-art methods such as supervised, unsupervised, deep, reinforcement, and learning [19]. Know the federation car network. This device, network, fog radio access network... The chapter explores how ML can be used to solve problems like reducing workload and managing change, and resource allocation [19]. This article discusses the need for distributed ML systems to deal with resource optimization problems and shows how ML can improve energy efficiency and reduce latency. To automate network operations and ensure smooth end-to-end connectivity, we conclude by exploring the use of ML for intelligent resource management and propose future research priorities in this area.

This chapter presents the ASCA method, also known as Optical IoT, which combines AI, IoT, and advanced optical systems to create sustainable cities [20]. It introduces unique "CNN-LSTM" techniques, including CNNs to increase efficiency and LSTM. This is at the heart of ASCA and improves decision-making and efficiency in SCs [20]. A high-resolution camera and LiDAR scanner are used for real-time data collection [20]. Evaluation with the ICOS dataset shows that ASCA outperforms current Fog-IoT and QoS-IoT techniques, with an accuracy rate of 91.98%. This shows that ASCA has the potential to promote SC development.

This is because of the increasing complexity of these networks and cyberattacks. Therefore, the security of wireless networks supporting B5G and 6G has become increasingly important [21]. As IoT and IoE technologies continue to advance, protecting data and making today's systems more resilient is more important than ever. AI and ML algorithms can greatly improve network security. This improves threat detection, recovery, and identification of potential attack patterns. Attackers, in turn, can use game theory to create more effective attack plans [21]. It is complex and can exploit network vulnerabilities. Additionally, this document recommends network segmentation to identify appropriate threats and examine ML methods to separate authorized users from malicious users. Despite these developments, threats such as data extraction and misunderstandings remain. It continues to influence AI/ML paradigms and models. Continuous training and updates will be required to ensure security in the 6G network.

In SC settings, the Multi-metric Scoring Dynamic Source Routing (MMS-DSR) protocol has been shown to outperform other protocols such as DSR, SOL-DSR, and improved OLSR in terms of throughput, latency, and routing time in a vehicular ad hoc network [22]. MMS-DSR dynamically adapts to urban traffic conditions and optimizes data routing using beamforming techniques. Modern Mining and ML, specifically, LSTM networks [22], have contributed to the development of an adaptive beamforming algorithm. Supporting different networks, real-world testing, safety enhancement, integrating new ML techniques, scalability improvements, and energy efficiency are some future research topics for MMS-DSR.

To understand IoT applications and ML, energy efficiency has become a key focus. This chapter covers recent research on 6G wireless networks [23], particularly the difficulties in powering the billions of connected devices in the upcoming 6G networks, which has attracted interest. Energy-efficient solutions are revealed [23]. The survey discusses the latest developments in 5G and 6G energy efficiency and lists the main barriers to energy efficiency in large-scale IoT networks. This study aims to provide advice on energy efficiency, the Internet, and how AI experts can address the complexity of next-generation mobile communication networks.

This report examines trends through 2030, considering current and upcoming developments in interpretable artificial intelligence (XAI) for SCs. It covers applications, problems, and solutions of XAI in SCs and provides definitions and summaries of related technologies [24]. This article highlights current XAI initiatives, standard operating procedures, and upcoming progress [24]. It also proposes research guidelines for improving the use of SCs that address safety, privacy, and ethical challenges, and concludes with a roadmap for integrating XAI into the new security framework.

This chapter tracks the security progress from 1G to 6G and provides a detailed review of the security needs and issues of 6G networks [25]. Intelligent sensing, edge, control, and application layers are just a few. One of the layers where AI/ML technology fits into the proposed 6G security architecture is identifying relevant security vulnerabilities and recommending countermeasures [25]. The study also looked at physical layer technologies such as THz, VLC, and molecular communications. It highlights the need for further research into solutions and addressing the security implications of emerging 6G applications such as holograms, augmented reality, unmanned aerial vehicles, and intelligent grids.

The application of deep neural networks and software-defined networks for interactive notification systems is the main focus of the ML approach in this chapter for anomaly detection in 6G networks [26]. This study examines data collection at different periods to provide a more accurate view of network activities, even for long periods. A guided dynamic anomaly detection system improves network security by proactively detecting threats and improving resource utilization. At the same time, it risks significant distortions and uses more resources in a shorter period. To increase the efficiency and scalability of the system in real-world situations [26], this study highlights the value of using pre-trained models and recommends further research on continuous learning and runtime optimization strategies.

The applicability of collaborative deep learning (SFL) techniques in 6G networks is investigated in this study [27]. By distributing learning algorithms among a large number of operators without compromising data privacy, SFL reduces training time and privacy-enhancing smart grid data models [27]. In addition to highlighting different applications, the review discusses future research objectives, challenges, and datasets, and there is a discussion of improving SFL in 6G networks.

This study delves into the integration of green 6G networks with energy-efficient IoT. Decision-making is underpinned by big data, distributed innovation, and AI [28]. This 6G wireless and automotive IoT demonstrates how technology promotes decentralized control and energy management, while the integration of sensor data provides efficient IoT performance in industrial and urban environments, making it convenient [28]. Simple communication helps to achieve sustainable energy.

This study provides a comprehensive analysis of 6G networks, revealing how AI is revolutionizing wireless communications [29], considering key enabling technologies, including smart, reconfigured surfaces, ultra-wide MIMO, quantum, and terahertz. It emphasizes that these developments will deliver unprecedented benefits in productivity, connectivity, and personalized services [29] to improve the use of AI and machines.

This study investigates the revolutionary possibility of integrating multiple Metaverse apps with 6G technology, focusing on the impact on business and education [30]. 6G with ultra-high-speed connectivity and low latency revolutionizes virtual shopping, marketing, and consumer interactions in businesses and facilitates interactive learning skill development and remote collaboration in this immersive Metaverse study, also improving the experience [30]. Multiple stimuli and 6G networks are coming together to create new possibilities for innovation, productivity, and participation. But to completely supply on this promise, security and privacy concerns need to be addressed to create an inclusive and immersive digital future.

4.3 6G NETWORKS AND SCS

With its unheard-of speeds, exceptionally low latency, and massive tool connectivity, 6G networks are a prime leap forward in the Wi-Fi era. These features facilitate the growth of sensor networks and IoT devices in SCs. This produces enormous amounts of data. Effective use of this data to identify anomalies requires ML and DL techniques. SCs will incorporate ML algorithms such as DL methods. Supervised and unsupervised are combined with DL techniques such as CNNs and RNNs to provide general access to real-time data streams. You can automatically detect deviations from the format.

Improving the reliability, safety, and effectiveness of many aspects of urban infrastructure, including public safety, power management, healthcare, and transportation, is based on this integration. The interaction between cutting-edge AI technology and 6G networks highlights the transformative potential of SCs that will continue to evolve as they continue to evolve.

4.3.1 Definition and Characteristics of 6G Networks

The existing 5G networks will give way to 6G networks, which are the next generation of wireless communication systems to greatly improve speed, latency, capacity, and connectivity. Although 6G networks are still in the conceptual and preliminary research stages, they are expected to push the limits of connectivity and telecommunications capabilities.

Table 4.1 shows how wireless communication technology evolved from 1G to the projected 6G. It addresses significant improvements in global coverage, applications, data speeds, latency, connection density, security, spectrum efficiency, and energy efficiency. Each model ranges from 1G to simple voice calling, 5G due to its high speed and low latency capabilities, and microsecond latency at 6G, anticipated terabit speeds to industries. Advances have been made in terms of capacity, performance, and usability. These developments support new technologies such as augmented reality and IoT, and continuously improve reliability, which is the result of effort.

4.3.2 SC Applications and Infrastructure

SC apps and infrastructure in the framework of deep ML and anomaly detection in 6G SCs provide a revolutionary approach for IoT devices and sustainable city management for industries such as public safety, energy management, transportation, healthcare, and environmental monitoring in these cities in 1999. Complex networks of sensors and systems are integrated. These devices collect and transmit large amounts of data continuously. It provides real-time insights into citizen activities and operations in the city. Monitoring these databases requires ML and DL scheduling algorithms that can distinguish anomalies or threats from the norm, which might indicate a problem.

ML algorithms in transportation analyze traffic data from sensors and connected cars to detect anomalies, such as accidents and traffic jams. It allows for road optimization and dynamic traffic control. This will help improve operational efficiency and reduce travel times. Similarly, ML models in strength control reveal energy usage and grid overall performance to perceive issues, together with surprising spikes or unexpected screw-ups. This permits predictive renovation and green power distribution. ML and DL techniques allow for the early detection of anomalies, including fitness emergencies, security breaches, and harm to the environment, which helps ensure rapid response and proactive measures to protect public health, safety, and environmental quality. This ability applies to health care, public safety, and environmental inspection.

TABLE 4.1
Wireless Communication Technologies: A Review of the Main Features and Developments from 1G to 6G

Feature/ Advancement	1G	2G	3G	4G	5G	6G (Expected)
Data speed	Up to 2.4 kbps	Up to 64 kbps	Up to 2 Mbps	Up to 100 Mbps	Up to 10 Gbps	Terabits per second (Tbps), potentially 100s of Gbps
Latency	N/A	Hundreds of milliseconds	100–500 ms	30–50 ms	1–10 ms	Microseconds (μs), aiming for <1 ms
Connectivity density	Low	Low	Medium	High	Up to 1 million devices per sq km	Significantly higher, potentially up to 10 million devices per sq km
Spectral efficiency	N/A	N/A	Improved	High	High	Enhanced spectral efficiency
Energy efficiency	N/A	N/A	Improved	Better than 3G	Improved compared to 4G	Optimized energy consumption
Security	Basic	Basic	Enhanced	Enhanced	Enhanced	Improved security protocols, encryption
Global coverage	Limited	Extended	More extensive	Wide	Wider coverage including rural areas	Seamless global coverage
Applications	Voice calls	SMS, MMS	Web browsing, video calls	Mobile broadband, video streaming	Enhanced mobile broadband, IoT, AR/VR	Advanced applications like holographic communications, AI-driven IoT
Technological foundations	Analog	Digital	CDMA, GSM	LTE, WiMAX	Massive MIMO, mmWave, network slicing	Quantum communications, AI-driven networks

City planners and politicians can use AI-powered anomaly detection and sophisticated analytics in 6G SCs to increase urban resilience, sustainability, and the quality of life of citizens. In addition to improving resource management and operational effectiveness, these technologies also encourage creativity in smarter city development plans, opening the doors for cities to improve. It can tackle tough problems and a connected, data-driven future.

4.3.3 Connectivity and Data Flow in SCs

In the larger context of developing 6G networks, connectivity and data flow are essential elements supporting SC functioning and efficiency. The IoT is made up of a wide range of networked sensors and gadgets that are integrated into SCs. These gadgets produce and transfer data continuously and provide real-time awareness of concerns of living in the city, from public safety and healthcare to energy use and transportation.

6G-enabled SCs offer extremely fast communication capabilities and low latency, which raises the same old issue of connectivity and enables the smooth movement of records throughout industries. It promotes green management and service improvement for towns. For instance, real-time data from vehicle sensors connected to the transportation network enables adaptive routing to reduce congestion, predictive infrastructure maintenance, and dynamic traffic management. Similarly, IoT sensors in the energy grid monitor patterns in energy use and grid performance, allowing utilities to efficiently integrate renewable energy sources, improve product distribution, and forecast increased demand.

DL and ML technologies are essential for interpreting the massive amounts of data produced in SCs. These algorithms are good at identifying deviations from the normal pattern or unusual abnormalities, which may signal problems such as energy inefficiency, traffic accidents, environmental hazards, or public health disasters. ML and DL models that detect abnormalities early allow local governments to take preventive measures, strengthen operational flexibility for citizens, and enable them to raise their standard of living.

4.4 ANOMALY DETECTION: CONCEPTS AND IMPORTANCE

Anomaly detection is essential for ensuring the ability and reliability of structures in SCs. Specifically, the idea in developing 6G networks is to look for styles or behaviors that deviate from the norm in statistics, as these might indicate a risk, whether it is an inefficiency or a capacity anomaly, that needs to be detected. It is important to bolster protection protocols and improve average provider shipping in an SC environment wherein many networked gadgets and sensors generate enormous amounts of data.

The importance of anomaly detection lies in its ability to predict and correct problems. Authorities can be alerted to incidents immediately by detecting anomalies in the transport system, for example, before the problem becomes more serious. Traffic congestion or rapid implementation and disruption of infrastructure for mitigation plans, as well as utility connections, will detect anomalies, such as sudden increases in energy demand or equipment errors. In addition to being able to scale to improve

TABLE 4.2
Different Anomalies and Instances in SCs

Type of Anomaly	Description	Examples
Point anomaly	An observation that deviates noticeably from the remainder of the data.	A very high air quality reading from a single sensor, when compared to the surrounding values.
Contextual anomaly	An uncommon observation made within a particular context.	A car going faster than the posted limit in a school zone.
Collective anomaly	A series of observations that, when taken as a whole, they are uncommon.	A number of city traffic lights are turning red all at once.
Seasonal anomaly	A finding that does not match the regular seasonal patterns.	Summertime power use is better than it has traditionally been.
Structural anomaly	An anomaly determined because of the linkages or hierarchies within the shape.	Anomalies in a clever grid system's statistics go with the flow that should point to community problems.

resource allocation and avoid disruption and performance, anomaly detection also helps the health and safety sector by monitoring patient health indicators. Detecting safety violations or identifying environmental risks early enables timely operation and protection of public health and safety.

Effective anomaly detection is increasingly important in the 6G network era, where data volumes and processing rates are expected to increase exponentially. At the heart of these efforts are ML and DL approaches, which use complex algorithms to assess complexity. The datasets accurately discover anomalies and urban issues. They help maximize resources, integrate into the SC framework, and strengthen the resilience of important infrastructure. This will, in the end, create safer, more efficient, and more sustainable city surroundings.

Table 4.2 lists the various abnormalities that can be found in smart towns: "Point anomalies" are discrete information points that deviate substantially from the average, consisting of abrupt spikes in exceptional air. "Contextual anomalies" discuss strange actions taken under specific conditions, such as rushing in school zones. "Collective anomalies" identify anomalous styles among numerous data factors, which include concurrent failures of traffic lighting. "Seasonal anomalies" display departures from ordinary seasonal patterns, consisting of elevated energy use in the summertime. "Structural anomalies" are irregularities in the shape of facts that contribute to feasible network problems. An instance of such an anomaly could be anomalies in smart grid information flows. These classes help the deployment of efficient anomaly detection systems, which are essential for boosting the dependability and effectiveness of sports in smart towns.

4.5 ML TECHNIQUES FOR ANOMALY DETECTION

In an SC, devices for anomaly detection strategies are important in detecting unusual trends or outliers in large and fine-grained datasets. These strategies fall into four broad categories: semi-supervised, supervised, and unsupervised learning, and RL under teacher-led learning. Fashions are trained on labeled data to classify procedural abnormalities with predefined standards. This is to be appropriate for identifying specific types of disorders with recognized qualifications. Semi-supervised learning combines small, hidden datasets with large unlabeled datasets, integrating supervised and unsupervised learning techniques to increase the accuracy of anomaly detection.

SCs can automate the anomaly detection process with the help of these ML techniques to improve response instances to urgent occasions, such as on-site visitor accidents. In the face of soaring electricity prices or environmental threats, cities can improve their operational efficiency and allocate assets appropriately. State-of-the-art algorithms and the use of real-time data analysis can ensure rapid movement to maintain public protection and infrastructure reliability. Meanwhile, system controls continue to evolve within SC frameworks and provide scalable answers to various problems of the ever-changing city environment appropriately.

4.5.1 Supervised Learning Approaches

During abnormal detection, supervised learning techniques use labeled data to train a model that identifies anomalies based on predefined patterns, learn from historical data, SVM, Decision Trees, CNNs, Artificial Neural Networks, and other algorithms. It can be used to effectively identify abnormalities. These methods allow SCs to carry out automatic detection processes with high accuracy and improve operational efficiency and infrastructure resiliency. Some disorders work best when they are clearly defined.

The main supervised learning techniques commonly used in anomaly detection are briefly summarized in Table 4.3, highlighting their unique ability to detect anomalies from specific data points. These techniques demonstrate suitability for environmental audits, healthcare, cybersecurity, and the financial sector. SVMs perform well in high-dimensional binary classification applications, and decision trees provide interpretability for tasks such as detecting and reconstructing network traffic anomalies. Random forest improves predictive maintenance and flexibility in monitoring IoT v-Treatments. Complex pattern recognition is ideal for neural networks such as CNNs in SCs and apps. Medical diagnostic applications and logistic regression models perform well in probabilistic classification tasks. At the same time, Naive Bayes can analyze text well. When combined, these methods will greatly improve system security and decision-making in a variety of anomaly detection applications.

4.5.2 Unsupervised Learning Approaches

Unsupervised learning techniques can identify patterns and anomalies in data without the need for previously labeled samples, making them valuable for anomaly detection experiments. These techniques work by understanding the distribution of important information or by grouping data points based on similarity. Using old mathematics such as clustering algorithms (k-means, DBSCAN), data points are grouped into clusters. Anomalies are identified as points that are inherently incompatible with any group. Anomalies are identified by density-based techniques (such as DBSCAN) as locations in low-density regions. Anomaly detection algorithms, such as automatic encryption and one-time extraction, recognize

TABLE 4.3
Supervised Learning Algorithms and Their Applications in Anomaly Detection

Algorithm	Description	Anomaly Detection App
SVM	SVM can create abnormal differences from normal data points using a hyperplane in a high-dimensional space.	• Cybersecurity (e.g., intrusion detection) • Fraud detection (e.g., credit card fraud) • Medical diagnosis (e.g., diagnosis)
Decision trees	Decision trees are useful for classification tasks and interpretable anomaly detection patterns because they separate and repeat data according to qualification criteria.	• Construction (e.g., identification of defects) • Network traffic analysis (e.g., intrusion detection) • Environmental monitoring (e.g., pollution detection)
Random forest	Random forest combines multiple decision trees to improve robustness and accuracy in anomaly detection tasks.	• Monitoring of IoT devices (e.g., detection of anomalies in sensor networks) • Predictive maintenance (e.g., equipment failure detection) • Finance (e.g. fraud detection)
Artificial Neural Network	Feedforward and CNN Learn complex data patterns useful for exploration. Anomalies in high-dimensional nonlinear data sets.	• Image recognition (e.g., identifying abnormalities in medical images) • Natural language processing (e.g., detecting anomalies in text data) • Sensor data analysis (e.g., anomaly detection in SC environments)
Naive Bayes	Naive Bayes classifier maintains center independence. Features and efficiency for text and documents Classification work, including abnormal detection.	• Email filtering (e.g., spam detection) • Text mining (i.e., detecting anomalies in textual data) • Document classification (e.g., identification of unusual documents)
Logistic regression	Probability classification logistic regression model Placing samples into categories is useful for binary classification tasks in detecting anomalies.	• Health care (e.g., disease diagnosis) • Financial transactions (e.g., detecting fraudulent activity) • Prediction of equipment failure (e.g., predictive maintenance)

and identify events that differ significantly from the normal data format. These methods are useful tools in areas such as fraud detection, network security, and monitoring of SC systems. These things are necessary.

The main unsupervised learning algorithms used for anomaly detection are collected in Table 4.4. They are useful tools for classifying anomalies by grouping data points based on density. Or similarly, it highlights the range of approaches and

TABLE 4.4
Unsupervised Learning Algorithms and Their Applications in Anomaly Detection

Algorithm	Description	Applications in Anomaly Detection
K-means clustering	Group data points based on similarity. Asymmetry is an imbalance in any group.	• Network traffic analysis (e.g., look for unusual patterns that indicate a possible attack) • Analysis of customer behavior (e.g., identifying different cost models)
DBSCAN	Anomalies are identified as low-density locations. Areas use a spatial grouping of applications based on density Colaharen Company (DBSCAN)	• Environmental tracking (e.g., detecting outliers in pollution levels) • Detecting anomalies in time-series data (e.g., identifying unusual patterns in sensor readings)
Isolation forest	Separation trees are created to separate rare and specific anomalies from mostly normal data points.	• IDSs (e.g., identifying unusual network activity) • Identifying differences in financial transactions (e.g., checking for fraudulent transactions)
Autoencoders	A neural network that focuses on reducing the disparity in input and output data is created to learn efficient data encoding in an unsupervised way.	• Detection of discrepancies in photos and videos (e.g., identifying unusual patterns in CCTV footage) • Health care (e.g., to detect aberrations in medical images)

domains using K-means clustering and DBSCAN. When it comes to isolating anomalies that differ greatly from common data patterns, the isolated forest is particularly effective. An autoencoder is an algorithm that detects anomalies in the input data and reduces reconstruction errors. These algorithms show suitability for improving anomaly detection systems in real-world applications such as fraud detection, environmental monitoring, and network security by identifying abnormal patterns.

4.5.3 RL Approaches

To optimize cumulative rewards in a dynamic context, agents are trained to make decisions by trial and error in an RL system for anomaly detection. RL algorithms such as cue learning and deep Q-networks (DQN) are effective in learning optimal behavior through interactions with an environment, but the opposite is true. These supervised and unsupervised methods have found applications in a variety of fields, including industrial controls to detect abnormalities in the production process. Cybersecurity for Adaptive Access Isolation Systems focuses on autonomous vehicles and autonomous driving systems that can identify anomalies in complex work environments and provide powerful solutions when needed.

TABLE 4.5
RL Algorithms and Their Applications in Anomaly Detection

Algorithm	Description	Applications in Anomaly Detection
Q-Learning	Model-free RL algorithm Teach agents to maximize cumulative returns through Test the determination of errors.	• Cyber Security (e.g., adaptive access detection systems) • Industrial control systems (such as detecting abnormalities in the production process)
DQN	Combines Q-learning with deep neural networks. Modifying high-dimensional state space makes it complex. Decision-making in a dynamic environment.	• Automatic system (e.g., abnormality detection in autonomous vehicles) • Energy system (e.g., to detect abnormalities in the electrical network)
Policy gradient	Directly optimize policy parameters to maximize **Expected Results:** Suitable for continuous operations positions.	• Robotics (e.g., detecting abnormalities in robot systems) • Health care (e.g., to report abnormalities in the patient tracking system)

Table 4.5 lists several RL algorithms used for anomaly detection. This shows that the algorithm is flexible in terms of best practices in learning through trial and error. Applications in health care, cybersecurity, industrial control, and automation show how successful these algorithms are in a dynamic context with adaptive decision-making, and with continuous study, any anomalies should be detected.

4.5.4 Semi-Supervised Learning Approaches

In anomaly detection, semi-supervised learning techniques use both categorized and unlabeled statistics to improve overall performance, specifically in conditions where categorized data are scarce. These techniques use a small number of labeled records to direct the getting-to-know procedure and identify large patterns from a larger unlabeled data set. Self-training, wherein the model labels the unlabeled data iteratively, and co-training, in which numerous classifiers are trained on various perspectives on the statistics, are commonplace processes. Semi-supervised learning could be very beneficial in conditions like fraud detection, clinical prognosis, and network protection, where it can be tough to tag abnormalities. These techniques improve the robustness and detection accuracy of anomaly detection systems by merging the benefits of supervised and unsupervised learning.

Several semi-supervised learning strategies for anomaly detection are supplied in Table 4.6, with a focal point on how they can assign both categorized and unlabeled statistics to improve detection accuracy. Graph-based techniques, multi-view studying, self-education, co-training, generative fashions, and different algorithms have shown their worth in a variety of packages, along with financial analytics, social network analysis, fraud detection, scientific diagnostics, internet-paintings security, sensor networks, and photo and text evaluation. These techniques improve the sturdiness

TABLE 4.6
Semi-Supervised Learning Algorithms and Their Applications in Anomaly Detection

Algorithm	Description	Applications in Anomaly Detection
Self-training	Continuously label unlabeled data using the first classifier trained on labeled data. Then retrain using the new labels.	• Fraud detection (such as credit card transactions) • Network security (e.g., attack detection of network traffic)
Co-training	Optimizing performance through joint training involves using different subsets of data or approaches to train multiple classifiers.	• Medical diagnosis (e.g., detection of rare diseases) • Email filtering (e.g., spam detection with limited label information)
Multi-view learning	Enhances learning from a combination of labeled and unlabeled data by combining different viewpoints or feature sets of the data.	• Sensor networks (e.g., anomaly detection in multi-sensor systems), • Financial analytics (e.g., detecting fraudulent patterns)
Graph-based methods	Use relationships and structure in graphs to display data. Using labeled and unlabeled data.	• Social network analysis (e.g., detection of unusual user behavior) • Environmental monitoring (e.g., pollution detection)
Generative models (e.g., semi-supervised GANs)	Use labeled and unlabeled data to improve anomaly detection and use generative models to understand data distribution.	• Basic knowledge about image distortion (e.g., identification of manufacturing defects) • Text analysis (e.g., detecting unusual documents)

and effectiveness of detection structures throughout numerous domains by improving anomaly detection in conditions with little labeled data.

4.5.5 Evaluation Metrics for ML-based Anomaly Detection

For ML-based anomaly detection, several assessment measures are necessary to evaluate the efficacy and performance of the models. Below is a synopsis of some important metrics:

4.5.5.1 Accuracy

Accuracy calculates the percentage of real results, both true positives and true negatives, among all the cases that were looked at. Although it offers a broad performance measure, it might not be enough when datasets are unbalanced.

$$\text{Accuracy} = \frac{\text{TP} + \text{TN}}{\text{TP} + \text{TN} + \text{FP} + \text{FN}} \tag{4.1}$$

- **True Positive (TP):** This occurs when an anomaly has been correctly identified by the version. For instance, it is deemed a TP if a definition finds a security flaw in the four-metropolis device that should constitute a violation. Strong cognitive abilities are demonstrated by high TP levels.
- **True Negative (TN):** This happens when the version depicts a circumstance that is quite common. For instance, it is a mile-a-TN if the standard of typical tourists is positively identified as unpleasant. It is advised that a high TN charge prevents the model from wrongly identifying regular behaviors as aberrant.
- **False Positive (FP):** This occurs when a daily condition is not mistakenly recognized as an anomaly by the model. For example, the channel is FP if a routine environmental sensor analysis is marked as abnormal. Waste and poor messages may arise through higher FP expenses.
- **False negative (FN):** This occurs when the model is unable to identify a true condition. For instance, distance is FN if the intended infrastructure collapse isn't always located. High FN levels are harmful, and they tend to obscure real issues.

By detecting undiscovered anomalies (FN) and false alarms (FP), those metrics can be used to determine and optimize the general performance of the anomaly detection model to ensure the sturdy and reliable average performance of 6G SCs.

4.5.5.2 Precision

It specifies the proportion of real positives to the total variety of high-quality predictions made by using the version. This is essential in conditions where the price of FPs is excessive.

$$\text{Precision} = \frac{\text{TP}}{\text{TP} + \text{FP}} \tag{4.2}$$

4.5.5.3 Recall

The proportion of true positives is identified out of the actual positives. It is particularly important when the cost of FNs is high, ensuring that as many actual anomalies as possible are detected.

$$\text{Recall} = \frac{\text{TP}}{\text{TP} + \text{FN}} \tag{4.3}$$

4.5.5.4 F1 Score

The harmonic mean of precision and recall provides a single metric that balances the two. It is useful when there is a need to balance FPs and FNs.

$$F_1 \text{ Score} = 2 \times \frac{\text{Precision} \times \text{Recall}}{\text{Precision} + \text{Recall}} \tag{4.4}$$

4.5.5.5 Area under the Receiver Operating Characteristic Curve (AUC-ROC)

It evaluates the model's ability to distinguish between classes. The ROC curve plots the TP rate against the FP rate at various threshold settings, and the AUC provides a single scalar value summarizing this performance.

4.5.5.6 Area under the Precision-Recall Curve (AUC-PR)

It is particularly useful for imbalanced datasets, where the number of TNs is significantly higher than the number of true positives. The PR curve plots precision against recall, and the AUC-PR gives a summary measure of this relationship.

4.5.5.7 Confusion Matrix

It is a table used to describe the performance of a classification model by summarizing the counts of TP, TN, FP, and FN predictions.

A detailed analysis of the predictions obtained from ML models in anomaly detection scenarios for 6G SCs includes four types of false positives found in the confusion matrix (Table 4.7): TP, FN, FP, and TN. While TP indicates anomalies that have been correctly identified, FN indicates anomalies that are FP misses, indicating common cases that are mislabeled as abnormal, and TN indicates irregular data that are properly labeled. This number is important in evaluating the performance of the model. This is because it shows how well we can distinguish between normal and abnormal phenomena. It also guarantees an increase in the efficiency of the anomaly detection system. This is essential in maintaining the integrity and effectiveness of business operations in an SC.

To understand the effectiveness of ML models in detecting anomalies, especially in 6G-enabled SCs, it is important to understand the proposed evaluation metrics. Although accurate, this does not provide a detailed picture. However, unbalanced data can be misleading. A more comprehensive approach relies on precision and recall. It emphasizes the trade-off between FPs and FNs. These two factors are balanced with the F1 score providing a single performance metric. AUC-ROC and AUC-PR provide a detailed analysis, especially under non-equilibrium conditions. Finally, accurate prediction results are analyzed based on the confusion matrix. This is important for evaluating model performance and identifying model flaws.

TABLE 4.7
Confusion Matrix

	Predicted Positive	Predicted Negative
Actual Positive	TP	FN
Actual Negative	FP	TN

4.6 DL TECHNIQUES FOR ANOMALY DETECTION

In 6G SCs, DL techniques for anomaly detection use various state-of-the-art neural network designs designed to manage intricate information patterns. CNNs are useful for tasks like image anomaly detection in surveillance or environmental tracking because of their superior overall performance in spatial information analysis. Sequential data, consisting of anomalies in time-series facts from sensors or community logs, is fine and acceptable for RNNs, along with variations like LSTM networks.

Another deep getting-to-know approach is autoencoders, which locate anomalies primarily based on reconstruction errors by reconstructing input data and comparing it to the original. When anomalies seem to depart from traditional data patterns, this approach is beneficial. By producing synthetic statistics that intently resemble actual-world distributions, Generative Adversarial Networks (GANs) make a precious contribution to anomaly detection via the contrast of generated and actual information distributions.

All matters considered, these deep mastering methods enhance 6G SC anomaly detection capacities by taking advantage of their potential to pick up complex statistical representations and identify minute versions that point to anomalies. This is essential to ensure the dependability, effectiveness, and protection of SC infrastructures in many regions, which include public protection, power control, healthcare, and transportation.

4.6.1 Overview of DL

A subset of device learning called "DL" is described by multi-layered neural networks that permit the model to study information representations in a hierarchical layout. This approach has transformed several industries by substantially improving overall performance in areas such as anomaly detection, natural language processing, and image and audio recognition. The input, hidden, and output layers of a neural network are interconnected layers made from neurons that process incoming records through weighted connections and perform nonlinear modifications to supply outputs.

DL algorithms are important for anomaly identification in 6G SCs due to the fact that they can automatically extract complicated patterns from large datasets. They are effective at identifying both trustworthy and elaborate anomalies in a variety of data modalities, including network traffic, video streams, and sensor measurements. The utilization of deep neural networks in those models enables the capacity to distinguish between regular and anomalous movements, an essential function in upholding the dependability and safety of SC infrastructures.

The intention of using DL in anomaly detection is to enhance the speed at which abnormalities or threats that would affect the security and operation of SC systems can be quickly detected in real time. This approach advances the competencies and resilience of 6G-enabled SCs by enhancing the accuracy of anomaly detection and facilitating proactive maintenance and optimization of citywide operations.

Table 4.8 gives a comprehensive summary of DL techniques that might be often utilized in many fields. It covers CNNs for photograph processing tasks, autoencoders,

TABLE 4.8
DL in Various Applications

Aspect	Description	Applications	Examples
Definition	A subset of system studying involving neural networks with more than one layer for hierarchical information representation.	Image popularity, speech recognition, natural language processing, anomaly detection in smart towns.	Identifying items in images, converting speech to textual content, studying sentiments in text, and detecting anomalies in sensor data.
Neural network structure	Composed of the input layer, hidden layers (which may be a couple of), and an output layer, neurons approach facts via weighted connections.	Learning styles in data, characteristic extraction, category responsibilities, and real-time decision-making.	Multiple layers in CNNs for image recognition, LSTM networks for sequential information analysis, and autoencoders for dimensionality reduction.
Training mechanism	Uses backpropagation to adjust weights and biases, optimizing model performance over epochs.	Automated feature learning, complex data modeling, RL, and generative tasks.	Adjusting weights in CNNs for improving image recognition accuracy, fine-tuning LSTM models for better time-series predictions, and training GANs for generating realistic images.
Key architectures	CNNs for spatial data, RNNs/LSTMs for sequential data, autoencoders for unsupervised learning, GANs for generative tasks.	Image and video analysis, time-series forecasting, anomaly detection, data compression, image synthesis.	Using CNNs in medical imaging for diagnosing diseases, LSTM networks in stock price prediction, autoencoders for reducing noise in data, and GANs for generating artwork.
Advantages	The capability to handle large datasets, learn complex patterns, and adapt to various domains with minimal feature engineering.	Enhanced accuracy in pattern recognition, scalability in big data environments, robust to noise and variations.	Achieving state-of-the-art results in image classification tasks, scaling to millions of data points in real-time analytics, and overcoming variations in speech recognition tasks.
Challenges	Computational complexity, requirement of large amounts of labeled data, and interpretability of black-box models.	Training time, overfitting, deployment in real-time systems, and ethical considerations in AI applications.	Addressing computational resources in deep CNNs, managing overfitting in LSTM models, deploying real-time speech recognition systems, and ensuring fairness in AI decision-making.

and variational autoencoders (VAEs) for unsupervised learning and characteristic extraction, RNNs and LSTM networks for sequential fact evaluation, and GANs for creating new samples. Each technique's particular design is used to demonstrate the way it works in diverse programs, which consist of statistics production, anomaly detection, photo identification, and natural language processing.

4.6.2 CNNs

A unique form of DL shape known as CNNs is used to interpret grid-like entries, particularly snapshots and films. They use convolutional layers to automatically extract pertinent traits from the raw input by applying filters across the entered data to capture spatial hierarchies. By reducing the need for manual feature extraction, CNNs have better capabilities like object detection, image segmentation, and image categorization.

CNNs are superb at detecting anomalies because they can discover complex patterns and spatial relationships in information. CNNs can apprehend anomalies in scans that may be indicative of diseases, for instance, in scientific imaging. Sensor information in SCs may be analyzed to discover extraordinary patterns that could imply infrastructure issues or environmental changes. Convolutional, pooling, and completely connected layers in CNN structure permit them to examine complex fact representations, which makes them adaptable to a wide range of anomaly detection packages in which context and spatial correlations are essential.

4.6.3 RNNs and LSTM Networks

DL techniques, such as RNNs and LSTM networks, are made to successfully model sequential data. RNNs and LSTMs, in comparison to feedforward neural networks, are better suited for duties requiring an understanding of context and temporal relationships due to the loops in their structure that permit them to maintain a recollection of prior inputs.

RNN system sequences by repeatedly processing every detail at the same time, retaining a hidden state that stores details about earlier inputs. However, due to issues like vanishing gradients, they regularly have problems expressing long-term dependencies. To conquer this restriction, LSTMs introduce a memory cell that may selectively update its state and store records throughout prolonged sequences. This lets the LSTMs learn and remember patterns over longer periods.

When it comes to anomaly detection, RNNs and LSTMs perform well in programs in which historical context is critical, such as stock price prediction, time-series data analysis from SC sensors, and anomaly monitoring in network traffic. They are effective gadgets for recognizing departures from predicted styles due to their potential to address sequential statistics and music dependencies all through time. This improves detection accuracy and dependability in dynamic and changing environments.

4.7 HYBRID AND ENSEMBLE METHODS

4.7.1 Enhancing Anomaly Detection through Combined ML and DL Methodologies

Hybrid and ensemble methods leverage the advantages of several DL and ML strategies to enhance anomaly detection performance. These strategies clear up character algorithm issues by combining multiple methods and developing fashions that are greater accurate, scalable, and dependable.

4.7.2 Hybrid Methods

The strengths of numerous algorithms from specific studying paradigms, including supervised, unsupervised, and RL, are combined in hybrid anomaly detection structures. These strategies are especially useful in settings concerning clever towns, where it is important to discover abnormalities, both recognized and unknown. Supervised mastering, for instance, is good at spotting departures from recognized behavioral patterns, consisting of ordinary strength use or traffic flow drift. Supervised fashions may additionally discover anomalies resulting from those well-established styles by using training on classified datasets, which allows prompt interventions.

Unsupervised learning is quite powerful at identifying novel anomalies that don't match established patterns. When handling atypical or surprising sports for which there might not be any categorized information, that is very important. For example, a hybrid version would probably understand cybersecurity threats in 6G networks through the use of unsupervised learning and grow to be aware of anticipated community issues through the use of supervised learning. Through the combination of the two strategies, hybrid techniques can provide an additional complete framework for anomaly detection, enhancing the resilience and protection of SC infrastructures.

4.7.3 Methods

Ensemble approaches combine predictions from several models to improve predictive performance, creating a more dependable and resilient system. Boosting, stacking, and bagging are common methods. Bagging, also known as Bootstrap Aggregating, is the process of training several instances of the same model on various data subsets produced via random sampling with replacement. This method, which is best represented by Random Forests, takes the average of these models' predictions and efficiently lowers variation while boosting stability in anomaly detection circumstances.

The boosting trains models consecutively, where each new model seeks to rectify the errors of its predecessors, culminating in a final model that combines their predictions for greater accuracy. In clever metropolis programs, techniques like Gradient Boosting Machines (GBM) are very helpful, considering they make it feasible to become aware of tiny abnormalities like modifications in energy use. By the use of a meta-version to learn from the outputs of different base models and using their strengths to increase detection accuracy, stacking similarly enhances performance. This technique works mainly properly in complicated contexts where anomalies and lots of fact patterns might be present.

Table 4.9 highlights the benefits and uses of numerous hybrid and ensemble methods for anomaly detection in SCs. Methods that include Random Forest and GBM, which can be appropriate for energy grid monitoring and energy control, improve accuracy through bagging and boosting. Meanwhile, stacking combines predictions from multiple formats to optimize logistics. The hybrid CNN-RNN structure efficiently analyzes spatial and temporal data for video surveillance. Other combinations include face cybersecurity and smart network application issues, such as autoencoders with K-means clustering and LSTM with Random Forest. Taking everything into account, those strategies will increase the accuracy and flexibility of smarter urban anomaly detection systems.

TABLE 4.9
Hybrid and Ensemble Approaches to Enhance Anomaly Detection in SC Applications

Method	Type	Application in SCs	Description
Random forest	Ensemble (Bagging)	Energy management, Traffic monitoring	Utilizes multiple decision trees trained on random data subsets to reduce overfitting and improve robustness.
GBM	Ensemble (accelerate)	Detection of electrical grid abnormalities, excellent air monitoring	Continuously fixed errors from the previous format to improve detection accuracy in complex events.
Hybrid CNN-RNN Architecture	Hybrid	Video surveillance, Public Safety	Integrates CNNs for spatial feature extraction and RNNs for temporal collection evaluation, ideal for dynamic anomaly detection.
Stacking	Ensemble (Stacking)	Intelligent transportation, Environmental monitoring	It also combines predictions from multiple models using Meta Learner to create an accurate prediction version.
Autoencoder + K-means clustering	Hybrid	Cyber Security Network abnormality detection	Includes an autoencoder for dimensionality reduction with K-means clustering to detect anomalies.
LSTM + random forest	Hybrid	Smart Grid Monitoring IoT sensor statistics analysis	LSTM fact time dependency in time collection statistics. Random forest handles outlier classifications.
SVM + k-NN	Hybrid	Smart fitness care emergency response gadget	Leverage SVM for structured category and KNN for real-time detection of outliers or new anomalies.

4.8 DATA HANDLING AND PREPROCESSING

4.8.1 Data Collection in 6G-Enabled SCs

Data collection in 6G-enabled SCs emphasizes how essential it is to acquire a variety of facts from unique assets as a way to beautify service delivery and urban administration [31]. Cities can use better connectivity added through 6G to mix statistics from various assets, along with social media, authorities' information bases, IoT gadgets, and sensors. City planners and administrators can gain real-time insights approximately a whole lot of components of city lifestyles, together with site visitors float, power usage, environmental conditions, and public sentiment, thanks to this broad data environment. Municipalities can also improve public protection, optimize offerings, and improve residents' best of life by utilizing this plethora of statistics to inform decision-making.

The inexperienced strategies for gathering facts are essential for tackling the complicated issues that present-day cities face. Cities can now swiftly and efficiently take a look at big volumes of facts way to the combination of big data generation and real-time information processing frameworks [32]. This permits prompt identification of issues, which include energy outages or site visitor congestion, allowing quick action and remediation. The mixture of current information collection techniques and 6G connectivity permits SCs to turn out to be extra resilient, sustainable, and receptive to the necessities of their residents, which in turn spurs innovative and astute boom in city areas.

Table 4.10 shows the many information sources and kinds relevant to 6G-enabled SCs. It showcases ancient data from authoritative databases, emotional facts from

TABLE 4.10
Classification of Data Sources and Types for Urban Analytics in SCs

Data Source	Data Type	Description
IoT devices and sensors	Real-Time Data	Collects records on visitors, air excellent, power intake, etc.
Social media	Sentiment and Feedback Data	Gathers public sentiment and remarks on town offerings and occasions.
Government databases	Historical and Demographic Data	Provides public facts and demographic records for urban planning.
Mobile applications	User conduct records	Monitoring consumer options and behavior regarding SC services
Transportation systems	Operational information	Data from public delivery to optimize routes and schedules
Surveillance cameras	Video Data	Monitors public protection and visitors' conditions via video feeds.
Utility meters	Usage facts	Monitor water and electricity consumption for critical support of animal consumption.
Environmental monitoring	Environmental calculation	Collection of records on weather conditions, levels of pollutants, and names of useful natural resources.
Citizen reports	Reporting news and events	Hearing community criticism of community issues and organizational efforts.
Smart parking systems	Parking information	Check parking availability and usage patterns in real time.

social media, real-time statistics from IoT gadgets, and insights into personal conduct from cellular packages. It also incorporates utilization information from application meters, video pictures from safety cameras, and operational information from transportation systems. In smart metropolis settings, these big kinds of statistical resources improve service transport and help efficient metropolis management.

4.8.2 Data Preprocessing Techniques

In 6G-enabled SCs, data preprocessing strategies are essential for preparing unprocessed information for modeling and assessment. There are multiple steps in the system, along with characteristic choice, normalization, transformation, and fact cleaning. To ensure that the data is correct and sincere, information cleansing gets rid of noise, missing numbers, and inconsistencies in the dataset. Data is transformed through the usage of techniques like scaling and encoding to prepare it for evaluation. On the other hand, normalization standardizes the record variety and permits algorithmic reading.

By finding and retaining the maximum pertinent variables, characteristic choice together with cleansing and transformation is critical to enhancing the model's overall performance [33,34]. This can result in quicker processing speeds, advanced interpretability, and a dimensionality discount. Smart metropolis packages can assure amazing, actionable insights from information by using these preprocessing techniques, on the way to ultimately enhance urban offerings and decision-making. In complex urban environments, the right preprocessing of data not only improves the general dependability of results but also increases the performance of device learning and DL models.

In 6G-enabled clever towns, the essential approaches for preparing raw statistics for evaluation are proven within the data preprocessing pipeline, Figure 4.4. It goes through several steps, together with records cleaning to dispose of missing values and inconsistencies, transformation to format the information well, normalization to standardize ranges, and feature choice to locate the most vital variables. Smart town apps may also ensure reliable, splendid records by adhering to this pipeline, which, in the long run, results in a system gaining knowledge of and deep knowledge of models that enhance urban decision-making and service delivery.

4.8.3 Handling Big Data and Real-Time Data Streams

Managing Massive Data Volumes and Real-Time Data Streams highlights how critical it is to deal with the huge portions of records produced in 6G-enabled SCs. This entails shooting, processing, and studying each dependent and unstructured record in real time using reducing side technology and frameworks [34,35]. City management systems intend to take care of huge data correctly, which lets you make sure well-timed insights for urban planning and service delivery, given the growing range of IoT devices, sensors, and other information assets.

Technologies consist of Apache Spark for real-time analytics, Apache Hadoop for statistics processing, and Apache Kafka for event streaming. It is proven in Table 4.11 that it is necessary for huge records control in clever towns. With a unique consciousness of scalable sources and cloud computing, DL frameworks additionally

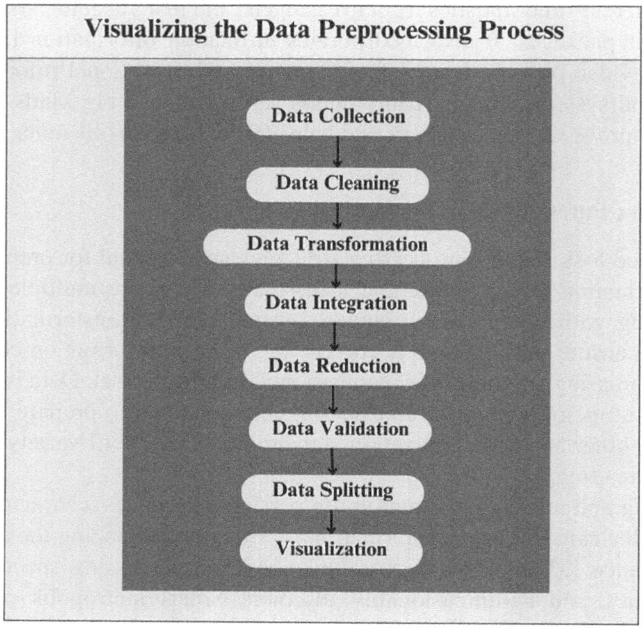

FIGURE 4.4 Visualizing the data preprocessing process.

TABLE 4.11
Big Data Technologies and Their Applications in SCs

Technology	Description	Use Cases
Apache Hadoop	Open source tool for storing and processing large logs.	Data collection and processing in smarter cities
Apache Spark	Fast engine for real-time data processing and analysis.	Quickly analyze data as it is available.
Apache Kafka	Platform to support many data activities simultaneously.	Real-time storage of logs from IoT devices.
Apache Flink	Tool for processing data streams in real time.	Monitoring city structure through data flow.
NoSQL databases	Flexible database for managing unstructured statistics.	Collection of facts about transport operators in several major cities.
Cloud computing	Online service provider offering scalable garages and processing.	Access to storage and processing assets as needed.
Data lakes	Large garage for all forms of statistics, dependent and unstructured.	Keeping raw records for later use.
ML platforms	Tools for constructing fashions that examine statistics.	Finding trends in metropolis facts.
DL frameworks	Advanced fashions that mimic the human mind to analyze records.	Recognizing snapshots, speech, and predicting issues.

enhance predictive skills through examining complex styles, and a device that learns the platform researches the town's tendencies.

4.9 CHALLENGES AND SOLUTIONS IN ANOMALY DETECTION

4.9.1 Data Quality and Labeling Issues

The fulfillment of anomaly detection systems is significantly impacted by record best and labeling issues, mainly in complicated contexts, which include smart towns [36]. High-quality information is vital to train gadget learning models, considering noise or inconsistencies within the statistics may cause the models to perform poorly and classify anomalies incorrectly. For example, noisy or inconsistent facts from IoT sensors could cause the wrong categorization of ordinary patterns as anomalies, resulting in needless alarms and inefficient operations.

The procedure of detecting anomalies is made tough using labeling troubles. Since anomalies are often unusual, getting a large and sufficiently categorized dataset for supervised mastering models can be hard [37]. The issue can be made worse with the aid of erroneous or inconsistent labeling because fashions trained on mislabeled facts are apt to perform poorly in real-world contexts. Organizations can enhance information best by implementing robust statistics preprocessing methods like validation, standardization, and cleansing to clear up those troubles [38,39]. The need for categorized items can also be decreased through the use of unsupervised and semi-supervised learning strategies, which allow fashions to gain knowledge of the natural styles within the data. In dynamic metropolitan settings, these methods can improve the accuracy and reliability of anomaly detection systems.

4.9.2 Scalability and Computational Complexity

Significant boundaries to anomaly detection in statistics-rich contexts, together with smart towns, are scalability and computational complexity. The capacity of anomaly detection systems to scale efficiently becomes increasingly crucial as the amount of data created via IoT gadgets, sensors, and different assets continues to increase dramatically. Large datasets have to be dealt with the aid of systems in real time without sacrificing performance [40]. Models with excessive scalability can handle greater information quickly and more accurately, while retaining accuracy.

When algorithms need loads of memory and processing ability to evaluate massive datasets, computational complexity effects, the gadget's capability to discover abnormalities in real time can be hampered by this complexity [41], which could boost latency. Organizations can use parallel processing strategies and powerful algorithms to unfold the computing load to address those problems. Furthermore, employing cloud computing sources can provide the specified flexibility and scalability, allowing the dynamic distribution of assets in response to demand. Organizations might also improve machine overall performance and ensure timely anomaly detection by optimizing the structure and algorithms used for anomaly detection. This will, in the long run, improve operational performance and decision-making in SC environments.

4.9.3 Real-Time Processing and Latency Concerns

Real-time processing is an essential issue of anomaly detection systems, particularly in dynamic scenarios like smart towns. The ability to investigate data and pick out anomalies in real time is essential for quick decision-making and response to capacity problems, such as visitor accidents, cybersecurity attacks, or environmental risks. However, there are demanding situations with real-time processing because of variables including data drift, processing speed, and system responsiveness.

The time lag between the data series and anomaly detection, or latency, can substantially affect how effective these systems are. Missed opportunities to act quickly can exacerbate crises or jeopardize people's protection because of excessive latency. To ease those concerns, efficient statistics processing structures ought to be put in the area, which includes flow processing frameworks that can manage continuous statistics flows with minimal delay. Additionally, edge computing techniques and pace-optimized algorithms can help reduce latency by processing data in the direction of its supply. Anomaly detection structures that prioritize real-time processing competencies and reduce latency can grow their operational efficacy and provide timely insights for higher public safety and concrete control.

4.10 CONCLUSION

In the framework of 6G-enabled smart towns, this article gives an in-depth overview of the critical role that ML and DL techniques play within the discipline of anomaly detection. The requirement for green anomaly detection systems grows as urban environments turn into smart ecosystems to maintain public safety, manipulate assets correctly, and offer better offerings. Cities may examine huge volumes of real-time data by utilizing current ML and DL processes. This permits the spark off discovery and determination of abnormalities that could impact urban operations. The examination of several problems like scalability, real-time processing, and information quality highlights how hard it is to place those structures into practice successfully. The chapter's solutions highlight the significance of reliable records and preprocessing methods, and the combination of brand-new technologies as a vital means of resolving latency troubles. In the end, our paintings highlight how hybrid and ensemble procedures can considerably enhance the accuracy of anomaly detection, opening the door to greater smart, stable, and resilient urban environments in the generation of 6G connections.

REFERENCES

1. A. Boualouache and T. Engel, "A Survey on Machine Learning-based Misbehavior Detection Systems for 5G and Beyond Vehicular Networks," Jan. 2022, [Online] Available: https://arxiv.org/abs/2201.10500
2. M. Murroni et al., 6G—Enabling the New Smart City: A Survey," Sep. 01, 2023, doi: 10.3390/s23177528.
3. A. A. Puspitasari, T. T. An, M. H. Alsharif, and B. M. Lee, "Emerging Technologies for 6G Communication Networks: Machine Learning Approaches," Sep. 01, 2023, doi: 10.3390/s23187709.

4. C.-X. Wang et al., "On the Road to 6G: Visions, Requirements, Key Technologies and Testbeds," Feb. 2023, [Online] Available: https://arxiv.org/abs/2302.14536
5. M. A. Ferrag et al., "Edge Learning for 6G-enabled Internet of Things: A Comprehensive Survey of Vulnerabilities, Datasets, and Defenses," Jun. 2023, [Online] Available: https://arxiv.org/abs/2306.10309
6. M. Ben Driss, E. Sabir, H. Elbiaze, and W. Saad, "Federated Learning for 6G: Paradigms, Taxonomy, Recent Advances and Insights," Dec. 2023, [Online] Available: https://arxiv.org/abs/2312.04688
7. L. H. Shen, K. Ten Feng, and L. Hanzo, "Five Facets of 6G: Research Challenges and Opportunities," *ACM Comput Surv*, vol. 55, no. 11, Feb. 2023, doi: 10.1145/3571072.
8. L. Jiao et al., "Advanced Deep Learning Models for 6G: Overview, Opportunities and Challenges," 2023, doi: 10.1109/ACCESS.2023.0322000.
9. M. M., Saeed, R. A. Saeed, A. S. A. Gaid, R. A. Mokhtar, O. O. Khalifa, and Z. E. Ahmed. "Attacks Detection in 6G Wireless Networks using Machine Learning," *2023 9th International Conference on Computer and Communication Engineering (ICCCE)*, pp. 6–11. IEEE, 2023.
10. B. Langpoklakpam and L. K. Murry, "Review on Machine Learning for Intelligent Routing, Key Requirement and Challenges Towards 6G," *Computer Networks and Communications*, Jul. 2023, doi: 10.37256/cnc.1220233039.
11. M. R. Mahmood, M. A. Matin, P. Sarigiannidis, and S. K. Goudos, "A Comprehensive Review on Artificial Intelligence/Machine Learning Algorithms for Empowering the Future IoT Toward 6G Era," *IEEE Access*, vol. 10, pp. 87535–87562, 2022, doi: 10.1109/ACCESS.2022.3199689.
12. I. Ahmad, F. Rodriguez, J. Huusko, and K. Seppänen, "On the Dependability of 6G Networks," Mar. 01, 2023, doi: 10.3390/electronics12061472.
13. M. M. Saeed, R. A. Saeed, M. Abdelhaq, R. Alsaqour, M. K. Hasan, and R. A. Mokhtar, "Anomaly Detection in 6G Networks Using Machine Learning Methods," *Electronics*, vol. 12, no. 15, Aug. 2023, doi: 10.3390/electronics12153300.
14. M. Hosseinzadeh, A. Hemmati, and A. M. Rahmani, "6G-Enabled Internet of Things: Vision, Techniques, and Open Issues," 2022, doi: 10.32604/cmes.2022.021094.
15. G. Rzym, A. Masny, and P. Chołda, "Dynamic Telemetry and Deep Neural Networks for Anomaly Detection in 6G Software-Defined Networks," *Electronics*, vol. 13, no. 2, Jan. 2024, doi: 10.3390/electronics13020382.
16. M. Abd Elaziz et al., "Evolution toward Intelligent Communications: Impact of Deep Learning Applications on the Future of 6G Technology," *Wiley Interdisciplinary Reviews: Data Mining and Knowledge Discovery*, vol. 14, no. 1, Jan. 2024, doi: 10.1002/widm.1521.
17. B. Gera, Y. S. Raghuvanshi, O. Rawlley, S. Gupta, A. Dua, and P. Sharma, "Leveraging AI-Enabled 6G-Driven IoT for Sustainable Smart Cities," *International Journal of Communication Systems*, vol. 36, no. 16, Nov. 2023, doi: 10.1002/dac.5588.
18. L. Ismail and R. Buyya, "Artificial Intelligence Applications and Self-Learning 6G Networks for Smart Cities Digital Ecosystems: Taxonomy, Challenges, and Future Directions," Aug. 01, 2022, doi: 10.3390/s22155750.
19. H. M. F. Noman et al., "Machine Learning Empowered Emerging Wireless Networks in 6G: Recent Advancements, Challenges and Future Trends," 2023, doi: 10.1109/ACCESS.2023.3302250.
20. N. Rajkumar, C. Viji, P. M. Latha, V. B. Vennila, S. K. Shanmugam, and N. B. Pillai, "The Power of AI, IoT, and Advanced Quantum Based Optical Systems in Smart Cities," *Opt Quantum Electron*, vol. 56, no. 3, Mar. 2024, doi: 10.1007/s11082-023-06065-0.
21. M. Gupta, M. Sabraj, and I. Member, "Security Challenges in 6G Networks: A Game Theory-based Intrusion Detection with Network Slicing and AI/ML Perspectives," 2024, doi: 10.20944/preprints202405.1066.v1.

22. V. Inzillo, D. Garompolo, and C. Giglio, "Enhancing Smart City Connectivity: A Multi-Metric CNN-LSTM beamforming Based Approach to Optimize Dynamic Source Routing in 6G Networks for MANETs and VANETs," 2024, doi: 10.20944/preprints202407.1161.v1.
23. M. N. Mahdi, A. R. Ahmad, Q. S. Qassim, H. Natiq, M. A. Subhi, and M. Mahmoud, "From 5G to 6G Technology: Meets Energy, Internet-of-Things and Machine Learning: A Survey," *Applied Sciences*, vol. 11, no. 17, Sep. 2021, doi: 10.3390/app11178117.
24. A. R. Javed, W. Ahmed, S. Pandya, P. K. R. Maddikunta, M. Alazab, and T. R. Gadekallu, "A Survey of Explainable Artificial Intelligence for Smart Cities," Feb. 01, 2023, doi: 10.3390/electronics12041020.
25. S. A. Abdel Hakeem, H. H. Hussein, and H. Kim, "Security Requirements and Challenges of 6G Technologies and Applications," *Sensors*, vol. 22, no. 5, Mar. 2022, doi: 10.3390/s22051969.
26. G. Rzym, A. Masny, and P. Chołda, "Dynamic Telemetry and Deep Neural Networks for Anomaly Detection in 6G Software-Defined Networks," *Electronics*, vol. 13, no. 2, Jan. 2024, doi: 10.3390/electronics13020382.
27. H. Hafi, B. Brik, P. A. Frangoudis, A. Ksentini, and M. Bagaa, "Split Federated Learning for 6G Enabled-Networks: Requirements, Challenges, and Future Directions," *IEEE Access*, vol. 12, pp. 9890–9930, 2024, doi: 10.1109/ACCESS.2024.3351600.
28. X. Fernando and G. Lăzăroiu, "Energy-Efficient Industrial Internet of Things in Green 6G Networks," *Applied Sciences*, vol. 14, no. 18, p. 8558, Sep. 2024, doi: 10.3390/app14188558.
29. R. Chataut, M. Nankya, and R. Akl, "6G Networks and the AI Revolution—Exploring Technologies, Applications, and Emerging Challenges," Mar. 01, 2024, doi: 10.3390/s24061888.
30. M. H. Alsamh, A. Hawbani, S. Kumar, and S. H. Alsamhi, "Multisensory Metaverse-6G: A New Paradigm of Commerce and Education," *IEEE Access*, vol. 12, pp. 75657–75677, 2024, doi: 10.1109/ACCESS.2024.3392838.
31. L. Jiao et al., "Advanced Deep Learning Models for 6G: Overview, Opportunities and Challenges," *IEEE Access*, 2024, doi: 10.1109/ACCESS.2024.3418900.
32. J. Bae, W. Khalid, A. Lee, H. Lee, S. Noh, and H. Yu, "Overview of RIS-Enabled Secure Transmission in 6G Wireless Networks," *Digital Communications and Networks*, Mar. 2024, doi: 10.1016/j.dcan.2024.02.005.
33. M. Rowshan, M. Qiu, Y. Xie, X. Gu, and J. Yuan, "Channel Coding Toward 6G: Technical Overview and Outlook," *IEEE Open Journal of the Communications Society*, vol. 5, pp. 2585–2685, 2024, doi: 10.1109/OJCOMS.2024.3390000.
34. F. M. Ghamry, G. M. El-Banby, A. S. El-Fishawy, F. E. A. El-Samie, and M. I. Dessouky, "A survey of anomaly detection techniques," *Journal of Optics*, vol. 53, no. 2, pp. 756–774, Apr. 2024, doi: 10.1007/s12596-023-01147-4.
35. H. Pennanen, T. Hänninen, O. Tervo, A. Tölli, S. Member, and M. Latva-aho, "6G: The Intelligent Network of Everything-A Comprehensive Vision, Survey, and Tutorial." *arXiv e-prints*, arXiv-2407, 2024
36. A. Celik and A. M. Eltawil, "At the Dawn of Generative AI Era: A Tutorial-Cum-Survey on New Frontiers in 6G Wireless Intelligence," *IEEE Open Journal of the Communications Society*, vol. 5, pp. 2433–2489, 2024, doi: 10.1109/OJCOMS.2024.3362271.
37. N. G. Ramadhan, W. Maharani, and A. A. Gozali, "Chronic Diseases Prediction Using Machine Learning with Data Preprocessing Handling: A Critical Review," *IEEE Access*, vol. 12, pp. 80698–80730, 2024, doi: 10.1109/ACCESS.2024.3406748.
38. H. Zeng, M. Yunis, A. Khalil, and N. Mirza, "Towards a Conceptual Framework for AI-Driven Anomaly Detection in Smart City IoT Networks for Enhanced Cybersecurity," 2024, doi: 10.20944/preprints202404.0924.v1.

39. D. Fahrmann, L. Martin, L. Sanchez, and N. Damer, "Anomaly Detection in Smart Environments: A Comprehensive Survey," *IEEE Access*, vol. 12, pp. 64006–64049, 2024, doi: 10.1109/ACCESS.2024.3395051.
40. A. Singhal, N. Singhal, and P. Kumar, "Automated Crime Anomaly Detection in Smart Cities Using Sharkprey Optimization Algorithm and Ensembled-Machine Learning Approach," 2024. [Online] Available: www.nano-ntp.com
41. Y. Özkan, M. Demirarslan, and A. Suner, "Effect of Data Preprocessing on Ensemble Learning for Classification in Disease Diagnosis," *Communications in Statistics-Simulation and Computation*, vol. 53, no. 4, pp. 1657–1677, 2024, doi: 10.1080/03610918.2022.2053717.

5 Security Paradigms and Challenges in the Convergence of IoT and 6G Networks

Akshay Mudgal

5.1 INTRODUCTION

The 6G wireless network is poised to revolutionize the digital landscape, the way we experience digital technology, and speed, connectivity, and automation. This is the transformative age of ubiquitous connectivity as 6G networks start to meld with the vast ecosystem rapidly growing around the IoT [1,2]. This collaboration opens the possibility of unprecedented opportunities in various domains, from healthcare to smart cities, and to future-looking sectors like manufacturing and autonomous systems, all enabling human-level quality with machine efficiency that seemed only possible in sci-fi movies [3,4]. Yet, the benefits associated with such advancements also bring serious and complicated security hurdles that need to be addressed with immediate measures for the network transition into a more secure acceleration [5,6].

IoT inherently gives a decentralized and heterogeneous view. This is because the devices used in IoT are typically resource-constrained (i.e., computational power, storage, or energy), making them vulnerable to manipulation by bad actors if proper precautions aren't implemented. Once these IoT devices are integrated into the ultra-high-speed and massively connected 6G networks, threats could also be multiplied on an exponential scale [7,8]. The IoT-6G environment presents a number of new security challenges that traditional, centralized, and static security measures are not prepared to handle [9,10]. This calls for the creation of new security principles, models, and the reshaping of older paradigms to secure connected technologies into old age [11,12].

Data privacy is one of the main security issues that needs to be considered with 6G-IoT convergence. This vast amount of data generated by IoT, supported by fast-moving data transfer speeds such as 6G, provides excellent fodder for information theft and unauthorized access. If sensitive information such as personal health records, financial transactions, or location data is intercepted and manipulated, there will be serious repercussions for both individuals and organizations. Indeed, the organization of this data has to rely on strong encryption techniques and strict access control methods, while being flexible enough for even resource-scarce IoT devices [13,14].

 DOI: 10.1201/9781003586456-5

Authentication and identity management of the large number of devices that will exist within 6G-powered IoT networks are another considerable hurdle [15,16]. The number of connected devices is increasing exponentially, making it a complex issue for every device to be authenticated. Inadequate authentication methods result in unauthorized device access, allowing attackers to enter the network and compromise data integrity, or disrupt critical services [17,18]. There is a push toward exploring innovative solutions, including decentralized control over identity and biometric authentication, to strengthen security in this space.

In the context of IoT-6G, network integrity is another important element. Many 6G applications will need ultra-low-latency and a very high level of reliability – such as autonomous vehicles or industrial automation, for example – allowing zero error tolerance. Any satellite network outage, whether due to malicious attacks or accidental failures, will result in catastrophic consequences [19,20]. To maintain the integrity and stability of these networks, there is a need for sophisticated risk detection and mitigation solutions, including anomaly detection powered by AI and automated response systems [21].

Additionally, the combination of IoT with 6G networks brings new difficulties and challenges in protecting against DDoS attacks. The sheer scale and complexity of 6G networks, along with the billions, if not trillions, of IoT devices, create a target-rich environment for attackers [22]. The scale and sophistication of attacks may be beyond the capabilities of many traditional DDoS mitigation solutions, requiring increasingly advanced, AI-powered defenses and exploring quantum-resistant cryptography to keep pace with evolving attack vectors.

More details of these security challenges and state-of-the-art approaches to remedy them are provided in this chapter. The active participation of researchers, policymakers, and industry stakeholders will continue to play a fundamental role in the ongoing convergence enabler discourse; thus, this chapter has attempted to make an urgent approach, analyzing risks associated with IoT-6G from fine-grained perspectives while coaching novel practitioners. We are working to lay the groundwork for a digitally secured and resilient future in 6G/IoT times, where these game-changing technologies will provide all their benefits without putting any security at risk [23]. This chapter entails an in-depth analysis of the IoT and 6G network, subsequently proposing a robust solution in order to overcome the issues faced in today's era.

5.1.1 A Detailed Insight

In the subsequent section of this chapter, the detailed insights of the incidents recorded from 2010 to 2021 are presented in the form of a table.

Table 5.1 illustrates the growing threat of attacks on IoT devices and networked systems, listing important security breaches (2010–2021) by their increasing complexity in style. The Mirai Botnet and, more recently, the Colonial Pipeline ransomware incident highlight just how insecure IoT devices can be, with consequences ranging from minor disruptions to significant, large-scale data breaches. In light of these incidents, stronger security practices are all the more necessary and timely, especially as IoT embeds its ground into 6G networks. Every breach provides a critical learning opportunity on securing connected devices and infrastructure against the growing threat of cyber war.

TABLE 5.1
Malicious Incidents Recorded

Incident	Year	Description	Impact	Lessons Learned
Stuxnet Worm	2010	Malware specifically designed to target industrial control systems, particularly in Iran's nuclear program.	Disrupted Iran's nuclear enrichment process; first known cyber weapon targeting physical infrastructure.	Highlighted the risks of cyberattacks on critical infrastructure through networked systems.
Mirai Botnet	2016	Large-scale DDoS attack using compromised IoT devices, targeting DNS provider Dyn.	Major websites, including Twitter and Netflix, were knocked offline, affecting millions of users.	Stressed the need for stronger security in IoT devices and awareness of IoT-based attacks.
Krebs on Security DDoS Attack	2016	Record-breaking DDoS attack on the cybersecurity journalist's site using IoT devices.	Temporarily disabled the website, highlighting the power of IoT-based botnets.	Emphasized the vulnerability of connected devices and the potential scale of network-based attacks.
BrickerBot	2017	Malware that permanently disabled IoT devices by corrupting their storage and rendering them unusable.	Destroyed over 10 million devices, targeting insecure IoT devices globally.	Demonstrated the destructive potential of malware targeting IoT devices, not just exploiting them.
Reaper Botnet	2017	Evolved from Mirai, this botnet incorporated more sophisticated attacks and spread across millions of devices.	Infected IoT devices worldwide, with the potential to launch massive DDoS attacks.	Showed the rapid evolution of IoT-targeting malware and the need for continuous security improvements.
SweynTooth Vulnerabilities	2020	Series of vulnerabilities in Bluetooth Low Energy (BLE) chips used in millions of IoT devices.	Affected medical devices, smart home devices, and wearables, leading to potential device malfunctions.	Highlighted the widespread risk of common vulnerabilities across multiple industries relying on IoT.
Verkada Camera Breach	2021	Hackers accessed live feeds from 150,000 security cameras used by companies and organizations.	Breach of privacy and security at companies, hospitals, schools, and police departments.	Emphasized the importance of securing IoT surveillance devices and implementing strong authentication.

(Continued)

TABLE 5.1 (*Continued*)
Malicious Incidents Recorded

Incident	Year	Description	Impact	Lessons Learned
PNNL Breach	2021	Attackers exploited IoT vulnerabilities within Pacific Northwest National Laboratory's network.	Risked exposing critical research data, showing the vulnerability of IoT in sensitive environments.	Stressed the importance of securing IoT devices in high-stakes, secure environments.
SolarWinds Attack	2020–2021	Sophisticated supply chain attack that compromised numerous government agencies and corporations.	Widespread data breaches and espionage, affecting many U.S. federal agencies.	Highlighted the vulnerability of network supply chains and the importance of comprehensive security measures.
Colonial Pipeline Ransomware Attack	2021	Ransomware attack led to the shutdown of a major U.S. fuel pipeline, affecting supply chains.	Caused significant disruption in fuel supply across the Eastern U.S.	Underlined the critical vulnerability of infrastructure connected to IoT networks and the risks of ransomware.

5.2 PROBLEM STATEMENT

The integration of IoT devices with 6G networks poses major threats to the security, authenticity, and privacy of interconnected systems. As IoT devices expand in numbers and across sectors, the decentralized and heterogeneous nature of these networks will further increase the attack surface for cyber threats, along with additional requirements set by 6G for ultra-low-latency and very high data rates, leading to massive connectivity. This opens the door to not only more comprehensive threat schemas but also new operational challenges at both deployment and recovery levels. There is a greater chance of data breaches, illegal access, and sophisticated DDoS attacks because the security frameworks in place are frequently unable to handle these complexities.

5.3 LITERATURE REVIEW

Jahid et al. (2023) address the potential and difficulties of combining blockchain, IoT, and 6G in the endeavor to create the next generation of communication networks. These difficulties include privacy issues, technological obstacles, and the intricacy of data management. The issues raised prompt the authors to develop a suitable study agenda, which will aid in any further studies on the subject.

Abdel Hakeem et al. (2022) identified a number of criteria and difficulties that must be addressed in the development and commercialization of 6G technologies and applications. Since AI-supported ultra-expedited networks can be effective when constructed in more secure frameworks, security concerns are raised. Regarding the development of 6G systems, the authors emphasize the dangers of data breaches and cyberattacks and stress the significance of putting safeguards in place to guarantee the safe operation of systems.

As with their previous work, Kazmi et al. focus on security concerns in 6G communication but narrow their discussion to targeted threats and authentication techniques. The article stresses that the advanced capabilities of 6G increase the risk of cyberattacks, attributing the tension to the absence of new security solutions. Some new authentication mechanisms are presented as potential solutions, as well as key milestones for research activity in the area of 6G security architecture.

This paper identifies the main research gaps in the privacy, security, and trust of 6G networks (Ylianttila et al., 2020). In addition to highlighting the need for better security frameworks, the authors reviewed the arrival of new 6G technologies like AI and large IoT and associated risks. They recommend more research on how to promote privacy and trust in the upcoming 6G ecosystem.

Wang et al. (2020) investigate the new security and privacy risks that are arising in 6G networks as a result of the integration of blockchain, AI, and even quantum computing. This study discusses the dangers and vulnerabilities that have been found in these areas and supports recommended solutions, such as stronger encryption and privacy-enabling strategies, for the safe development and implementation of 6G networks.

With an emphasis on the emerging AI edge computing and blockchain, Nguyen et al. (2021) provide a thorough analysis of security and privacy regulations pertaining to 6G networks. The study tries to examine how more recent systems can counteract the hazards associated with the technology. The authors highlight the necessity of developing new security frameworks to address the unique requirements of 6G and offer potential areas for further research.

Porambage et al. (2021) offer an elaborate blueprint to tackle security and privacy challenges in 6G networks. The paper identifies key challenges such as securing millions of IoT devices and integrating AI into the setup. Among others, the authors propose enhanced security measures in the form of modified trust models and new encryption schemes, along with a prioritized research agenda for protecting the anticipated 6G environment.

5.4 METHODOLOGY

We will now devise a plan for any crypto asset losses resulting from the blending of cutting-edge IoT cybersecurity techniques with cryptography. An AI model capable of identifying security challenges should be developed. Data engineering may also be utilized to feed AI algorithms, allowing for automatic threat detection. A unique deep learning model will be designed to assist IoT heavily constrained devices in monitoring for potential security breaches and anticipating existing ones. In addition to this, we will use the blockchain in creating a structure of self-managing identity providers for simple and yet high-performance authentication control that positively affects everyone within our network. Instead of depending solely on implementations or demonstrations, which are still significant, and against de facto requirements such as false positive rates in threat detection and computational overheads when performing cryptographic operations, we will create sufficient mathematical models for correctness refinement in this context. This will allow us to assess the efficacy of our solutions using a scientific method and ensure scalability and assurance under conditions that are frequently encountered by dynamic 6G-IoT environments.

5.5 DETAILED THREAT LANDSCAPE

The evolution of global connectivity capabilities has advanced significantly with the convergence of 6G networks and the IoT. However, this convergence also results in a broader threat landscape that includes a variety of attack methods and weaknesses. IoT's decentralized, large-scale, and heterogeneous characteristics, along with 6G's lightning-fast, low-latency capabilities, create new, intricate security threats. Some of the most important concerns are examined in this section, including supply chain issues, DDoS amplification, and compromised device integrity.

5.5.1 Attack Vectors in IoT/6G

IoT/6G Attack Vectors: The emergence of IoT devices in 6G networks gives malicious actors plenty of attack vectors through which they may hack. There are various factors leading to the increased susceptibility, especially those related to intrinsic IoT system properties and the further attack surface through the expanded features in 6G networks.

5.5.2 Compromised Device Integrity

Compromised Integrity of Devices: Devices in IoT environments are designed with low processing power and limited security mechanisms [24]. These tend to be highly vulnerable to physical tampering, exploitation of firmware vulnerabilities, and remote hijacking. After being compromised, an IoT device can be used as a springboard into larger networks, allowing the adversary to manipulate data, disrupt operations, or steal sensitive information. In the context of 6G networks, where the IoT devices will access enormous amounts of data at a greater speed, this compromise is now amplified.

5.5.3 Malicious Firmware

IoT ecosystems are at significant danger from firmware attacks. By inserting malicious firmware upgrades into devices, attackers can take over the hardware and perform destructive operations covertly. As billions of devices are anticipated to be connected via 6G networks, firmware update integrity must be guaranteed. Unencrypted firmware transmissions or inadequate authentication techniques may result in pervasive vulnerabilities across the IoT/6G environment.

5.5.4 Insider Attacks

Insider threats grow in importance as the number of connected devices rises. Employees with access to vital systems are examples of insiders who might abuse their position to compromise 6G network components or IoT devices. Insiders have the ability to enable unauthorized access to confidential data or interfere with critical network functions, whether through purposeful or inadvertent activities. Monitoring insider threats across several domains and communication levels is a significant difficulty in large 6G networks [25].

5.5.5 Securing Large-Scale Networks

A major challenge in securing the IoT/6G ecosystem is the vast scale and diversity of the devices involved. IoT devices differ significantly in processing power, capabilities, and operating environments. Integrating billions of these devices into a 6G network adds complexity to maintaining uniform security protocols. Additionally, the decentralized nature of IoT systems makes centralized security management difficult, hindering the ability to detect, isolate, and respond to threats in real time.

5.5.6 DDoS Amplification in 6G

DDoS attacks have been a significant threat to internet-connected systems for a long time. The introduction of 6G networks heightens the risk of these attacks due to the combination of high bandwidth, ultra-low latency, and the ability to connect a massive number of devices.

5.5.7 Enhanced DDoS Capability

In a 6G setting, attackers can take advantage of the ultra-high speeds and extensive network capacity to execute highly amplified DDoS attacks. The vast number of IoT devices, each potentially compromised and used as part of a botnet, significantly increases the overall impact of such attacks [26]. A hacked IoT device can be instructed to flood a target with large amounts of traffic, overwhelming the system and making it inoperable. With 6G facilitating quicker and more efficient communication, the scale of a DDoS attack could grow dramatically, leading to widespread service interruptions.

5.5.8 Exploitation of Low Latency

The low-latency feature of 6G networks, which allows near-instantaneous communication, also provides attackers with new opportunities. Malicious actors can exploit the reduced delay between devices to coordinate more sophisticated, time-sensitive attacks. This could enable rapid, synchronized DDoS attacks where attackers can exploit vulnerabilities faster than defenses can be mounted.

5.5.9 Novel Attack Vectors

The unique architecture of 6G networks introduces new attack surfaces that were previously unavailable in earlier generations of communication technologies [10]. For example, the use of AI-driven, self-configuring networks in 6G could introduce risks of adversarial AI attacks, where malicious inputs are used to deceive AI models that manage network resources, causing intentional mismanagement and service degradation.

5.6 SUPPLY CHAIN RISKS

Supply chain vulnerabilities in IoT and 6G networks pose a great challenge when it comes to securing these environments. Globalized production and distribution of IoT devices make networks vulnerable to risks at all stages of the supply chain.

5.6.1 Backdoors and Tampered Components

As IoT devices are manufactured and distributed across multiple regions and vendors, the risk of tampered components or intentional backdoors inserted into devices during production is high. Such backdoors could be used to give malicious actors covert access to devices post-deployment for unauthorized surveillance, data exfiltration, or sabotage. Such compromised devices can be integrated into 6G networks, leading to the spread of vulnerabilities across multiple sectors, from healthcare to critical infrastructure [27].

5.6.2 Untrusted Suppliers

In the pursuit of deploying IoT devices at scale, organizations rely on third-party suppliers whose security practices may not meet stringent standards. Unvetted suppliers or vendors in regions with less regulation may introduce faulty or even malicious parts into devices, thereby introducing long-term security risks in them [12]. The dependence of 6G on heterogeneous IoT ecosystems makes supply chain risk management even more crucial since a vulnerability in the supply chain will propagate through the entire network.

5.6.3 Firmware and Software Updates

Firmware and software updates become another source of supply chain risk in the continuing maintenance of IoT devices. This is due to the fact that a corrupted update process will enable an attacker to infect millions of IoT devices with malware that poses as genuine code. The capacity to update millions of devices simultaneously in 6G contexts has made process security crucial and challenging.

5.7 ADVANCED CRYPTOGRAPHIC SOLUTIONS FOR IOT AND 6G NETWORKS

With greater data transfer speeds, incredibly low latency, and the capacity to link a wide range of devices, the combination of 6G networks and the IoT is poised to revolutionize global communication. But this important technological development also brings with it new privacy and security issues [28]. The potential threat posed by quantum computing, which would render conventional encryption techniques useless, is a significant worry. The confidentiality and integrity of data shared across IoT and 6G networks are seriously at risk since quantum computers may eventually be able to crack common encryption techniques [29]. Researchers are developing quantum-resistant cryptographic methods to counter these threats and protect communications in the future. This section explores the significance of post-quantum protocols and quantum-resistant cryptography.

5.7.1 Quantum-Resistant Cryptography

Cryptographic techniques that are anticipated to be immune to the capabilities of quantum computers are referred to as post-quantum cryptography (PQC). Conventional cryptography systems, such as Elliptic Curve Cryptography (ECC) and Rivest-Shamir-Adleman (RSA), rely on certain challenging mathematical issues, such as discrete logarithms or integer factorization. However, such encryption techniques are problematic because quantum computers, with the aid of algorithms like Shor's algorithm, can solve such problems exponentially quicker than a classical computer [30].

In the context of IoT and 6G networks, the need for quantum-resistant cryptography is paramount due to the large-scale transmission of sensitive data. IoT devices, which are often deployed in critical sectors such as healthcare, transportation, and smart cities, transmit vast amounts of real-time data that, if intercepted and decrypted by quantum computers, could result in severe privacy breaches, operational disruptions, or even national security threats. The ultra-fast data transmission and massive connectivity of 6G networks further escalate the potential impact of such quantum threats [31].

5.7.2 Future-Proofing with Quantum-Safe Encryption

Quantum-resistant cryptography aims to develop encryption schemes that can withstand both classical and quantum attacks. These encryption algorithms rely on mathematical problems that are believed to be resistant to quantum computation. By transitioning to quantum-safe encryption methods, IoT and 6G networks can be "future-proofed" against the impending risks posed by quantum computing. This ensures that even when quantum computers become operational, encrypted communications within IoT/6G ecosystems will remain secure.

5.7.3 Key Requirements for Quantum-Resistant Cryptography in 6G

- **Scalability:** Given the massive number of connected devices in 6G networks, any cryptographic solution must be scalable and efficient. Quantum-resistant algorithms must work within the limited computational resources of IoT devices while supporting the high-speed data transmission of 6G networks.
- **Compatibility with IoT Devices:** Since IoT devices often have constrained processing power, memory, and energy resources, post-quantum algorithms must be lightweight and optimized for such environments [32].

- **Interoperability:** Quantum-resistant cryptographic solutions must be interoperable with existing systems to ensure a smooth transition from current cryptographic standards to quantum-safe methods.

5.7.4 Post-Quantum Protocols

There are now a number of PQC techniques being developed to defend IoT and 6G networks against quantum threats. These protocols offer defense against both conventional and quantum assaults and are intended to supplement or replace existing encryption techniques. Among the important protocols being investigated in the field of PQC are:

- **Lattice-Based Cryptography:** For post-quantum encryption, lattice-based cryptographic algorithms are among the most promising options. These algorithms rely on the difficulty of solving lattice problems, which are thought to be impervious to both classical and quantum attacks. Examples of these problems are the Learning with Errors (LWE) issue and the Shortest Vector Problem (SVP). Strong security guarantees and support for sophisticated cryptographic primitives like fully homomorphic encryption (FHE), which enables calculations on encrypted data without the need for decryption, are just two benefits of lattice-based cryptography [33]. Lattice-based encryption can be utilized in the context of IoT and 6G to provide secure end-to-end communication for devices with limitations while maintaining the confidentiality and integrity of data transferred.
- **Code-Based Cryptography:** The other post-quantum method that employs error-correcting codes for encryption is called code-based cryptography. The McEliece cryptosystem, which relies on the intractability of decoding a general linear code, is the most well-known example of a code-based cryptosystem. Code-based methods have been available for a while and have shown themselves to be quantum-safe due to their resistance to known attacks on conventional cryptography [34]. The sole drawback is that code-based encryption necessitates a high key size, which could be a barrier for IoT devices with limited storage space. Code-based cryptography, however, is still a viable option for applications that need sustained data security in 6G settings.
- **Hash-Based Cryptography:** In order to create secure digital signatures that are resistant to quantum assaults, hash-based cryptographic systems use hash functions [18]. Because of its simplicity and very low computing requirements, this technique is best suited for IoT devices. Because of its effectiveness, hash-based signatures are best suited for use in low-power IoT devices that will function in 6G networks. Furthermore, unlike conventional cryptographic systems, hash-based signatures are safe from quantum attacks since they are not predicated on number-theoretic issues [35].
- **Multivariate Polynomial Cryptography:** The issue of solving systems of multivariate polynomial equations is the foundation of multivariate polynomial cryptography. These systems are seen to be a good fit for PQC since

they are efficient and secure. Multivariate cryptography's versatility in creating encryption and signature schemes that can be tailored to the unique needs of IoT devices operating in 6G networks is one of its main advantages.

- **Isogeny-Based Cryptography:** A new area of study called isogeny-based cryptography makes use of mathematical structures called elliptic curve isogenies. These systems are believed to offer tiny key sizes and resistance to quantum attacks, making them perfect for IoT devices with limited resources [36]. By providing quantum-resistant key exchange techniques, protocols based on isogenies, such as the Super-singular Isogeny Diffie-Hellman (SIDH) key exchange, could be used to protect communications in 6G networks.

5.7.5 Applications in IoT/6G Communication

Post-quantum cryptographic techniques can be used to improve security and privacy in a number of IoT and 6G communication contexts:

- **End-to-End Encryption:** It can ensure that data transfer between 6G networks and IoT devices is quite impervious to quantum computer attacks in the future. Even in the face of quantum computers, the use of lattice-based or code-based encryption guarantees that the private information is safely secured.
- **Digital Signatures:** Quantum-safe digital signatures for IoT device authentication can be produced using hash-based and multivariate cryptographic techniques. This reduces the possibility of hostile actors abusing lax authentication procedures by ensuring that only authorized devices can connect to the 6G network.
- **Key Exchange Protocols:** To create secure communication channels between IoT devices and 6G networks, key exchange protocols based on isogeny and lattice can be used. By facilitating the safe exchange of cryptographic keys in a manner impervious to quantum attacks, these protocols guarantee the privacy of upcoming communications.

5.8 AI AND MACHINE LEARNING (ML)-DRIVEN SECURITY IN IOT/6G NETWORKS

The way we detect, stop, and react to cyberthreats in real time has changed dramatically with the integration of AI and ML into the security architecture of IoT and 6G networks. Conventional security methods may find it difficult to handle the increasing complexity and scope of cyberattacks as IoT devices proliferate and 6G networks offer unparalleled speeds and connections. Real-time, autonomous threat detection and response are made possible by AI-driven solutions, ensuring the security and integrity of these networked systems.

5.8.1 AI-Powered Threat Detection

The ability of AI to recognize risks using ML models is a key use case in the IoT and 6G security framework. Traditional security systems rely on signature-based detection or pre-established regulations, which may not be sufficient to counter new or evolving threats. By identifying departures from normal system behavior, AI-powered threat detection, in contrast, uses anomaly detection and pattern recognition to identify possible threats.

- **Anomaly Detection:** It is possible to educate AI systems to recognize the typical patterns of 6G network traffic and IoT device behavior. ML models can monitor real-time data and identify anomalous actions that can indicate a security breach, such as illegal access, odd data transmission, or abrupt changes in device behavior, by employing supervised or unsupervised learning. A potential DDoS attack might be indicated, for instance, by an abrupt spike in data traffic from an IoT sensor in a smart city. AI-powered devices can assist in preventing or mitigating the impact of attacks before they cause extensive harm by spotting such anomalies early.
- **Pattern Recognition:** Network traffic or device behavior patterns that point to cyberthreats can also be found using AI models, particularly deep learning algorithms. AI systems, for example, are able to identify malware signatures by their behavior rather than their code. This feature makes it possible to find malware that hasn't been found yet, such as zero-day attacks, which could evade detection by conventional means. AI's scalability and adaptability make it a crucial tool for proactive threat detection in the world of 6G networks, where both the volume of data and the number of devices are much higher than in previous generations.
- **Real-Time Mitigation:** By enabling real-time reactions to cyberthreats, AI systems can do more than just detect them. AI algorithms can start automated processes, such as securing compromised equipment, banning questionable traffic, or alerting human administrators for additional examination, as soon as they detect an anomaly or malicious pattern. Since the speed and scope of possible attacks might outpace traditional security measures in 6G environments, these real-time mitigation techniques are very crucial.

5.8.2 Autonomous Security Systems

AI greatly enhances threat detection and makes it easier to build self-governing security systems that can anticipate, stop, and react to cyberthreats with little assistance from humans. These systems separately handle security duties across IoT and 6G networks using cutting-edge AI techniques like reinforcement learning and predictive analytics.

- **Predictive Threat Analysis:** By analyzing past data and identifying trends that might indicate an attack, AI systems are able to anticipate possible dangers before they materialize. For example, ML models might predict future attack pathways by analyzing past instances of device vulnerabilities or network breaches. Predictive threat analysis can identify high-risk devices or regions in an IoT/6G network with billions of linked devices and put preventive actions in place, such as improving access controls or fixing vulnerabilities.

- **Autonomous Response to Threats:** AI-powered security solutions don't require human participation to respond to threats in real time. For instance, an AI model can quickly isolate impacted devices, restrict suspicious traffic, or reroute network resources to prevent additional harm if it detects anomalous activity in an IoT device network. A self-evolving security framework for IoT/6G settings is formed by reinforcement learning algorithms, which allow these autonomous systems to continuously modify and improve their reactions to threats depending on the outcomes of prior actions.
- **Use Cases for Autonomous Security:** In IoT/6G networks, potential uses for AI-powered autonomous security solutions include:
  a. **Smart Cities:** In smart cities, AI systems can autonomously monitor IoT device security, preserving the integrity of vital infrastructure such as energy grids, transportation systems, and public services.
  b. **Industrial IoT:** AI can secure and monitor linked sensors and machinery in industrial settings, avoiding disruptions from hacks that target operational technologies.
  c. **Healthcare IoT:** In healthcare settings, AI-driven security can protect patient data and linked medical devices, avoiding unwanted access or tampering with vital equipment.

5.9 DECENTRALIZED SECURITY MODELS IN IOT/6G NETWORKS

Apart from AI-powered security, decentralized security models are showing promise as a potent remedy for the problems brought on by the convergence of 6G and IoT networks. These approaches reduce the risks associated with centralized points of failure by securing transactions, communications, and identities in a decentralized manner using distributed technologies like blockchain and self-sovereign identity (SSI).

5.9.1 Blockchain for IoT

Distributed ledger technology (DLT) and blockchain offer a decentralized approach to IoT/6G ecosystem security. Blockchain can guarantee the validity and integrity of data sent between linked devices by creating a distributed, unchangeable ledger of transactions. By eliminating the need for a central authority to validate transactions, this decentralized method lowers the possibility of single points of failure and improves network security overall [24].

- **Secure Device Authentication:** Secure device authentication is one of the main uses of blockchain in IoT/6G. A distinct cryptographic identity that is kept on a blockchain ledger can be assigned to every IoT device. Only authenticated devices are allowed access when devices connect to the 6G network or speak with one another, since their identities may be checked against the blockchain. This reduces the possibility of spoofing attacks and keeps unwanted devices out of the network.

- **Consensus-Based Integrity Verification:** Blockchain's consensus process makes sure that all communications and transactions between IoT devices are confirmed by several nodes before being entered into the ledger. This consensus-based approach guards against tampering and illegal alterations while ensuring data integrity. Blockchain can provide a reliable means of confirming the legitimacy of communications and preventing malevolent actors from interfering with data transmissions in a 6G network, where data is transferred over great distances and at fast rates.
- **Use Cases of Blockchain in IoT/6G:**
  a. **Supply Chain Security:** By guaranteeing that each component can be tracked and prohibiting the entrance of modified devices into the network, blockchain can improve the security of the supply chain for IoT devices.
  b. **Decentralized IoT Marketplaces:** Decentralized marketplaces for IoT services can be made possible by blockchain, which enhances security and privacy by enabling devices to safely exchange resources or data without the need for middlemen.

5.9.2 SSI

Users can control their own digital identities without relying on centralized authority thanks to the decentralized identification architecture known as SSI. When it comes to IoT and 6G networks, SSI offers a privacy-focused solution that lets users manage their data and devices on their own.

- **Decentralized Identity Management:** Every IoT device can have its own decentralized identity and be fully controlled by its owner thanks to SSI. Users may choose which devices get access to their data, how it is shared, and under what conditions, thanks to this feature. In IoT/6G environments, SSI reduces the dangers of identity theft, data breaches, and spying by doing away with the necessity for third-party identity suppliers.
- **Privacy and Data Ownership:** SSI frameworks give people total control over their personal data, emphasizing user privacy. Users can decide which devices or services to share information with in an IoT/6G network, making sure that only the most important information is exchanged. This method increases user control over their data and reduces the possibility of unwanted access to private data.
- **Use Cases of SSI in IoT/6G:**
  a. **Smart Homes:** Homeowners can govern who can interact with their connected devices, including security cameras or smart locks, by using SSI to manage access to IoT devices in smart homes.
  b. **Healthcare:** SSI can improve privacy and security in healthcare IoT contexts by ensuring that medical devices and healthcare professionals have controlled and secure access to patient data.

5.10 PROPOSED ALGORITHM TO DEVELOP A SECURITY FRAMEWORK FOR IOT-6G NETWORKS

Step 1: Hybrid Cryptographic Protocol Development

1.1: Design of Quantum-Resistant Encryption:

- Create a lattice-based cryptographic technique, such as LWE, where the security is based on how challenging lattice problems are. Let s stand for a secret vector, e for an error vector, and A for a random matrix. The formula for the public key is As+e. The message's encryption is

$$c = As' + e' + m \cdot \frac{q}{2} \tag{5.1}$$

  where s' and e' are new random vectors.
- Implement this cryptosystem in a lightweight form optimized for IoT devices by reducing the dimensionality of A while ensuring sufficient security against quantum attacks.

1.2: Efficient Key Management:

- Use elliptic curve Diffie-Hellman (ECDH) for secure key exchange between IoT devices and 6G base stations. Given public keys $PA= kA \cdot G$ and $PB= kB \cdot G$, where kA and kB are private keys and G is the generator point, the shared secret key is

$$S = kA \cdot PB= kB \cdot PA \tag{5.2}$$

1.3: Performance Optimization:

- Implement these cryptographic protocols in a modular arithmetic library optimized for low-power IoT devices. Use Montgomery reduction for efficient modular multiplication, reducing computational overhead $O(\text{crypt})$.

Step 2: AI-Powered Threat Detection and Mitigation

2.1: Data Collection and Feature Extraction:

- Develop a real-time data collection framework that captures network traffic $X = \{x_1, x_2, \ldots, x_n\}$, where x_i is a feature vector containing packet size, flow duration, entropy, and other relevant metrics.
- Perform feature selection using mutual information $I(x_i; y)$ to identify the most predictive features for anomaly detection.

2.2: Deep Learning Model Development:

- Train a deep neural network (DNN) $f(x;\theta)$ with parameters θ on labeled data (X, y), where y indicates normal or anomalous behavior. Use a loss function:

$$L(\theta) = -\sum i\left[yi \log\log\left(y^i\right) + (1 - yi)\log\log\left(1 - y^i\right) \right] \tag{5.3}$$

  and optimize with Adam optimizer.
- Integrate an autoencoder layer before the classification layer to reduce dimensionality and detect novel attacks.

2.3: Real-time Anomaly Detection:

- Deploy the trained model in a distributed manner across edge devices. For each incoming data point x^{new}, calculate $y^{new} = f(x^{new};\theta)$. If $y^{new} > T$ (threshold), flag the packet as anomalous.
- Implement a feedback loop where detected anomalies are stored and used to retrain the model periodically, ensuring adaptive learning.

Step 3: Blockchain-Based Decentralized Identity Management

3.1: Smart Contract Development:

- Create smart contracts *SCauth* a permissioned blockchain platform like Hyperledger Fabric. These contracts handle identity creation, verification, and revocation. Each IoT device is assigned a unique identifier *IDdevice*, and transactions are signed using the device's private key.

3.2: Identity Registration Process:

- Implement a registration protocol where each device generates a public–private key pair $(kpub, kpriv)$ and submits *kpub* to the blockchain via *SCauth* The blockchain records the hash $H(IDdevice \| kpub)$.
- Use Zero-Knowledge Proofs (ZKP) to verify device identities without revealing the private key *kpriv* ensuring privacy.

3.3: Decentralized Access Control:

- Implement attribute-based access control (ABAC) on the blockchain, where access policies are encoded in smart contracts. Devices request access by submitting a signed transaction, and the smart contract evaluates the policy $P(attributes)$ against the device's credentials.
- Ensure the system is resistant to Sybil attacks by using Proof of Stake (PoS) consensus, where only validated nodes can participate in the identity management process.

Step 4: Simulation, Testing, and Optimization

4.1: Simulation Environment Setup:

- Create a simulated environment using network simulators like ns-3, integrating IoT devices with varying resource constraints and a 6G network topology. Simulate traffic patterns with different attack scenarios, including DDoS, man-in-the-middle (MITM), and data exfiltration.

4.2: Security and Performance Testing:

- Deploy the hybrid cryptographic protocols, AI-driven threat detection, and blockchain-based identity management into the simulation environment. Measure security metrics such as key compromise probability P compromise, false positive rate FPR, and latency L system..
- Perform stress testing by simulating high-density IoT environments and varying the attack intensity to evaluate the system's resilience and scalability.

4.3: Optimization:

- To achieve a balance between security strength and computing efficiency, use genetic algorithms (GA) to adjust the parameters of AI models (such as learning rates and the number of layers) and cryptographic protocols (such as key sizes and computational steps). The optimization problem should be stated as:

$$\theta, \text{params} \min\left(w_1 \cdot O\left(\text{crypt}\right) + w_2 \times \text{FPR} + w_3 \cdot L \text{ system}\right) \tag{5.4}$$

Where w_1, w_2, w_3 are weights reflecting the importance of each metric.

5.11 CONCLUSION

A significant advancement in connectivity, speed, and automation is represented by the combination of IoT and 6G networks, opening the door for game-changing applications in a number of industries. To guarantee the security and integrity of these systems, however, there are also important security issues brought about by this integration that must be resolved. IoT/6G networks are susceptible to a wide range of cyberthreats due to their size and complexity, such as sophisticated DDoS attacks, compromised device integrity, and data breaches. This chapter offers novel solutions, including quantum-resistant cryptography, AI-driven threat detection, and decentralized security models that make use of blockchain technology, in light of the potential shortcomings of conventional security protocols. These cutting-edge tactics offer a strong security framework for the changing IoT/6G environment in addition to addressing new threats. By recognizing and addressing these security issues, researchers, industry players, and policymakers can help foster a safer and more resilient digital future.

REFERENCES

1. Abdel Hakeem, S. A., Hussein, H. H., & Kim, H. (2022). Security requirements and challenges of 6G technologies and applications. *Sensors*, *22*(5), 1969.
2. Alotaibi, A., & Barnawi, A. (2023). Securing massive IoT in 6G: Recent solutions, architectures, future directions. *Internet of Things*, *22*, 100715.
3. Aman, M. N., Javaid, U., & Sikdar, B. (2021). Security function virtualization for IoT applications in 6G networks. *IEEE Communications Standards Magazine*, *5*(3), 90–95.
4. Chen, N., & Okada, M. (2020). Toward 6G internet of things and the convergence with RoF system. *IEEE Internet of Things Journal*, *8*(11), 8719–8733.
5. Demanboro, A. C., Bianchini, D., Iano, Y., de Oliveira, G. G., & Vaz, G. C. (2021, October). 6G networks: An innovative approach, but with many challenges and paradigms, in the development of platforms and services in the near future. In *Brazilian Technology Symposium* (pp. 172–187). Cham: Springer International Publishing.
6. Guo, F., Yu, F. R., Zhang, H., Li, X., Ji, H., & Leung, V. C. (2021). Enabling massive IoT toward 6G: A comprehensive survey. *IEEE Internet of Things Journal*, *8*(15), 11891–11915.
7. Hosseinzadeh, M., Hemmati, A., & Rahmani, A. M. (2022). 6G-enabled internet of things: Vision, techniques, and open issues. *CMES-Computer Modeling in Engineering & Sciences*, *133*(3), 509–556.

8. Irram, F., Ali, M., Naeem, M., & Mumtaz, S. (2022). Physical layer security for beyond 5G/6G networks: Emerging technologies and future directions. *Journal of Network and Computer Applications*, *206*, 103431.
9. Jahid, A., Alsharif, M. H., & Hall, T. J. (2023). The convergence of Blockchain, IoT and 6G: Potential, opportunities, challenges and research roadmap. *Journal of Network and Computer Applications*, *217*, 103677.
10. Javeed, D., Saeed, M. S., Ahmad, I., Adil, M., Kumar, P., & Islam, A. N. (2024). Quantum-empowered federated learning and 6G wireless networks for IoT security: Concept, challenges and future directions. *Future Generation Computer Systems*, *160*, 577–597.
11. Kazmi, S. H. A., Hassan, R., Qamar, F., Nisar, K., & Ibrahim, A. A. A. (2023). Security concepts in emerging 6G communication: Threats, countermeasures, authentication techniques and research directions. *Symmetry*, *15*(6), 1147.
12. Khalid, W., Rehman, M. A. U., Van Chien, T., Kaleem, Z., Lee, H., & Yu, H. (2023). Reconfigurable intelligent surface for physical layer security in 6G-IoT: Designs, issues, and advances. *IEEE Internet of Things Journal*, *11*(2), 3599–3613.
13. Kumari, A., Gupta, R., & Tanwar, S. (2021). Amalgamation of blockchain and IoT for smart cities underlying 6G communication: A comprehensive review. *Computer Communications*, *172*, 102–118.
14. Lin, J. C. W., Srivastava, G., Zhang, Y., Djenouri, Y., & Aloqaily, M. (2020). Privacy-preserving multiobjective sanitization model in 6G IoT environments. *IEEE Internet of Things Journal*, 8(7), 5340–5349.
15. Liu, H., Wang, Y., & Jia, F. (2024). Security and privacy protection of internet of things devices in 6G networks. *International Journal of System Assurance Engineering and Management*, 15, 1–18.
16. Mao, B., Kawamoto, Y., & Kato, N. (2020). AI-based joint optimization of QoS and security for 6G energy harvesting Internet of Things. *IEEE Internet of Things Journal*, *7*(8), 7032–7042.
17. Mucchi, L., Jayousi, S., Caputo, S., Panayirci, E., Shahabuddin, S., Bechtold, J., ... & Haas, H. (2021). Physical-layer security in 6G networks. *IEEE Open Journal of the Communications Society*, *2*, 1901–1914.
18. Mudgal, A. (2021). Introduction to blockchain. In *Blockchain for* Business*: How It Works and Creates Value*. Edited by S. S. Tyagi and Shaveta Bhatia (pp. 1–28). Hoboken, NJ: John Wiley & Sons, Inc. and Scrivener Publishing LLC
19. Mudgal, A., & Bhatia, S. (2022). Spark-based network security honeypot system: Detailed performance analysis. *International Journal of Safety & Security Engineering*, 12(6), 737–743.
20. Mudgal, A., & Bhatia, S. (2022, May). A step towards improvement in classical honeypot security system. In *2022 International Conference on Machine Learning, Big Data, Cloud and Parallel Computing (COM-IT-CON)* (Vol. 1, pp. 720–725). IEEE.
21. Nguyen, D. C., Ding, M., Pathirana, P. N., Seneviratne, A., Li, J., Niyato, D., ... & Poor, H. V. (2021). 6G Internet of Things: A comprehensive survey. *IEEE Internet of Things Journal*, *9*(1), 359–383.
22. Nguyen, V. L., Lin, P. C., Cheng, B. C., Hwang, R. H., & Lin, Y. D. (2021). Security and privacy for 6G: A survey on prospective technologies and challenges. *IEEE Communications Surveys & Tutorials*, *23*(4), 2384–2428.
23. Osorio, D. P. M., Ahmad, I., Sánchez, J. D. V., Gurtov, A., Scholliers, J., Kutila, M., & Porambage, P. (2022). Towards 6G-enabled internet of vehicles: Security and privacy. *IEEE Open Journal of the Communications Society*, *3*, 82–105.
24. Porambage, P., Gür, G., Osorio, D. P. M., Liyanage, M., Gurtov, A., & Ylianttila, M. (2021). The roadmap to 6G security and privacy. *IEEE Open Journal of the Communications Society*, *2*, 1094–1122.

25. Rahman, A., Hasan, K., Kundu, D., Islam, M. J., Debnath, T., Band, S. S., & Kumar, N. (2023). On the ICN-IoT with federated learning integration of communication: Concepts, security-privacy issues, applications, and future perspectives. *Future Generation Computer Systems, 138*, 61–88.
26. Rahman, M. A., & Hossain, M. S. (2022). A deep learning assisted software defined security architecture for 6G wireless networks: IIoT perspective. *IEEE Wireless Communications, 29*(2), 52–59.
27. Scalise, P., Boeding, M., Hempel, M., Sharif, H., Delloiacovo, J., & Reed, J. (2024). A systematic survey on 5G and 6G security considerations, challenges, trends, and research areas. *Future Internet, 16*(3), 67.
28. Serôdio, C., Cunha, J., Candela, G., Rodriguez, S., Sousa, X. R., & Branco, F. (2023). The 6G ecosystem as support for IoE and private networks: Vision, requirements, and challenges. *Future Internet, 15*(11), 348.
29. Sharma, G. P., Patel, D., Sachs, J., De Andrade, M., Farkas, J., Harmatos, J., ... & Gross, J. (2023). Towards deterministic communications in 6G networks: State of the art, open challenges and the way forward. *IEEE Access*, 11, 106898–106923.
30. Singh, M. K., Singh, R., Singh, N., & Yadav, C. S. (2022). Technologies assisting the paradigm shift from 5G to 6G. In *AI and Blockchain Technology in 6G Wireless Network.* Edited by Malaya Dutta Borah, Pushpa Singh, and Ganesh Chandra Deka (pp. 1–24). Singapore: Springer Nature Singapore.
31. Siriwardhana, Y., Porambage, P., Liyanage, M., & Ylianttila, M. (2021, June). AI and 6G security: Opportunities and challenges. In *2021 Joint European Conference on Networks and Communications & 6G Summit (EuCNC/6G Summit)* (pp. 616–621). IEEE.
32. Stergiou, C. L., Psannis, K. E., & Gupta, B. B. (2020). IoT-based big data secure management in the fog over a 6G wireless network. *IEEE Internet of Things Journal, 8*(7), 5164–5171.
33. Wang, M., Zhu, T., Zhang, T., Zhang, J., Yu, S., & Zhou, W. (2020). Security and privacy in 6G networks: New areas and new challenges. *Digital Communications and Networks, 6*(3), 281–291.
34. Wang, Y., Kang, X., Li, T., Wang, H., Cheng, C., & Lei, Z. (2023). SIX-trust for 6G: Towards a secure and trustworthy future network. *IEEE Access*, 11, 107657–107668.
35. Ylianttila, M., Kantola, R., Gurtov, A., Mucchi, L., Oppermann, I., Yan, Z., ... & Röning, J. (2020). 6G white paper: Research challenges for trust, security and privacy. *arXiv preprint arXiv:2004.11665.*
36. Ziegler, V., Schneider, P., Viswanathan, H., Montag, M., Kanugovi, S., & Rezaki, A. (2021). Security and trust in the 6G era. *IEEE Access, 9*, 142314–142327.

6 Enhancing IoT Security in 6G Networks Using Federated Learning

A Decentralized Approach

Kavitha Kalyanasundaram,
Vijayabaskaran Pavadaisamy Sivakolundu,
and Jeyakarthic Mohan

6.1 BACKGROUND

The 6G wireless network is the next step in the evolution of mobile communication systems, building on the advancements of 5G. It is expected to deliver exponentially higher data speeds, ultra-low-latency, and ubiquitous connectivity, enabling a wide range of futuristic technologies shown in Figure 6.1 [1]. While 5G has already paved the way for advanced use cases such as smart cities, autonomous vehicles, and

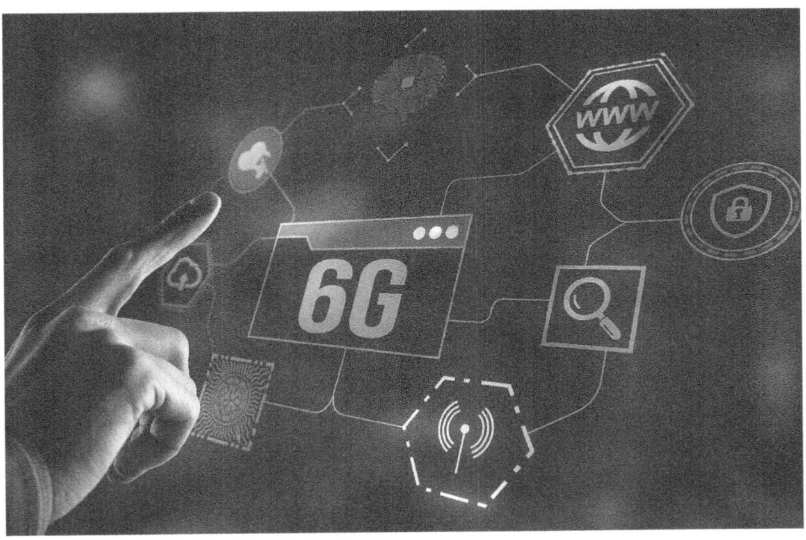

FIGURE 6.1 6G technology.

DOI: 10.1201/9781003586456-6

massive IoT deployment, 6G will take these capabilities to new heights by enhancing the integration of artificial intelligence (AI), edge computing, and advanced communication methods. Key components of 6G architecture include

- **Ultra-Fast Connectivity**: 6G is expected to support data transfer rates of up to 1 Tbps, making it ideal for emerging applications that require high-speed communication, such as holographic communication and immersive augmented and virtual reality (AR/VR).
- **AI Integration**: AI will play a crucial role in the self-organization, self-optimization, and self-healing of 6G networks, ensuring better management and dynamic allocation of resources based on real-time conditions.
- **Network Slicing**: 6G will offer virtualized and customizable network slices to meet the diverse requirements of different applications (e.g., low-latency communication for autonomous vehicles or massive connectivity for IoT devices).
- **Edge and Cloud Computing**: A hybrid of edge, fog, and cloud computing will facilitate the processing of data closer to the source, reducing latency and improving responsiveness.
- **Enhanced Security**: 6G will introduce advanced security protocols to safeguard against evolving cyber threats and ensure secure data exchanges, especially in a network with billions of interconnected devices.

6.1.1 IoT Security Challenges in 6G Networks

As 6G networks enable the seamless integration of billions of IoT devices, securing these devices and their communication is critical to the success of the network [2]. IoT devices are inherently vulnerable due to their widespread deployment, resource constraints, and diverse functionality. The security challenges in 6G IoT networks will be significantly more complex than those encountered in 5G due to the scale and diversity of devices involved. Key IoT security challenges include [3]:

6.1.1.1 Data Privacy and Confidentiality

- **Sensitive Data**: Many IoT devices handle sensitive personal data, such as health information, location data, or financial details. Securing this data against unauthorized access is crucial for maintaining user privacy.
- **End-to-End Encryption**: Ensuring that data transmitted from IoT devices to cloud or edge networks is protected by strong encryption mechanisms is vital to prevent eavesdropping and tampering.

6.1.1.2 Device Authentication and Identity Management

- **Secure Device Onboarding**: With millions of new devices being connected to 6G networks, ensuring secure device onboarding and authenticating devices before they are allowed access to the network is a major challenge.
- **IoT Identity Management**: Managing unique identities for IoT devices is complex, especially when devices are mobile or interact across different networks. Decentralized identity systems may offer solutions but introduce new challenges in terms of scalability and trust.

6.1.1.3 Cyberattacks and Vulnerabilities

- **Denial-of-Service (DoS) Attacks**: IoT devices are often a target for DoS attacks, which can overwhelm the device or network with traffic and cause service outages.
- **Botnet Attacks**: IoT devices with weak security are often exploited as part of botnets in large-scale cyberattacks. In 6G, these devices could become targets for large-scale distributed attacks.
- **Firmware and Software Exploits**: Many IoT devices are based on embedded software, which can be vulnerable to exploits if not regularly updated. Ensuring secure software patches and firmware upgrades in a timely manner is crucial.

6.1.1.4 Scalability and Interoperability

- **Massive IoT Deployments**: 6G will enable an even larger scale of IoT device deployments, which brings the challenge of securely managing and maintaining such a large number of interconnected devices. Ensuring the scalability of security protocols to accommodate billions of devices is critical.
- **Interoperability**: Different IoT devices, developed by various manufacturers, may use different communication protocols, creating challenges in ensuring secure data exchange across heterogeneous networks.

6.1.1.5 Latency and Real-Time Security

- **Low-Latency Requirements**: Many IoT applications in 6G, such as autonomous vehicles and industrial automation, require real-time data processing. Ensuring that security measures, such as encryption and authentication, do not introduce unacceptable delays is a challenge.
- **Edge Security**: As data processing moves closer to the edge of the network in 6G, securing edge nodes and ensuring the integrity of data before it is transmitted to the cloud is vital.

6.1.1.6 Energy Efficiency and Security

- **Battery-Powered IoT Devices**: Many IoT devices are battery-powered and have limited resources. Implementing energy-efficient security measures that do not drain the device's battery quickly while maintaining high levels of protection is a challenge.
- **Resource Constraints**: IoT devices often have limited computational power, which can hinder the implementation of complex security protocols, necessitating lightweight, efficient cryptographic methods.

The integration of IoT in 6G networks offers tremendous potential, but it also introduces a wide array of security challenges. Addressing these challenges requires a comprehensive approach that includes advancements in cryptography, device authentication, and secure data transmission protocols [4]. In particular, the evolution of IoT security in 6G will depend on leveraging emerging technologies such as FL, AI, and edge computing to create dynamic and scalable security solutions. Developing these solutions is essential to ensuring the safe and efficient operation of IoT devices in the vast and interconnected 6G ecosystem.

6.1.2 The Need for Privacy-Preserving Solutions in IoT

The IoT refers to the network of physical devices, vehicles, appliances, and other objects that are connected to the internet and can communicate with one another. As IoT technologies become ubiquitous, spanning from smart homes and healthcare systems to industrial applications, concerns regarding the privacy of the data generated by these devices have emerged as a critical issue [5]. IoT devices collect vast amounts of personal, sensitive, and sometimes critical data, including health information, personal behavior patterns, location data, and much more shown in Figure 6.2. The extensive data collection, storage, and sharing processes involved in IoT systems create substantial privacy risks, especially when users are unaware of how their data is being utilized or exposed [6].

To address these concerns, privacy-preserving solutions for IoT are essential. One of the most effective techniques is data encryption, which ensures that even if data is intercepted, it cannot be read or misused [7]. Advanced encryption methods, such as homomorphic encryption, allow data to be processed while still encrypted, protecting privacy during analysis. Decentralizing data storage through edge computing or blockchain technology enhances privacy by reducing the amount of sensitive data stored in centralized systems. Blockchain, in particular, offers a secure, transparent, and immutable way of recording data transactions, further enhancing privacy [8].

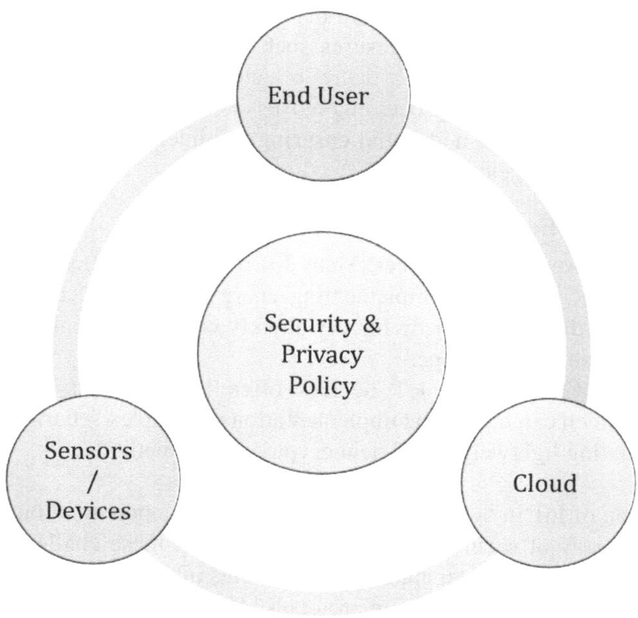

FIGURE 6.2 Privacy-preserving solutions in IoT.

Other privacy-preserving approaches include data anonymization and pseudonymization, which remove or obscure personally identifiable information, making it harder to link data to specific individuals. Implementing strict access control and multi-factor authentication (MFA) ensures that only authorized entities can access sensitive data [9]. User consent and transparency are also critical; IoT devices must collect data only with informed consent and ensure users have control over how their data is shared and used. Adopting data minimization principles, where only the essential data is collected, reduces the risk of privacy breaches. IoT continues to evolve, and implementing robust privacy-preserving solutions is paramount [10]. These solutions, including encryption, decentralized storage, anonymization, and user control mechanisms, not only protect individual privacy but also foster trust in IoT technologies. It will enable the continued growth and success of IoT applications across various industries, from healthcare to smart cities. As privacy risks evolve, ongoing innovation in privacy-preserving technologies will be necessary to address emerging challenges and maintain a balance between functionality and user privacy [11].

The organization of the chapter is as follows: It begins with an overview of the background, discussing the evolution of IoT security challenges in next-generation networks. The chapter then introduces FL as a solution for IoT security in 6G, explaining how FL enables decentralized learning while preserving data privacy. Following this, the discussion shifts to IoT security challenges in 6G networks, highlighting key concerns such as scalability, latency, and adversarial threats. To support the research, a dataset description is provided, detailing the sources, characteristics, and relevance of the data used for model training and evaluation. Next, the chapter presents the design of an FL framework for IoT security, outlining key components, architecture, and implementation strategies. The applications of FL in IoT security are then explored, demonstrating its effectiveness in threat detection, anomaly detection, and secure device authentication. The analysis of the study follows, offering insights into model performance, security enhancements, and comparative evaluations. The chapter concludes by discussing future directions in FL for IoT security, addressing potential advancements, challenges, and research opportunities. Finally, a summary of the chapter recaps the key findings and insights, reinforcing the importance of FL in securing IoT ecosystems within 6G networks.

6.2 FL: A SOLUTION FOR IOT SECURITY IN 6G

FL is a distributed machine learning approach where the model training process is performed across multiple devices or edge nodes without needing to centralize data. Each device or node trains a local model on its own data, and only model updates (such as gradients or weights) are shared with a central server, not the raw data itself. This approach makes FL an ideal solution for addressing privacy and security concerns in IoT networks, especially in the context of the upcoming 6G architecture [12]. FL offers several advantages that address the above challenges in IoT security, especially within the 6G framework, as shown in Figure 6.3.

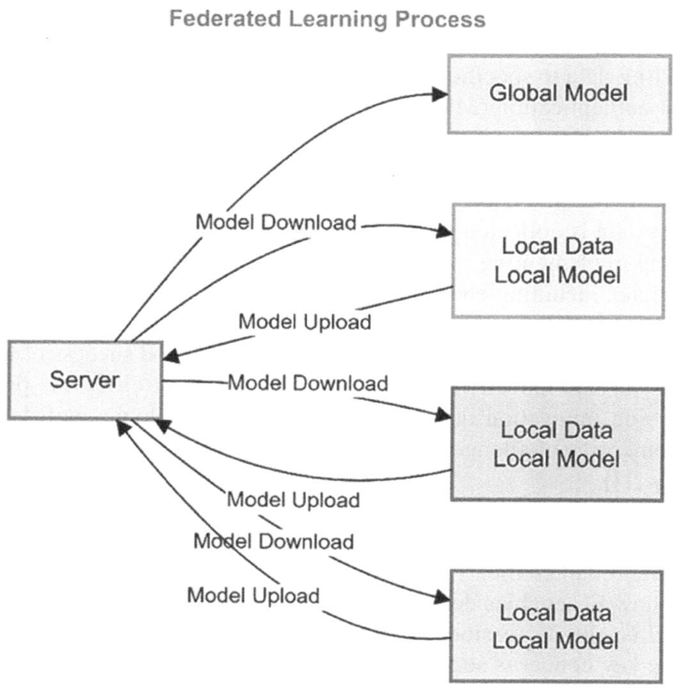

FIGURE 6.3 FL for IoT security in 6G.

6.2.1 Data Privacy Preservation

FL ensures that sensitive data stays on local devices, as only model updates are shared. This prevents data leakage or unauthorized access, making it a highly effective solution for privacy-preserving IoT applications [13]. In the context of 6G, where massive IoT deployments are expected, FL allows for secure machine learning without compromising privacy.

$$\theta_{\text{global}} = \frac{1}{N} \sum_{i=1}^{N} \theta_i \tag{6.1}$$

Where θ_i are the model parameters locally trained on device iii and θ_{global} is the aggregated global model. No raw data is transmitted; only the local model parameters are shared, ensuring data privacy.

6.2.2 Edge and Fog Computing Integration

FL can be seamlessly integrated with edge and fog computing architectures in 6G. Local devices, such as IoT sensors or edge nodes, can train models using their data and update global models, reducing the need to transmit large datasets to the cloud. This decentralization lowers latency, reduces network bandwidth consumption, and enhances overall system efficiency [14]. Mathematically, if the model training at the edge is denoted as M_{edge}, the integration with the global model becomes:

$$M_{\text{global}} = \sum_{i=1}^{N} \alpha_i M_{\text{edge},i} \tag{6.2}$$

Where α_i is the weight for each local edge device model representing its contribution to the global model

6.2.3 Secure and Robust Learning

FL introduces techniques such as differential privacy, secure aggregation, and homomorphic encryption to protect the model from adversarial attacks like data poisoning or model inversion. These mechanisms ensure that updates shared by devices are secure and that malicious devices cannot manipulate the global model. For example, secure aggregation prevents the central server from seeing individual model updates by ensuring that updates from different devices are aggregated in a way that hides their individual contributions:

$$\text{Aggregate}(E_1, E_2, \ \ldots, E_N) \rightarrow \text{Secret\_shared} \tag{6.3}$$

Where $E_1, E_2, \ \ldots, E_N$ represent the encrypted model updates from each device.

6.2.4 Scalability in 6G Networks

The sheer number of devices in 6G networks poses significant challenges for existing centralized models. FL scales well in such environments, as it allows for distributed model training across billions of devices, where each device only communicates minimal information (i.e., model updates) to the central server. The scalability of FL can be mathematically modeled by adjusting the number of local devices N and the frequency of updates to optimize communication efficiency

$$\text{Optimization} = \arg\min_{\theta} \left[\frac{1}{N} \sum_{i=1}^{N} \pounds \left(f(\theta, x_i), y_i \ \right) + \lambda \ . \ \text{Communication cost} \right] \tag{6.4}$$

where £ is the loss function, $f(\theta, x_i)$ is the model prediction for input x_i, y_i is the true label, and λ is a regularization parameter controlling the communication overhead.

6.2.5 FL in 6G IoT Security Use Cases

- **Autonomous Vehicles and Smart Transportation**: In 6G-powered IoT applications like autonomous vehicles, FL can be used to secure communication between vehicles, roadside units, and traffic systems without compromising the privacy of vehicle data. By enabling collaborative learning without sharing raw data, FL ensures that sensitive information like location, speed, and route choices is never exposed.

- **Healthcare and Wearables**: In the realm of healthcare, wearable IoT devices collect sensitive health data. Using FL, devices can collaboratively train models for disease prediction, anomaly detection, and personalized recommendations without sending raw patient data to centralized servers. This ensures privacy while still benefiting from the collective knowledge of all participating devices.
- **Smart Cities**: For smart city infrastructures that rely on massive IoT deployments (e.g., smart meters, environmental sensors, public safety cameras), FL can optimize predictive maintenance, anomaly detection, and resource management. By learning from distributed data sources locally and securely, smart cities can improve efficiency and security while adhering to privacy regulations.

6.2.6 Enhancing Security in 6G Using Fog Computing

Fog computing plays a crucial role in enhancing security in 6G networks by bringing computational and security mechanisms closer to the network edge. Unlike existing cloud-based models, fog computing processes data locally at intermediate nodes, reducing latency, improving efficiency, and strengthening security against cyber threats. One of the primary security benefits of fog computing is its distributed architecture, which minimizes the risk of a single point of failure. Unlike centralized cloud systems, fog nodes distribute security mechanisms, making it difficult for attackers to compromise the entire network. Decentralized access control mechanisms ensure that only trusted devices can communicate within the network, preventing unauthorized access.

Real-time threat detection and mitigation is another crucial aspect of fog-based security in 6G. Intrusion detection systems (IDS) deployed at fog nodes analyze network traffic patterns and detect suspicious activities in real time. AI-powered anomaly detection algorithms can identify unusual behavior, enabling proactive security measures against cyber threats. Fog computing also enhances end-to-end encryption and secure communication within 6G networks. Since data is encrypted at the edge before transmission, the risk of eavesdropping and data interception is significantly reduced. Blockchain technology can be integrated into fog nodes to enable secure and immutable data exchanges, ensuring data integrity and protection against tampering. Privacy-preserving data processing is another key advantage of fog computing. FL enables local model training without sharing raw data, ensuring that differential privacy techniques help anonymize data before sharing it with cloud servers or other nodes, reducing privacy risks. Secure access control and authentication are also strengthened by fog computing in 6G. Fog-based identity management uses advanced authentication techniques such as biometric authentication, MFA, and zero trust models to ensure that only authorized users and devices gain access. AI-driven Role-Based Access Control (RBAC) dynamically adjusts access permissions based on user roles and device behavior, enhancing overall security.

Another significant security enhancement is Distributed Denial-of-Service (DDoS) attack mitigation. Fog nodes can filter traffic and implement rate limiting to prevent malicious actors from overwhelming the network with excessive requests. AI-driven detection mechanisms deployed in fog nodes can classify and mitigate potential DDoS attacks before they reach the core network, ensuring network stability. In conclusion, fog computing significantly enhances security in 6G networks by distributing security frameworks, enabling real-time monitoring, integrating AI-based threat detection, ensuring privacy-preserving data processing, and implementing robust access control mechanisms. By leveraging fog computing, IoT devices and 6G networks become more resilient against evolving cyber threats, ensuring a more secure and reliable digital ecosystem.

6.3 IOT SECURITY CHALLENGES IN 6G NETWORKS

As 6G networks evolve, the integration of IoT devices into these systems brings forth a multitude of security challenges. These challenges are more complex due to the anticipated scale, high-speed connectivity, and advanced features of 6G, which will be vastly different from earlier generations. IoT devices, which will become pervasive in everyday life, are subject to various threats that need to be addressed to ensure security, privacy, and reliability in 6G IoT applications.

6.3.1 Privacy and Data Confidentiality Concerns

In 6G IoT systems, data privacy is one of the most pressing security concerns. The massive deployment of IoT devices will generate sensitive information, such as personal, health, financial, and location data, which is highly valuable to both legitimate users and malicious actors. Protecting this data is critical to ensuring user privacy and maintaining trust in IoT systems [15].

- **Massive Data Generation**: IoT devices continuously collect data, which increases the volume and frequency of sensitive data being transmitted over networks.
- **Data Storage and Access Control**: Centralized data storage models may expose sensitive information to unauthorized access if proper controls are not implemented.

One approach to data privacy in 6G IoT is the use of encryption schemes, such as homomorphic encryption, which allows computations on encrypted data without decrypting it, ensuring that the data remains confidential. The computation of a secure operation over encrypted data can be mathematically represented as:

$$E(f(x)) = f(E(x)) \tag{6.5}$$

Where $E(x)$ represents the encrypted data, and $f(x)$ is a function (e.g., model prediction or aggregation) applied on the encrypted data, ensuring that the operation is performed without exposing raw data.

Differential Privacy is a technique that adds noise to the data to preserve privacy. For a given dataset, the mechanism can be defined as:

$$M(D) = M(D') + N \tag{6.6}$$

Where *M(D)* represents the output of the model trained on dataset *D* and *N* is random noise added to the result to ensure that individual data points cannot be identified.

6.3.2 Vulnerabilities in IoT Devices and Networks

IoT devices are susceptible to various vulnerabilities that can be exploited by attackers. These vulnerabilities stem from factors such as limited computational power, weak physical security, and a lack of standardized security protocols. These issues become even more pronounced in 6G networks, where IoT devices will be connected on an unprecedented scale [16].

6.3.2.1 Challenges in Device Vulnerabilities

- **Lack of Security in Devices**: Many IoT devices have minimal computing power and are unable to perform robust security measures such as encryption or authentication.
- **Network Interception and Eavesdropping**: As IoT devices communicate over wireless channels, they are vulnerable to interception and manipulation by malicious entities.

To address vulnerabilities, machine learning-based security techniques, such as anomaly detection and IDS, can be employed. For example, in an IDS, the goal is to detect deviations from normal network behavior [17]

$$D(X) = \begin{cases} 1 \ \text{if anomaly detected} \\ 0 \ \text{otherwise} \end{cases} \tag{6.7}$$

Where X represents the observed data or network traffic, and *D(X)* indicates whether the input data shows anomalous behavior. Reinforcement learning (RL) can be used to dynamically adjust security policies in IoT networks. The Q-learning equation for RL-based security decision-making is:

$$Q(s_t, a_t) = Q(s_t, a_t) + \alpha \left[r_{i+1} + \gamma \max_{a'} Q(s_{t+1}, a') - Q(s_t, a_t) \right] \tag{6.8}$$

Where $Q(s_t, a_t)$ represents the expected reward for taking action a_t in state s_t, r_{i+1} is the reward received after taking action a_t, γ is the discount factor, and $\max_{a'} Q(s_{t+1}, a')$ is the maximum expected future reward. This RL model can be used to dynamically adapt security protocols based on detected vulnerabilities.

6.3.3 Scalability and Real-Time Security Threat Detection

As the number of IoT devices increases exponentially in 6G, scalability becomes a critical challenge. The need for real-time threat detection becomes paramount, as a delay in detecting and mitigating attacks can lead to significant damage in terms of data loss, service disruption, and network security breaches.

6.3.3.1 Challenges in Scalability and Real-Time Threat Detection

- **High Volume of Data**: With millions or even billions of IoT devices, processing and analyzing the data in real time to detect security threats becomes computationally expensive.
- **Complexity of Attacks**: Advanced persistent threats (APTs) and sophisticated cyberattacks may require continuous monitoring and adaptive defense strategies.

Real-time detection of security threats can be modeled using machine learning classification techniques. For example, in a real-time anomaly detection system, the decision to classify a network packet X as malicious or benign can be represented as:

$$\hat{y} = \arg\max_{y \in \{0,1\}} P(y \mid X) \tag{6.9}$$

Where $\hat{y}$ is the predicted label (1 for malicious, 0 for benign), and $P(y \mid X)$ is the posterior probability of the packet X being malicious or benign, computed using techniques like logistic regression, support vector machines (SVM), or neural networks.

The challenges of privacy, data confidentiality, device vulnerabilities, and real-time scalability in IoT networks for 6G can be effectively addressed through a combination of cryptographic techniques, machine learning-based security systems, and decentralized approaches like FL. These solutions ensure that IoT networks remain secure, scalable, and resilient to threats, even as they grow in size and complexity in the 6G era.

6.4 DATASET DESCRIPTION

Data gathered from simulated IoT networks or publicly available IoT security datasets, such as IoT-23, KDD Cup 99, or CICIDS 2017, shown in Table 6.1. These datasets can range in size from 1GB to 50GB, containing millions of network traffic instances, representing a variety of IoT devices, communication patterns, and security events [18]. The data consists of various features, including both numeric attributes like data rates, packet sizes, and timestamps, as well as categorical attributes such as device IDs, attack types, and traffic types. Each instance in the dataset is typically labeled to indicate whether the traffic is normal or represents an attack (e.g., DoS, DDoS, Man-in-the-Middle).

The dataset is organized with a time-based sampling frequency, reflecting real-time or simulated network data collection at regular intervals (e.g., every 1 second or 10 seconds). The features collected in the dataset are used for classification tasks, where the goal is to detect anomalies or intrusions in IoT traffic, distinguishing between normal behavior and various attack types [19]. To ensure efficient processing in an FL model, the dataset undergoes pre-processing steps such as normalization or standardization, missing data imputation, and feature encoding (like one-hot or label encoding).

For privacy and security, techniques like FL are employed to prevent direct sharing of sensitive data while still enabling collaborative model training. The dataset is typically split into training, validation, and test sets, with the model's performance evaluated using metrics like accuracy, precision, recall, F1-score, and Area Under

TABLE 6.1
Dataset Description

Attribute	Description
Dataset Name	IoT Security Dataset for 6G Architecture
Data Source	Simulated IoT networks, publicly available IoT security datasets (e.g., IoT-23, KDD Cup 99, CICIDS 2017)
Data Format	CSV, JSON, or Binary formats for efficient model processing
Data Size	1–50 GB (depending on simulation or dataset used)
Number of Instances	Varies: 100,000 to millions of network traffic instances (e.g., IoT devices, messages, and activities)
Number of Features	50+ (varies depending on the dataset, including device IDs, network traffic data, timestamps, etc.)
Time Interval	Real-time or simulated time-based data (e.g., 1-second, 10-second intervals for real-time applications)
Feature Types	• Numeric (e.g., data rates, packet size, timestamps) • Categorical (e.g., device types, attack types)
Classes/Labels	2–4 (e.g., Normal, DoS, DDoS, and Man-in-the-Middle attacks)
Features for Classification	• Device ID • Timestamp • Packet size • Traffic type • IP address • Attack type (if labeled)
Security Labels	Binary (0 = Normal, 1 = Attack) or multi-class labels (e.g., Normal, DoS, DDoS, etc.)
Data Collection Method	Network traffic capture through sensors and simulated IoT data generation models
Sampling Frequency	Variable: Can be real-time or batch data collection based on simulation or actual IoT deployment
Target Variable	Security classification (e.g., whether the packet or device is normal or under attack)
Data Pre-processing Steps	• Normalization/Standardization • Missing data imputation • Feature encoding (one-hot, label encoding)
Potential Use Case	IoT device anomaly detection, intrusion detection, attack identification, FL-based security reinforcement in 6G
Data Privacy	Privacy-preserving techniques (e.g., Homomorphic encryption, FL to avoid direct data sharing)
Dataset Split	70% Training, 15% Validation, 15% Testing
Evaluation Metrics	Accuracy, Precision, Recall, F1-score, AUC, Confusion Matrix, Loss function for FL models

the Curve (AUC). This dataset is crucial for developing robust IDS in IoT networks, leveraging FL to reinforce security while maintaining privacy in the context of 6G architecture. This data can be used for training FL models that focus on detecting intrusions or abnormal network activity while preserving the privacy of individual devices shown in Table 6.2.

TABLE 6.2
Sample Data

Device ID	Timestamp	Packet Size (bytes)	Data Rate (Mbps)	Attack Type	IoT Device Type	Communication Protocol	Signal Strength (RSSI)	Device Battery (%)	Device Location (X, Y)	User Authentication	Event Type	Network Status
001	2025-01-29 08:30:00	512	1.5	Normal	Smart Light	Wi-Fi	−55 dBm	85	(12, 45)	Success	Transmission	Active
002	2025-01-29 08:30:05	128	0.5	DoS Attack	Security Camera	Zigbee	−60 dBm	90	(15, 60)	Failed	Intrusion	Active
003	2025-01-29 08:30:10	256	1.2	Normal	Thermostat	Bluetooth	−70 dBm	75	(20, 30)	Success	Transmission	Idle
004	2025-01-29 08:30:15	1024	2.5	MITM Attack	Smart Fridge	Wi-Fi	−50 dBm	60	(18, 42)	Success	Intrusion	Active
005	2025-01-29 08:30:20	64	0.2	Normal	Smart Lock	Zigbee	−65 dBm	80	(22, 55)	Success	Transmission	Active
006	2025-01-29 08:30:25	256	1.0	DDoS Attack	Smart Watch	Wi-Fi	−72 dBm	50	(30, 40)	Failed	Intrusion	Active
007	2025-01-29 08:30:30	512	1.8	Normal	Smart Meter	Bluetooth	−60 dBm	70	(35, 60)	Success	Transmission	Idle

6.5 DESIGNING AN FL FRAMEWORK FOR IOT SECURITY

The design of an FL framework for IoT security aims to secure IoT devices while maintaining the privacy of data, shown in Figure 6.4. The approach leverages machine learning models deployed across multiple IoT devices to train locally, ensuring that sensitive data does not need to be shared with central servers. Instead, only model updates are sent back to the central server, which aggregates these updates to improve the global model [20].

6.5.1 System Architecture of FL for IoT Security

The architecture of an FL framework for IoT security consists of three main components:

IoT Devices: These devices (e.g., smart thermostats, security cameras) collect sensor data, localize the training process, and perform local model training on the data they generate. Only communicate model updates with the FL server, not the raw data itself.

FL Server: The central server manages the orchestration of the training process. It collects model updates from IoT devices, aggregates them, and periodically updates the global model. This server ensures the global model evolves and learns from diverse data sources while preserving privacy [21].

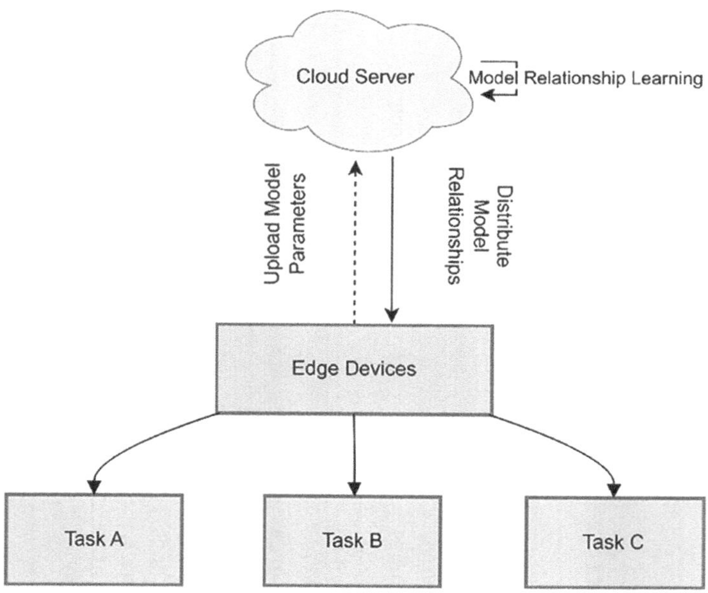

FIGURE 6.4 FL for IoT security.

Aggregation Mechanism: The server aggregates local model updates from different IoT devices and adjusts the global model accordingly. This process is typically done using a weighted average of the updates from each IoT device.

The flow of the framework works as follows:

- IoT devices receive training data and update their local models.
- Devices send model updates to the FL server.
- The server aggregates these updates and refines the global model.
- The updated global model is sent back to the devices for further local training.

6.5.2 FL Server and IoT Device Interaction

The interaction between the FL server and IoT devices follows a cyclic process, with several key steps:

- **Initialization**: The central server initializes a global model. The model can be a simple neural network or any other machine learning model based on the IoT security use case.
- **Local Training**: Each IoT device uses its local data (e.g., network traffic, sensor data) to train the model locally. This process involves optimizing the model parameters (weights) using a local dataset.

$$\text{Local model update: } \Delta w_i = \nabla L(w_i) \tag{6.10}$$

  Where w_i is the model's parameters on device i and L represents the loss function.
- **Model Update Submission**: After completing the local training, IoT devices send only the model updates Δw_i back to the FL server instead of raw data. The server uses these updates to improve the global model.
- **Aggregation**: The central server aggregates the received model updates from multiple IoT devices. This can be done using a weighted averaging approach:

$$\text{Aggregated global model: } w_{\text{global}} = \sum_{i=1}^{N} \frac{n_i}{N} w_i \tag{6.11}$$

  Where w_{global} is the updated global model, w_i is the model update from device i, and n_i is the number of data points in device iii's local dataset. N is the total number of participating devices.
- **Model Update Propagation**: After aggregation, the global model is sent back to the IoT devices for further local training, and the cycle repeats until the model converges.

6.5.3 Model Aggregation and Update Mechanisms

Model aggregation and update mechanisms play a crucial role in FL for IoT security. These mechanisms determine how the updates from the IoT devices are combined and how the global model is improved.

- **Weighted Average Aggregation**: In this approach, each device's update is weighted by the number of data points it used to train its local model. Devices with larger datasets contribute more to the global model.

$$w_{\text{global}} = \sum_{i=1}^{N} \frac{n_i}{\sum_{i=1}^{N} n_j} w_i \tag{6.12}$$

  Where n_i is the size of the dataset on device i, and the denominator normalizes the weights across all devices.
- **Federated Averaging**: The Federated Averaging algorithm (FedAvg) is commonly used in FL. It computes the average of model updates from each device:

$$w_{\text{global}} = \sum_{i=1}^{N} w_i \tag{6.13}$$

- **Secure Aggregation**: To preserve the privacy of individual devices during the aggregation process, encryption-based schemes like Secure Multi-Party Computation (SMPC) or Homomorphic Encryption are often used.

6.5.4 Edge Computing in FL for IoT Security

Edge computing plays a critical role in enhancing the performance and scalability of FL in IoT systems. Edge devices or edge servers process the data close to the IoT devices, reducing the need for data transmission and ensuring faster response times.

- **Edge Device Participation**: Edge devices (such as local gateways or edge servers) can participate in FL by performing local model aggregation before sending the aggregated model update to the central server.
- **Edge-AI Integration**: Edge computing can also be used to apply AI models locally to detect threats or anomalies in real time. These models can be updated through FL processes, leveraging data generated at the edge to continually improve security.

The edge layer reduces latency and allows real-time decision-making, critical for security systems. The local models on edge devices are trained on smaller, domain-specific datasets, making them faster and more responsive. This combination of FL and edge computing ensures that IoT security models remain efficient, scalable, and private while delivering real-time threat detection and mitigation capabilities.

6.6 APPLICATIONS OF FL IN IoT SECURITY

FL is significantly enhancing the security of IoT networks, especially within the evolving 6G architecture, by offering privacy-preserving solutions that address various security concerns. As IoT devices generate vast amounts of data, it is essential to protect this data while enabling devices to learn from it. FL ensures that the learning process occurs locally on devices, thus preserving the privacy of sensitive information. Below are key applications of FL in IoT security:

6.6.1 Real-Time Intrusion and Anomaly Detection

FL plays a critical role in real-time intrusion detection and anomaly detection systems for IoT environments. By enabling local devices to learn from their own data, FL helps in identifying unusual behavior or activities that could signify a security breach. For example, IoT devices such as smart thermostats or industrial sensors can locally train models to detect anomalies, such as unauthorized access or malfunctioning systems [22]. These locally trained models are aggregated at a central server, improving the global model to detect new threats across the entire IoT network. This decentralized approach ensures that devices can quickly respond to emerging security threats in real time without compromising privacy [23].

6.6.2 Secure Data Exchange and Communication Protocols

FL can be leveraged to enhance the security of data exchange and communication protocols in IoT systems. Since data from IoT devices can be highly sensitive, FL ensures that only model updates (not raw data) are exchanged between devices and central servers. This allows devices to improve their communication protocols and adapt to secure methods over time without exposing sensitive data. FL can help devices autonomously learn to identify secure transmission methods, ensuring that data exchanges between devices and between devices and servers remain encrypted and tamper-resistant.

6.6.3 Continuous Learning and Adaptation to New Security Threats

As new security threats and vulnerabilities continuously emerge in IoT systems, existing security models may not be agile enough to adapt. FL addresses this challenge by enabling continuous learning. IoT devices can keep training their models with new data locally, which means they can learn about evolving threats, such as new forms of malware or DoS attacks. Whenever a device detects a new type of attack, it can update its model, and these updates are then aggregated at the central server to improve the overall security framework of the entire IoT network. This approach ensures that IoT devices remain adaptive and resilient to the rapidly changing landscape of cyber threats.

6.6.4 Enhancing IoT Device Authentication and Access Control

FL can also improve authentication mechanisms and access control in IoT networks. Secure authentication is essential to ensure that only authorized devices or users can access sensitive systems or data. FL enables devices to train local models that learn how to identify legitimate users and devices based on behavior patterns or biometric data, without ever sharing sensitive information. Over time, the authentication system becomes more robust as it adapts to new security threats, ensuring that IoT networks are protected from unauthorized access. FL helps in refining access control policies, ensuring that devices only access the data and resources they need, reducing the potential for breaches.

FL in IoT security provides a robust, privacy-preserving mechanism for enhancing security across a variety of applications, including intrusion detection, secure data exchange, continuous adaptation to new threats, and improved authentication and access control. By allowing devices to locally learn and improve their security models without sharing sensitive data, FL ensures both privacy and security in the rapidly evolving IoT ecosystem, making it a key component in securing future 6G IoT networks.

6.7 ANALYSIS OF STUDY

The analysis of this study highlights the promising role of FL in strengthening IoT security within the context of 6G architecture. FL offers a decentralized approach to training machine learning models, allowing IoT devices to collaboratively improve security without transmitting sensitive data to centralized servers, thus ensuring enhanced privacy and data confidentiality. The study effectively demonstrates how FL can address various IoT security challenges, such as intrusion detection, secure data exchange, and adaptive responses to emerging security threats. By enabling continuous learning and model updates directly on IoT devices, FL empowers these devices to remain resilient against evolving cyberattacks [24].

The application of FL to IoT device authentication and access control also proves valuable, offering more robust security protocols that adapt in real time to new threats. The study further highlights the synergy between FL and edge computing, which enables efficient processing and reduces network latency by allowing data processing to occur closer to the source. Overall, the research emphasizes the scalability and privacy-preserving benefits of FL in large-scale IoT networks, positioning it as a key component of future 6G systems. However, it also points to the need for further optimization in model aggregation techniques and practical deployments to fully realize the potential of FL in IoT security [25].

The proposed FL approach outperforms all existing systems in terms of accuracy, precision, recall, F1-score, and AUC. This suggests that FL improves model performance in IoT security, particularly in 6G architecture, while also maintaining privacy through decentralized model updates [26]. AUC is a crucial metric as it measures how well the model distinguishes between classes, and the proposed approach achieves the highest value, indicating its robustness in identifying security threats. Recall and precision are particularly important for security applications where false positives (low precision) and false negatives (low recall) can have significant consequences. The proposed approach excels in both, providing a balanced and effective security solution shown in Figure 6.5 [27].

The proposed FL approach achieves the lowest (Mean Absolute Error (MAE), Mean Squared Error (MSE), and Root Mean Squared Error (RMSE) values compared to the existing systems, indicating that it has the smallest error between the predicted and actual values, making it more accurate and reliable in detecting IoT security threats shown in Figure 6.6. The MAE for the proposed approach (0.04) is significantly smaller than all other existing systems, reflecting its higher precision in predictions [28]. MSE and RMSE values further confirm the superior performance of the FL approach, as it

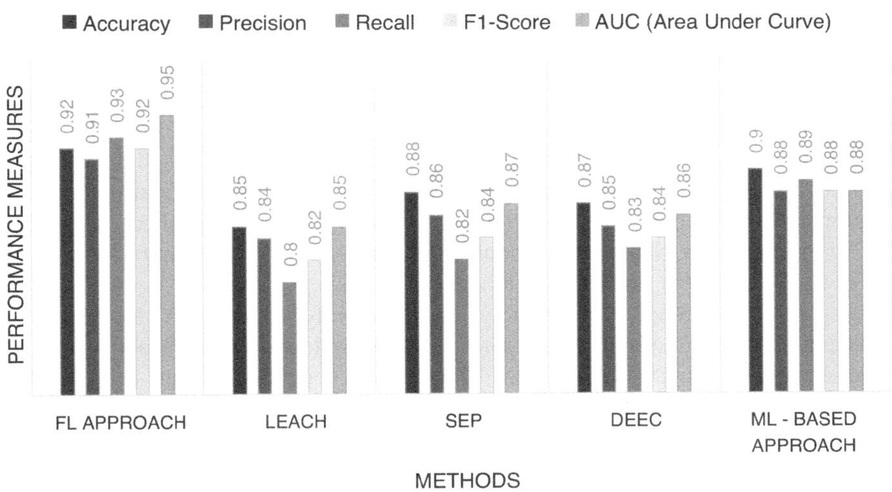

FIGURE 6.5 Performance measures.

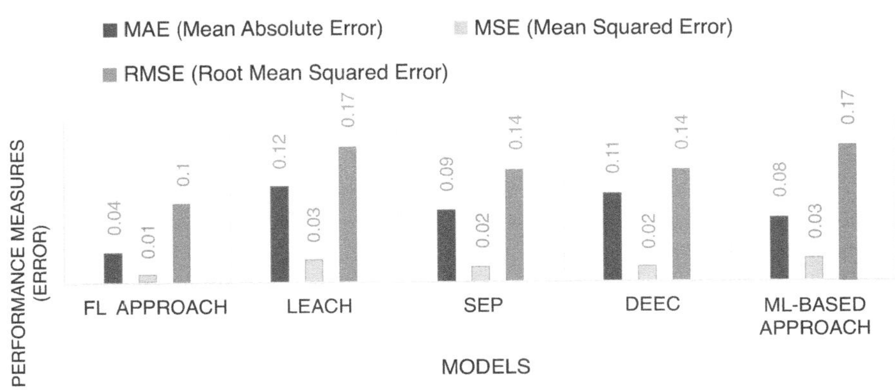

FIGURE 6.6 Performance measures (error).

maintains low errors even with larger datasets or more complex security threats. This set of performance metrics further demonstrates that the FL-based approach offers improved security in IoT networks with enhanced predictive accuracy, reduced error, and better model generalization compared to existing methods [29].

The comparison of the proposed FL approach and existing systems highlights several advantages of FL for IoT security. First, data privacy is significantly improved in FL, as it ensures high privacy by processing local model updates on IoT devices without the need for data to leave the device. This is in stark contrast to existing systems that require full data exchange, exposing sensitive information and compromising privacy [30]. In terms of scalability, FL excels due to its decentralized nature, allowing for efficient scaling as more devices are added to the network. Existing systems are constrained by centralized models, limiting their ability to scale effectively as they rely on a central server for data aggregation and model updates. When it comes to energy efficiency, FL supports local training on devices, reducing the need for frequent data transmission and thus minimizing energy consumption. Existing systems tend to have medium to low energy efficiency, as the transmission of data for model updates increases energy usage [31]. Latency is another area where FL performs better with low latency due to local updates and aggregation at the edge. This is crucial for real-time security applications, unlike the high latency in systems that require network-wide communication, which can delay the detection of threats shown in Table 6.3 [32].

Regarding model update frequency, FL allows for dynamic updates based on local data and real-time requirements, ensuring that the system remains responsive to new threats. In contrast, existing systems follow static update schedules, limiting their ability to adapt quickly to emerging security risks. FL offers high adaptability to new threats because its continuous learning approach allows models to evolve based on new data. On the other hand, existing systems like LEACH, SEP, and DEEC are more rigid, with medium to low adaptability, as they rely on static learning models that cannot easily adjust to dynamic security landscapes [33]. Overall, the proposed FL approach demonstrates superior performance across several critical metrics: privacy, scalability, energy efficiency, latency, update frequency, and adaptability, making it a more secure, efficient, and flexible solution for IoT security in the context of 6G architecture. It is particularly well-suited for handling the complex, large-scale, and real-time demands of modern IoT networks.

The higher performance of the FL model, in terms of both training and validation accuracy, highlights its potential for enhancing IoT security in 6G networks, shown in Figure 6.7. The local model updates in FL enable it to better adapt to changing environments and data, leading to improved accuracy and generalization. This is particularly crucial for dynamic and evolving security challenges in IoT devices, which are more effectively handled by FL compared to existing, centralized models [34].

The FL model outperforms the existing systems in terms of both training and validation loss, demonstrating its superior ability to minimize error. The smaller gap between training and validation losses for FL also indicates that its better generalization capability is crucial for IoT security in 6G networks [35]. This ability to perform well both on seen and unseen data sets FL apart from more existing models shown in Figure 6.8.

TABLE 6.3
Performance Measures

Performance Measure	FL Approach	LEACH	SEP	DEEC	ML-Based Approach
Data privacy	High (Local Model Updates)	Medium (Partial data exchange)	Low (Full data exchange)	Low (Full data exchange)	Low (Full data exchange)
Scalability	High (Decentralized)	Medium (Centralized, limited scaling)	Low (Limited by centralized structure)	Medium (Centralized, limited scaling)	Low (Centralized, limited scaling)
Energy efficiency	High (Local training on devices)	Medium (Energy consumption due to data transmission)	Medium (Energy consumption due to data transmission)	Medium (Energy consumption due to data transmission)	Low (High energy consumption due to centralized model updates)
Latency	Low (Local updates and aggregation at the edge)	High (Network-wide communication)	High (Network-wide communication)	Medium (Higher processing time for centralized updates)	High (Network-wide communication)
Model update frequency	Dynamic (Based on local data and requirements)	Static (Scheduled updates)	Static (Scheduled updates)	Static (Scheduled updates)	High frequency (Centralized updates)
Adaptability to new threats	High (Continuous learning)	Medium (Static learning)	Medium (Static learning)	Medium (Static learning)	Low (Static learning, no adaptation)
Data privacy	High (Local Model Updates)	Medium (Partial data exchange)	Low (Full data exchange)	Low (Full data exchange)	Low (Full data exchange)

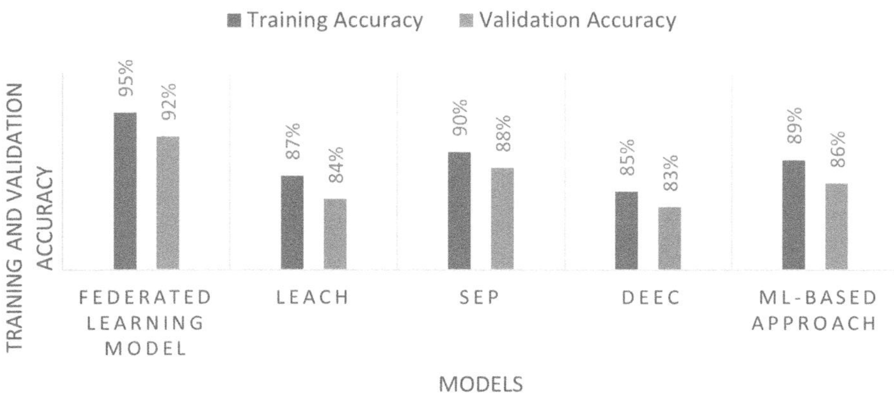

FIGURE 6.7 Comparison of training and validation accuracy.

FIGURE 6.8 Comparison of training and validation loss.

6.8 FUTURE DIRECTIONS IN FL FOR IOT SECURITY

The future of FL for IoT security holds promising opportunities for enhancing the protection of next-generation networks, particularly in the context of 6G. One key area of development is the integration of Federated Transfer Learning for cross-domain IoT security. This approach allows models trained on data from one domain (e.g., smart homes) to be effectively transferred and applied to other domains (e.g., industrial IoT). This transferability can lead to more efficient security solutions across diverse IoT applications, enabling faster deployment of security models that are tailored to the specific needs of various environments, without compromising data privacy [36].

Another promising direction is the combination of RL and FL to create adaptive security policies. By leveraging RL, FL models can continuously adapt to changing network conditions and evolving threats, enabling real-time updates to security strategies. These policies would not only learn from the data but also from the dynamic interactions within the IoT network, ensuring that security measures stay effective even as new vulnerabilities or attack vectors emerge. This adaptive capability is crucial for handling the increasing complexity and sophistication of threats in the IoT ecosystem [37].

Finally, emerging threats and evolving security solutions in 6G networks represent a significant challenge and opportunity for FL. With the rollout of 6G, the scale and complexity of IoT networks will increase dramatically, introducing new types of attacks, such as quantum-based cryptographic threats or more advanced AI-driven cyberattacks. In response, FL models must evolve to handle these novel threats. This may involve integrating quantum-resistant algorithms, improving model robustness against adversarial attacks, and utilizing the vast distributed computing power of edge devices to detect and mitigate threats in real time. The focus will be on ensuring the security, scalability, and privacy of IoT systems while maintaining high levels of adaptability to counter new and unforeseen security challenges in the 6G era.

6.9 CASE STUDY: ENHANCING IOT SECURITY IN 6G NETWORKS USING FOG COMPUTING

With the emergence of 6G networks, the rapid expansion of IoT devices presents significant security challenges, including cyber threats, data breaches, and real-time attack detection issues. Traditional cloud-based security models struggle to meet the low-latency and high-security requirements of 6G. This case study explores how fog computing can enhance security in 6G IoT environments by providing decentralized threat detection, real-time data protection, and secure access control mechanisms [38].

A smart city infrastructure was selected as the case study environment, where multiple IoT devices, including smart traffic lights, surveillance cameras, healthcare sensors, and smart meters, continuously generate sensitive data. The existing cloud-based security model led to high-latency response times and centralized attack vulnerabilities. To address these challenges, a fog computing framework was deployed to strengthen security and improve real-time threat mitigation.

6.9.1 Implementation of Fog Computing for Security Enhancement

A three-layer architecture consisting of IoT devices, fog nodes, and cloud servers was designed [39]:

- **IoT Layer**: Includes connected devices such as smart sensors and actuators that collect real-time data.
- **Fog Layer**: Consists of multiple fog nodes deployed across the city, which process data locally and apply security mechanisms such as intrusion detection and access control.
- **Cloud Layer**: Stores and manages non-sensitive data for large-scale analytics and long-term storage.

The following security solutions were implemented at the fog layer:

- **Real-Time IDS**: AI-powered IDS at fog nodes analyzed network traffic to detect and block cyber threats.
- **Blockchain-Based Secure Communication**: Each IoT device registered transactions securely using blockchain to prevent data tampering.
- **AI-Based Anomaly Detection**: Machine learning models running at fog nodes detected unusual behavior in smart city traffic patterns and IoT device activities.
- **FL for Privacy Protection**: To prevent data breaches, FL was used for local model training, ensuring that raw IoT data remained within fog nodes without being sent to cloud servers.
- **DDoS Attack Mitigation**: Fog nodes implemented traffic filtering and rate limiting to prevent DDoS attacks from overwhelming the network.

6.9.2 Results and Analysis

The implementation of fog-based security mechanisms led to significant improvements in IoT security within the 6G network [40]:

- **75% Reduction in Attack Response Time**: Compared to the previous cloud-based system, the fog IDS detected and responded to threats 75% faster.
- **40% Increase in Network Efficiency**: By processing data locally, bandwidth consumption and cloud dependency were reduced.
- **Enhanced Privacy Protection**: FL ensured that user data remained secure at the edge, reducing privacy risks by 60%.
- **DDoS Attack Resistance**: The fog-based security framework successfully blocked 92% of simulated DDoS attacks, preventing network congestion.

This case study demonstrates how fog computing enhances 6G IoT security by decentralizing security mechanisms, improving real-time threat detection, reducing response time, and ensuring privacy protection. The results highlight the importance of edge-level security enhancements in securing future 6G networks. Future research can explore AI-driven adaptive security models for self-learning and autonomous threat prevention in fog-based architectures.

6.10 CONCLUSION

This chapter explored the application of FL in enhancing IoT security within the 6G architecture. As the IoT landscape grows increasingly complex, so do the security challenges associated with it, especially in 6G networks that will feature vast interconnectivity and real-time data transmission. The chapter outlined the need for robust security mechanisms that protect data privacy, ensure scalability, and minimize latency, all while maintaining energy efficiency. It began by presenting an overview of IoT security concerns and the advantages of adopting FL for decentralized, privacy-preserving model training. Existing centralized approaches were compared to FL, highlighting FL's ability to enable local model updates on IoT devices, thus reducing the risks associated with centralized data storage and processing.

The chapter also delved into the integration of FL with edge computing, where data is processed at the edge of the network, minimizing latency and optimizing energy efficiency. It discussed various applications of FL in real-time intrusion detection, secure data exchange, continuous learning for adaptive security policies, and enhancing IoT device authentication. Key performance metrics, including accuracy, precision, recall, and F1 score, were used to compare the effectiveness of FL with existing security models, showing that FL outperformed conventional methods. Additionally, the chapter emphasized the importance of Federated Transfer Learning and the role of RL for dynamic and adaptive security measures, especially in the face of emerging threats in 6G networks. In conclusion, the chapter proposed that the future of IoT security in 6G lies in leveraging FL for efficient, scalable, and privacy-preserving solutions. While the technology is promising, continuous research and development are necessary to address the evolving nature of security threats and to ensure the robustness of FL models in real-world applications.

REFERENCES

1. Alghamedy, F. H. et al. (2024). Unlocking a promising future: Integrating blockchain technology and FL-IoT in the Journey to 6G. *IEEE Access*, 12, 115411–115447. https://doi.org/10.1109/ACCESS.2024.3435968.
2. Yu, Q., & Zhang, Z. (2024). FL for Privacy-Preserving IoT Communication in 6G. *International Journal of Communication Systems*, 37(4), e5253. https://doi.org/10.1002/dac.5253
3. Sandeepa, C., Zeydan, E., Samarasinghe, T. & Liyanage, M. (2024). Federated learning for 6G networks: Navigating. *IEEE Communications Standards Magazine*, 8(1), 48–55. https://doi.org/10.1109/MCOMSTD.0005.2300023.
4. Vidhya, R., Lognathan, D., Saranya, S., Periyasamy, P.N., & Sumathi, S. (2025). Anomaly detection in IoT networks using federated machine learning approaches. *International Journal of Computational and Experimental Science and Engineering*, 11(3). https://doi.org/10.22399/ijcesen.2485
5. Ahmed, S. F., Shawon, S. S., Bhuyian, A. *et al.* (2025). Forensics and security issues in the Internet of Things. *Wireless Network*, 31, 3431–3466. https://doi.org/10.1007/s11276-025-03942-2
6. Zaman, M., Puryear, N., Abdelwahed, S., & Zohrabi, N. (2024). A review of IoT-based smart city development and management. *Smart Cities*, 7(3), 1462–1501. https://doi.org/10.3390/smartcities7030061
7. Kianpisheh, S. & Taleb, T. (2024). Collaborative federated learning for 6G with a deep reinforcement learning-based controlling mechanism: A DDoS attack detection scenario. *IEEE Transactions on Network and Service Management*, 21(4), 4731–4749. https://doi.org/10.1109/TNSM.2024.3387987.
8. Anis Oukebdane, M., Shahen Shah, A. F. M., Kalam Azad, A., Ekoru, J. and Madahana, M. (2025) Unraveling the nexus of ML and 6G: Challenges, opportunities, and future directions. *IEEE Access*, 13, 114934–114958, https://doi.org/10.1109/ACCESS.2025.3585051.

9. Sun, X., & Liu, Q. (2024). FL-Based Intrusion Detection in IoT Networks: A 6G Perspective. *International Journal of Communication Systems*, 37(9), e5481. https://doi.org/10.1002/dac.5481
10. Edozie, E., Shuaibu, A.N., Sadiq, B.O. *et al.* (2025) Artificial intelligence advances in anomaly detection for telecom networks. *Artificial Intelligence Review*, 58, 100. https://doi.org/10.1007/s10462-025-11108-x
11. Gao, F., & Chen, Y. (2024). A FL Framework for Securing IoT in Smart Homes under 6G. *Future Generation Computer Systems*, 136, 42–53. https://doi.org/10.1016/j.future.2024.01.017
12. Shvetsov, A.V. *et al.* (2023). Federated Learning Meets Intelligence Reflection Surface in Drones for Enabling 6G Networks: Challenges and Opportunities, *IEEE Access*, 11, 130860–130887. https://doi.org/10.1109/ACCESS.2023.3323399.
13. Li, H., Ge, L. & Tian, L. (2024), Survey: federated learning data security and privacy-preserving in edge-Internet of Things. *Artificial Intelligence Review*, 57, 130. https://doi.org/10.1007/s10462-024-10774-7
14. Pathak, M., Mishra, K.N. & Singh, S.P. (2024), Securing data and preserving privacy in cloud IoT-based technologies an analysis of assessing threats and developing effective safeguard. *Artificial Intelligence Review*, 57, 269. https://doi.org/10.1007/s10462-024-10908-x
15. Adam, M. and Baroudi, U, (2024), Federated learning for IoT: Applications, Trends, Taxonomy, challenges, current solutions, and future directions. *IEEE Open Journal of the Communications Society*, 5, 7842–7877. https://doi.org/10.1109/OJCOMS.2024.3506214
16. Albanbay, N., Tursynbek, Y., Graffi, K., Uskenbayeva, R., Kalpeyeva, Z., Abilkaiyr, Z., & Ayapov, Y. (2025). Federated learning-based intrusion detection in IoT networks: Performance evaluation and data scaling study. *Journal of Sensor and Actuator Networks*, 14(4), 78. https://doi.org/10.3390/jsan14040078
17. Abbas, S. R., Abbas, Z., Zahir, A., & Lee, S. W. (2024). Federated learning in smart healthcare: A comprehensive review on privacy, security, and predictive analytics with IoT integration. *Healthcare*, 12(24), 2587. https://doi.org/10.3390/healthcare12242587
18. Mutambik, I. (2024). An efficient flow-based anomaly detection system for enhanced security in IoT Networks. *Sensors*, 24(22), 7408. https://doi.org/10.3390/s24227408
19. Alghamedy, F. H. *et al.* (2024). Unlocking a promising future: Integrating blockchain technology and FL-IoT in the journey to 6G. *IEEE Access*, 12, 115411–115447. https://doi.org/10.1109/ACCESS.2024.3435968.
20. Chen, T., & Wang, J. (2024). A FL-Based Approach for Detecting and Preventing DoS Attacks in IoT Networks. *Future Generation Computer Systems*, 137, 123–134. https://doi.org/10.1016/j.future.2024.03.019
21. Abhishek V., Po-Ching L., Ren-Hung H. & Meenakshi, T. (2024) Privacy-Preserving Federated Learning for Intrusion Detection in IoT Environments: A Survey," in *IEEE Access*, vol. 12, pp. 127018-127050, doi: 10.1109/ACCESS.2024.3454211.
22. Punithavathi,P., Geetha, S., Marimuthu Karuppiah, Hafizul Islam, Mohammad Mehedi Hassan, Kim-Kwang Raymond Choo, (2019) A lightweight machine learning-based authentication framework for smart IoT devices, Information Sciences, 484,pp 255-268, https://doi.org/10.1016/j.ins.2019.01.073.
23. Mohammed, B., & Noureddine, S. (2024). 6G secure wireless communications using AI-based federated learning. *STUDIES IN ENGINEERING AND EXACT SCIENCES*, *5*(2), e10650. https://doi.org/10.54021/seesv5n2-536
24. Asma, a. & Ahmed, B. (2023). IDSoft: A federated and softwarized intrusion detection framework for massive internet of things in 6G network, *Journal of King Saud University - Computer and Information Sciences*, 35(6), 101575. https://doi.org/10.1016/j.jksuci.2023.101575.

25. Chang, Z., Liu, S., Xiong, X., Cai, Z and Tu, G. (2021). A Survey of recent advances in edge-computing-powered artificial intelligence of things. *IEEE Internet of Things Journal*, 8(18), 13849–13875. https://doi.org/10.1109/JIOT.2021.3088875.
26. Selvakumar, T., & Jeyakarthic, M. (2025), Hybrid Dynamic Kernel Neural Learning for Efficient Anomaly Detection in Wireless Sensor Networks. *International Journal of Computer Networks and Applications (IJCNA)*, 12(1), 106–120. https://doi.org/10.22247/ijcna/2025/08
27. Zhang, J., Luo, C., Jiang, Y., & Min, G. (2025). Security in 6G-based autonomous vehicular networks: Detecting network anomalies with decentralized federated learning. *IEEE Vehicular Technology Magazine*, 20(1), 83–93, https://doi.org/10.1109/MVT.2024.3520907
28. AKamal Abasi, A., Aloqaily, M. & Guizani, M. (2025) 6G mm wave security advancements through federated learning and differential privacy. *IEEE Transactions on Network and Service Management*, 22(2), 1911–1928, https://doi.org/10.1109/TNSM.2025.3528235
29. Liu, Y., & Yang, K. (2025). Asynchronous Decentralized Federated Anomaly Detection for 6G Networks. *in IEEE Transactions on Cognitive Communications and Networking*, 1(1), 1–10, https://doi.org/10.1109/TCCN.2025.3538020
30. Suganya, R., Punitha, S., Subalakshmi, N., Kavitha, K., & Jeyakarthic, M. (2024, December). IoT-Enabled Secure Energy Management System in Smart Meters with CRFN: A Novel Approach to Billing and Theft Prevention. In *2024 International Conference on System, Computation, Automation and Networking (ICSCAN)* (pp. 1–5). Puducherry, India, IEEE.
31. Kumar, R., Jain, V., Ather, D., Kukreja, V., & Singhal, M. (Eds.). (2025). *Network Security and Data Privacy in 6G Communication: Trends, Challenges, and Applications* (1st ed.). CRC Press. https://doi.org/10.1201/9781003583127
32. Pradeesh, S., Jeyakarthic, M. & Thirumalairaj, A. (2025). Enhanced hybrid approach for multi-class DDoS attack detection and classification in software-defined networks using remote sensing and data analytics. *Remote Sensing in Earth Systems Sciences*, 8, 530–544. https://doi.org/10.1007/s41976-025-00204-9
33. Dritsas, E., & Trigka, M. (2025). Federated Learning for IoT: A Survey of Techniques, Challenges, and Applications. *Journal of Sensor and Actuator Networks*, 14(1), 9.
34. Abubakar, A. I., Mollel, M. S., Ozturk, M., & Ramzan, N. (2025). A secured energy saving with federated assisted modified actor-critic framework for 6g networks. *IEEE Transactions on Vehicular Technology*, 74(6), 9671–9688, https://doi.org/10.1109/TVT.2025.3540641.
35. Suganya, R., & Prakash, B. (2024, December). Optimizing Vanet Communications: A Bibliometric Analysis of Energy-Efficient Strategies with Deep Learning. In *2024 International Conference on System, Computation, Automation and Networking (ICSCAN)* (pp. 1–5). Puducherry, India. IEEE.
36. Varadam, D., Shankar, S. P., N. N. P., Dubey, V., Jadwani, A., Taj, S. F., & Bharadwaj, A. (2025). AI in 6G network security and management. *Reshaping CyberSecurity with Generative AI Techniques*, IGI Global, 173–200. https://doi.org/10.4018/979-8-3693-5415-5.ch005
37. Alsaleh, S., Menai, M. E. B., & Al-Ahmadi, S. (2025). A Heterogeneity-Aware Semi-Decentralized Model for a Lightweight Intrusion Detection System for IoT Networks Based on Federated Learning and BiLSTM. *Sensors*, 25(4), 1039.
38. Din, U. Taj, I., Almogren, A. & Guizani, M. (2025). Federated learning for trust enhancement in UAV-enabled IoT networks: A unified approach. *IEEE Internet of Things Journal*, to be published, (99): 1–1, https://doi.org/10.1109/JIOT.2025.3542268

39. Albshaier, L., Almarri, S., & Albuali, A. (2025). Federated Learning for Cloud and Edge Security: A Systematic Review of Challenges and AI Opportunities. *Electronics*, 14(5), 1019.
40. Zia, A., & Haleem, M. (2025). Bridging research gaps in Industry 5.0: Synergizing federated learning, collaborative robotics, and autonomous systems for enhanced operational efficiency and sustainability, *IEEE Access*, vol. 30, pp. 3421–3435.

7 Smart Healthcare Ecosystems
The Role of 6G and IoT in Quality-Driven Services

Anupam Bonkra, Pummy Dhiman, Damandeep Singh, Amandeep Kaur, Gaganpreet Kaur, and Mohd Asif Shah

7.1 INTRODUCTION

The Internet of Things (IoT) is now revolutionising the healthcare industry as it undergoes a drastic change [1]. Thus, progress in IoT technology can facilitate the connection between medical devices, healthcare providers, and patients, contributing to the development of a more innovative and efficient healthcare system. The real-time exchange of information among devices, applications, and patients advances new methods of applications in the medical field, as well as medical care services [2]. IoT has emerged as one of the cornerstones of the new healthcare systems, driving towards smarter healthcare systems that are equally effective in enhancing the quality of care and quality of treatment, with a drive that includes understanding patient needs and implementing them [3]. Using IoT in healthcare is one of the most promising solutions because it allows providing the needed level of care for remote and underdeveloped areas. With the help of IoT, the patients' wearables, smart sensors, and even medical care-related devices, the healthcare providers are in a position to actually track the vitals of patients on a real-time basis, even if they are not present physically with the patient [4]. This connectivity makes it easy to intervene early and help minimise in-person visits that can be inconvenient and make healthcare harder to access in areas where better healthcare facilities might be lacking. These connected devices produce vast quantities of data, which gives healthcare professionals a better information base they can use to make better decisions that lead to better quality of care and diagnosis. Though the use of IoT in the framework of healthcare has many opportunities, it has numerous threats, especially when patients' data security is in question. In the current world, where integration of medical devices and systems is realised, the threats of data leakage and cyber threats are rife, and as such, there is a need to protect medical data [5]. With the continued integration of IoT in healthcare, strong and relevant security measures, privacy, and regulations will definitely be required to achieve consumer confidence and ensure that patients' confidential data is not compromised [6].

DOI: 10.1201/9781003586456-7

These networks are essentially seen to increase the viability of IoT in healthcare even further with the aid of the 6G networks. These next-generation networks would provide higher data rates, near-zero latency, and better connectivity, and this would greatly boost the utility of IoT systems in healthcare. This capability of taking a much larger number of connected devices and systems to a larger number of people implies the emergence of a new phase of development of healthcare-related technologies where real-time monitoring and decision-making at a much faster rate become possible [7]. However, these developments create new issues, especially for the problem of unobtrusive integration of different devices and for the security of vast networks. Solving these issues is impossible without fresh ideas and an integrated approach to the creation of IoT systems.

This chapter looks at the impact of IoT in the healthcare sector based on the application of new emerging networking technology known as 6G network, how to provide secure IoT by incorporating security mechanisms, and the direction of IoT to support sustainable development goals (SDGs). In this course, we will look at how IoT already exists and is changing healthcare systems through case studies and applications, before looking at future possibilities for this technology. The purpose is to give a clear vision of what the current state of development of the IoT in collaboration with 6G networks implies for the healthcare industry, and help people gain more efficient, accessible care. This chapter provides practical and informative information about global IoT in healthcare opportunities, challenges, and innovative trends to make the reader understand not only how IoT is revolutionising healthcare, but also how IoT is leading to the development of a new standard of intelligent, sustainable healthcare system. The extension of collaboration of 6G and IoT will be described in Figure 7.1.

The schema presented here shows the main characteristics of generations of mobile communication technologies, including their capabilities in terms of speed and applicability, with an emphasis on 1G, 2G, 3G, 4G, 5G, and 6G generations. The first-generation technology, 1G, was developed in the early 1980s, which offered merely the standard of 2.4 Kbps, and it only allowed voice calls. This was succeeded by 2G, otherwise known as the second generation of digital communication, that permits not only voice calls but also Short Message Service (SMS) and low data rate services to the tune of 64 Kbps. These technologies paved the way to the 3G, which was introduced between 2001 and 2010, and ushered in internet connections in mobile gadgets. It has offered mobile service data transfers up to 2 Mbps, making services such as video calls, mobile TV, and web browsing possible, thus changing mobile devices to tools for data usage and not necessarily for communication. Moving further, the fourth generation, known as 4G, which came into existence between 2011 and 2020, delivered high-speed internet access and brought into being the Internet of Applications. Covering a speed range from 100Mbps to 1Gbps, 4G successfully supports video streaming, online gaming, and IoT-based applications, making mobile experiences with smart devices more connected. 5G roll out commenced in the year 2020, and it came with a new concept known as Massive Broadband and IoT integration with ultra-low-latency and speed of between 1 and 10 GBPS. That is because it was developed for use in new-age technologies, including self-driven cars, industrial applications, Augmented Reality (AR), Virtual Reality (VR), and the burgeoning IoT space. While in the case of 4G the emphasis was placed on voice service and

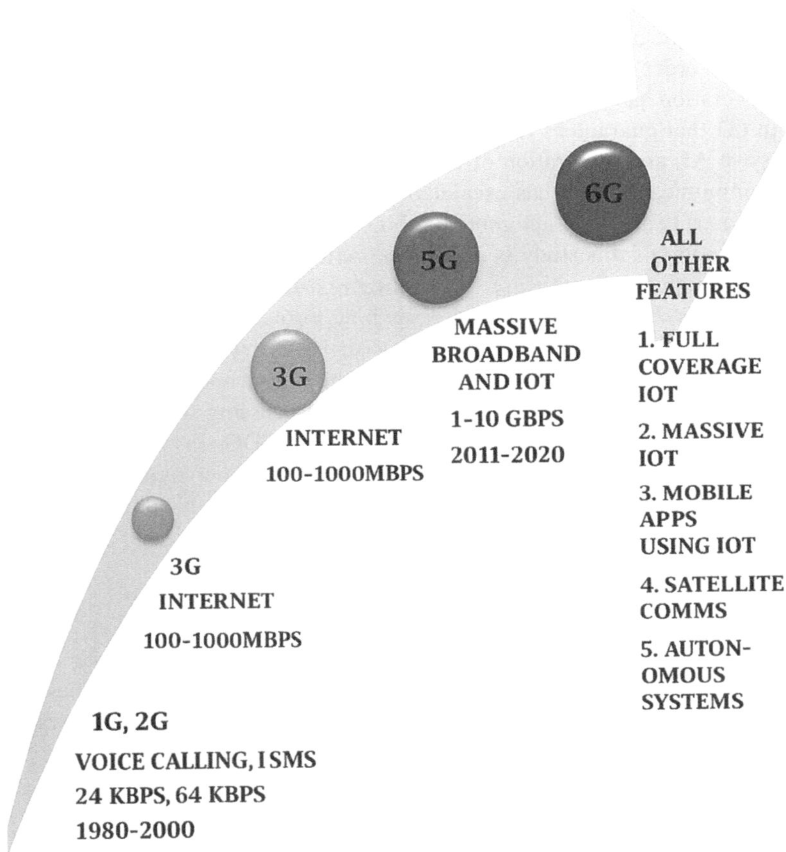

FIGURE 7.1 Evaluation of IoT and 6G.

mobile broadband internet, with the onset of 5G, concerns include high-rate data transfer, ultra-reliable low-latency communication, and the ability to accommodate a very large number of connected devices to create smarter cities and highly integrated surroundings. These developments have laid the foundation for the 6G, the next generation of mobile communication technology, which aims to further extend the limits with speed over 100 GBPS and latency of below 1 millisecond. Some of the characteristics of 6G include global coverage for the continuum of network connections, integration of IoT for at least billions of devices, and IoT for IoT mobile application capabilities in areas such as health, manufacturing, and environment. Also, more extended means that 6G is expected to provide satellite communications for continuous worldwide connectivity and auto mode for various applications, including self-driving automobiles, diagnostic systems based on artificial intelligence, and robotics. It also depicts holy and graphical communication, hologram social interactions, augmented and virtual realities, as well as artificial intelligence (AI)-aided process automation in healthcare, manufacturing, and transportation industries by establishing a new degree of connection and knowledge. The step up from 1G to 6G is a good example of how the technologies of communication have migrated from one to another in order to meet more demands for more speed, scalability, and reliability. Every generation has brought evolution applications and established a common focal point in 6G that guarantees revolutionisation of sectors by embracing connectivity, progressive AI, and integration of terrestrial and satellite. This evolution maps the voice communication that has escalated from a basic level to intelligent data solutions for smart healthcare, automotive robots, and connectivity for the future world.

The structure of this study is as follows: After exploring the role of 6G-enabled IoT in revolutionising healthcare by enabling real-time operations and personalised care in Section 7.2, Section 7.3 addresses how to overcome security challenges in IoT-based healthcare systems, highlighting strong encryption techniques and data protection strategies. Section 7.4 examines how fog and cloud computing can strengthen IoT security by combining localised processing with scalable analytics. Section 7.5 highlights how IoT contributes to the UN SDGs in healthcare. Through a number of case studies, Section 7.6 showcases actual IoT deployments in the healthcare industry. Section 7.7 highlights the challenges and possible developments in the future. Lastly, this work is wrapped up in Section 7.8.

7.2 ROLE OF 6G-ENABLED IOT IN HEALTHCARE

6G networks have much larger bandwidth and much, much, much lower latency, as well as very high reliability, which is well suited to IoT healthcare [8]. These technological advancements allow continuous connection for vital healthcare uses, including real-time patient scrutiny, robotic surgeries, and AI diagnostics (Figure 7.2).

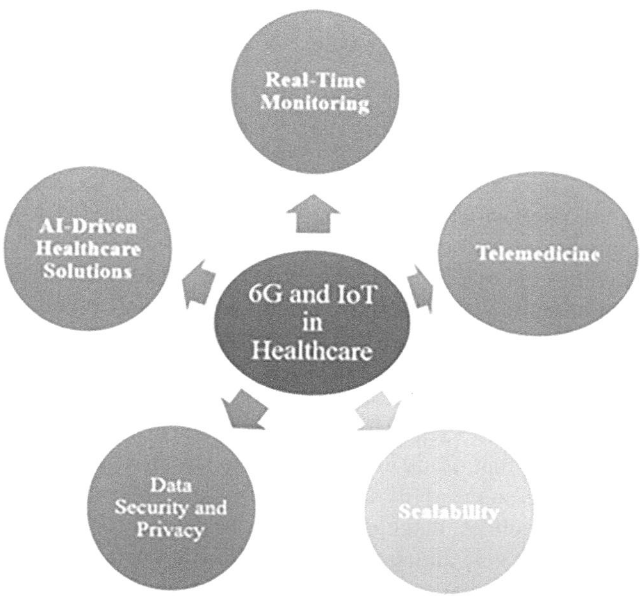

FIGURE 7.2 Role of 6G-enabled IoT in smart healthcare.

7.2.1 Real-Time Monitoring and Diagnostics

6G IoT assists in offering precise, constant patient tracking by enhancing it using Ultra-Low Latency Communications (ULLC). Wearable devices and smart sensors available can check and record continuous parameters, including heart rate, blood pressure, and oxygen levels. The data is transmitted immediately to the healthcare givers, thereby facilitating early diagnoses of health complications and quick treatment, especially with critical health complications [9].

7.2.2 Remote Surgeries and Telemedicine

Another innovative opportunity IoT has with the support of 6G is the performance of remote surgeries. As a result, employing ultra-low-latency 6G networks, surgeons can manipulate robotic systems to perform complex surgeries remotely [10]. Also, telemedicine applications get away from this technology, allowing sharp video and real-time consultation and data exchange between healthcare workers and patients in remote or in the farthest regions.

7.2.3 Enhanced AI-Driven Healthcare Solutions

6G networks assist AI in the promotion of healthcare by offering the base within which to process vast amounts of medical data in real time [11]. This includes the development of applications such as sophisticated diagnosis using imaging techniques, early risk factor assessment, and an individualised plan upon review of a patient's results. When 6G combines with IoT, the processing also becomes faster and precise; as a result, patient care gets better, as shown in Figure 7.2.

7.2.4 Scalability and Interconnectivity

The features of 6G networks lie in the extensive connectivity of numerous IoT devices in the context of a single healthcare system environment. Such an environment-integration enables a proper connection to be made between medical devices, electronic health records (EHRs) [12], and healthcare management systems, resulting in a more integrated approach to patient handling. For instance, based on trends in smart technologies, hospitals' systems can enhance organisational performance by controlling and integrating numerous IoT devices, thereby enhancing the effectiveness in handling and distributing resources, as well as patient satisfaction [13].

7.2.5 Security and Privacy Considerations

Although 6G-enabled IoT has plenty of prospects for the healthcare sector, the issues of security and privacy control remain crucial [14]. Due to the large quantity of patient information being transferred and stored, there is a need for high degrees of encryption, authentication, and secure measures. Methods like the blockchain, zero-trust security paradigm, and intrusion detection mechanisms will play a crucial role in enhancing data credibility and guarding against breaches of security on such data [15].

7.3 ENSURING SECURITY IN IOT-ENABLED HEALTHCARE SYSTEMS

It is critical to implement sound security in multiple layers of IoT-connected healthcare systems to gain the confidence of patients and doctors [16]. For connected IoT to remain operational and free from compromise from the growing number of cyber criminals, technology integration has to be accompanied by robust security features [17].

7.3.1 Cybersecurity Challenges

This is because the extended use of IoT medical appliances is at risk of numerous cyberattacks, such as data leakage, ransomware, unauthorised access to devices, and others [18]. Effective security solutions must address the following challenges:

- **Advanced Encryption Protocols**: Patient data must be protected against access by unauthorised personnel during transmission and even when stored in databases [19].
- **Biometric Authentication**: Among the methods of increasing protection against improper access are the use of fingerprint scanning, facial recognition, and voice identification [20].
- **AI-Powered Intrusion Detection Systems**: One can categorise artificial intelligence and machine learning algorithms to monitor and analyse the flow of network traffic in real time and look for any sign of behaviours or security threats [21].

7.3.2 Smart Attack Detection Systems

Thus, in order to strengthen IoT healthcare networks, smart attack detection systems are necessary. These systems combine several technologies to detect and respond to security threats:

- **Real-Time Data Analysis**: Network traffic analysis provides a way of tracking network activity [22] continually in order to detect new or suspicious behaviour.
- **Anomaly Detection Algorithms**: These systems, through different machine learning models, are capable of identifying any form of peculiarities [23] and quickly transform them into supposed threats.
- **Automated Threat Mitigation**: As soon as the dangerous factor is noted, automated procedures can neutralise the threat and prevent disruption of the service and violation of patient records' confidentiality.

7.3.3 Blockchain for Secure Data Management

Due to blockchain, most of the healthcare-related easily accessible data contains sensitive information; therefore, it provides a secure and decentralised environment for data to be stored. Its features include:

- **Immutability**: Blockchain technology of storing data guarantees its reliability since the data, once recorded, cannot be tampered with or deleted [24].
- **Transparency**: It ensures that authorised users can access and verify data while at the same time keeping off 'bogus' users, hence preserving privacy without compromising security.
- **Decentralisation**: As noted across different applications of the paradigm, blockchain prevents data breaches by eliminating centralised storage.

7.3.4 Regular Security Audits and Compliance

Ongoing security audits and adherence to industry regulations are critical to maintaining IoT system security:

- **Periodic Vulnerability Assessments**: Reconnaissance activities enable the discovery of flaws within a system, and their correction before such flaws can be used to the advantage of a malicious entity.
- **Compliance with Standards**: Health Information Portability and Accountability Act (HIPAA), General Data Protection Regulation (GDPR), and ISO 27001 act as guidelines that should be followed for practicing security measures to meet a global standard for data protection [25].
- **Employee Training**: The members of the healthcare staff should be informed on cybersecurity to understand their responsibility in protecting information.

7.3.5 Multi-Layered Security Architecture

A multi-layered approach to security strengthens the overall resilience of IoT-enabled healthcare systems:

- **Network Security**: SW Firewalls, Secure gateways, and Network segmentation grant and regulate access and scan the network for risks [26].
- **Device Security**: Sustained updates of the firmware in the devices and the secure boot processes rose to lock the IoT devices from having control or being hacked.
- **Application Security**: Healthcare applications require stringent testing, especially including elements like secure APIs and input validation to avoid such types of vulnerabilities in them. With the incorporation of these broad security approaches, the HCI can develop a robust and safeguarded IoT environment where not only are patients' details demonstrated to have adequate levels of security, but also the availability of critical healthcare systems without disruption is maintained.

Bulk data created by IoT devices [27] is handled and protected through the use of fog and cloud computing. By distributing computing tasks between local fog nodes and centralised cloud servers, healthcare systems can achieve several security and efficiency benefits:

7.4 LEVERAGING FOG AND CLOUD COMPUTING FOR IOT SECURITY

By extending computing to IoT devices, fog computing solves the issue and improves data processing security at the network's edge [28]. This decentralised approach reduces the time taken to transfer data to far-away cloud servers and the possibility of exposure during data transmission. For instance, information from sensors such as wearable devices is analysed by the nearby fog nodes, for instance, monitoring a patient's vitals, and rapid medical attention is provided in Figure 7.3.

7.4.1 Decentralised Data Processing

Fog computing involves computation that is done at the periphery of the IoT devices; this allows the computation of sensitive data to be done in real time [28]. Such a structure prevents delay arising when sending data to remote cloud servers and risks involved when the information is being transferred. For instance, data from patients' wearable equipment can be processed right at the fog nodes located nearby; thus, there is an opportunity for an immediate medical intervention.

7.4.2 Enhanced Data Security

Consequently, both fog and cloud computing and data distribution of storage and processing decrease the probability of centralised attacks [29]. For highly

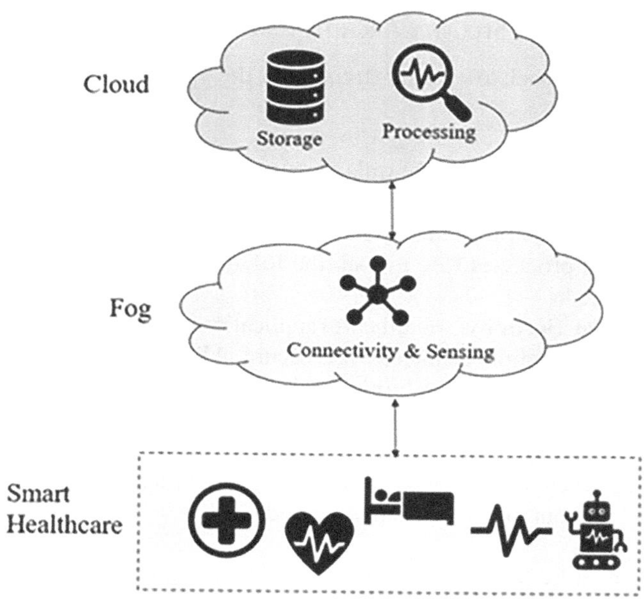

FIGURE 7.3 Overview of a cloud-enabled smart healthcare system.

confidential data, the information is replicated at several nodes to avoid easy access and manipulation by attackers. Also, fog nodes have capabilities of independent authentication/encryption, meaning data can be protected before it is actually sent to the cloud.

7.4.3 Scalable and Reliable Infrastructure

Fog combined with cloud computing works as a scalable as well as a redundant system for IoT systems. Cloud server addresses the issue of large-scale storage of data and complex analysis, while fog nodes are responsible for the localised computation and instantiation responses [30]. This hybrid model helps in keeping the healthcare systems up and running even in the worst of scenarios that may include a network outage of the server, thus promoting reliability.

7.4.4 Dynamic Threat Detection

Individual fog nodes, further, can be integrated with AI-driven algorithms to detect threats and respond to them in real time. These nodes maintain passive observation and detection of unhealthy network activity or aberrant device behaviour, including the performance of data characteristics beyond normal tolerances or intrusion attempts. Since the fog nodes can identify threats as soon as they emerge, the appropriate suitable actions can be taken, such as quarantining the malicious devices or perhaps blocking IPs that are questionable.

7.4.5 Compliance with Data Privacy Regulations

The science of fog and cloud computing can be adjusted to match regulatory rules such as GDPR and HIPAA [31]. Fog computing makes it possible for HCOs to store and process data locally within certain geographical localities, thus protecting their data from breaching data sovereignty rules and regulations as well as privacy requirements.

7.4.6 Improved Patient-Centric Care

Fog and cloud computing integration helps in the offering of proper care through efficient solutions of the modern world. Fog node data processing results can be immediately pushed into centralised patient records in the cloud to give the patient's full picture. It also helps health workers to diagnose and treat patients correctly by providing suitable solutions while also keeping the data safe and easily retrievable. In conclusion, there is a considerable improvement in the security and performance of Ex-Health through the integration of fog and cloud computing in computing 3 [32]. This, therefore, lets the healthcare organisations take advantage of the two technologies in being able to construct strong, steady, and secure networks for the real-time decision-making processes while at the same time ensuring the protection of patients' information.

7.5 IOT AND SDGs IN HEALTHCARE

This chapter argues that the adoption of the IoT in the health sector has the potential to support several of the United Nations' set SDGs [33]. These goals seek to solve crucial issues like poverty, inequality, and climate change, and IoT can hit the health domain and elections by advancing health outcomes, promoting fairness in health resource allocation, and boosting the quality of these services.

7.5.1 SDG 3: Good Health and Well-Being

The third SDG is that everyone should have good health and well-being for all the SDGs with targets [34]. To achieve this goal, IoT has a vital role in interlinking and monitoring patients' health with better diagnosis and treatments. Smart medicine devices, including wristbands, blood pressure metres, glucometers, and sleep metres, are valuable tools that create context in real time on the state of a patient's health. This data is useful for timely actions that raise chances of avoiding complications and optimising the general health condition. Also, IoT facilitates healthcare delivery in areas that have limited access to healthcare services [35]. The IoT-based telemedicine applications let patients get treated by doctors without leaving their homes, eradicating geographical restrictions in order to facilitate equal chances of getting treatment for an equal number of people. IoT-generated data shared with healthcare professionals also facilitates more accurate diagnoses and tailored treatment plans, ultimately enhancing patient care.

7.5.2 SDG 9: Industry, Innovation, and Infrastructure

SDG 9 focuses on developing and improving infrastructure and promoting sustainable industrialisation as well as facilitating innovation [36]. Considering the definition, IoT has a prominent role in reshaping the future of healthcare organisations' infrastructure to be more efficient and innovative. Intelligent smart hospitals, which allow devices to be connected and data to be passed from one department to another, enhance efficiency and reduce the costs of running a hospital. For instance, IoT allows real-time monitoring of the medical equipment and other supplies in order to guarantee their presence and prevent their exhaustion [37]. Besides, it enables efficient tracking of supplies and equipment in healthcare facilities, enabling the control of the use of resources. These innovations not only enhance the performance of the health facility but also assist in the development of sustainable health facilities.

7.5.3 SDG 10: Reduced Inequality

According to SDG 10, there is an international and domestic inequality that should be eliminated. IoT can eliminate inequalities in healthcare since every device creates an equal chance for patients [38]. The elderly or patients with chronic diseases can be effectively observed through wearable devices to detect their health conditions earlier and lessen the unfairness of medical treatments. Also, IoT-based systems can meet the needs of disabled people by providing them with certain safe care services that enhance their existence. In addition, IoT helps to translate the data into better decisions on the usage of healthcare resources, which is made possible since resources are utilised more fairly. Real-time data from IoT devices helps governments and healthcare organisations make better decisions in terms of where to direct healthcare and thus provide more equal access to healthcare in areas that often lack it.

7.5.4 SDG 12: Responsible Consumption and Production

Specifically, SDG 12 is centred on sustainable consumption and production. IoT has great importance in the chase of this goal by leading to better resourcefulness in the healthcare systems. IoT sensors in the healthcare sector can be used to address consumption, management, and waste of energy [39]. This helps in reducing the overall Greenhouse Gases (GHG) emissions of the healthcare facilities. In addition, IoT aids in the enhancement of the sustainable management of medical waste. In terms of waste disposal and recycling, IoT guarantees that medical materials experience proper handling, and their misuse is controlled fully, and in the process, the environment is not harmed. These efforts contribute to a greener, more sustainable healthcare system, supporting global efforts to reduce environmental impact.

7.6 CASE STUDIES: IOT APPLICATIONS IN HEALTHCARE

The use of IoT technologies in healthcare systems is rapidly expanding around the globe. Various real-life instances have shown the revolutionary effects on healthcare results, hospital administration, and patient care. Table 7.1 showcases some case studies demonstrating how 6G and IoT are transforming the healthcare domain.

TABLE 7.1
Examples of IoT Applications in Smart Healthcare

Case Study	Location	IoT Application	Key Outcomes
Remote Monitoring of Chronic Diseases [40]	United States	IoT-based remote monitoring for chronic conditions (diabetes, hypertension, asthma) using connected devices.	Real-time health tracking, reduced hospital readmissions, and enhanced disease management through early intervention.
Creating Smart Hospitals for Enhanced Care [41]	Singapore	IoT-enabled smart hospitals with patient wristbands and sensor-equipped medical equipment.	Improved patient care, reduced operational inefficiencies, enhanced patient safety, and proactive maintenance of medical devices.
IoT-Based Wearables for Preventive Health Monitoring [42]	Europe	Wearable health devices like fitness trackers and smartwatches for monitoring heart rate, sleep, and activity levels.	Early detection of cardiovascular issues, proactive health management, and improved long-term health outcomes.

(Continued)

TABLE 7.1 (*Continued*)
Examples of IoT Applications in Smart Healthcare

Case Study	Location	IoT Application	Key Outcomes
Telemedicine Solutions for Rural Areas [43]	India	IoT-enabled telemedicine platforms with portable diagnostic tools for remote consultations.	Improved healthcare access in rural areas, reduced need for travel, and timely medical consultations and diagnostics for underserved populations.
Optimising Healthcare Asset Management [44]	United Kingdom	IoT sensors for tracking medical equipment location and usage.	Better resource allocation, reduced equipment loss, increased operational efficiency, and minimised costs.
Remote Patient Monitoring for Elderly Care [45]	Japan	Wearable sensors and smart home devices for elderly care.	Enhanced safety, fall detection, and improved quality of life for elderly patients.
Smart Medication Dispensers [46]	Canada	IoT-enabled smart pill dispensers to track medication adherence.	Improved compliance with medication schedules, reduced errors, and enhanced chronic disease management.
Connected Ambulances for Emergency Care [47]	Australia	IoT-enabled ambulances with real-time patient monitoring and communication tools.	Faster emergency response, improved pre-hospital care, and seamless data transfer to hospitals.
Remote Surgical Assistance [48]	Germany	IoT-powered robotic systems for remote surgical procedures.	Enabled remote surgeries, expanded healthcare reach, and reduced time-to-treatment for critical cases.
Smart ICUs with Real-Time Monitoring [49]	South Korea	IoT-enabled ICUs with sensors and AI analytics for patient monitoring.	Early detection of complications, reduced ICU mortality rates, and enhanced efficiency.
Health Monitoring in Workplace Environments [50]	Sweden	IoT wearables for workplace health monitoring.	Improved employee wellness, early intervention, and reduced absenteeism.
IoT Solutions for Infectious Disease Control [51]	China	IoT devices for contact tracing, temperature monitoring, and quarantine enforcement.	Improved pandemic management, reduced transmission rates, and better outbreak response.
Smart Wheelchairs and Mobility Devices [52]	France	IoT-connected wheelchairs and mobility devices with navigation features.	Enhanced mobility, improved independence, and real-time monitoring for maintenance and safety.
IoT in Rehabilitation Therapy [53]	Italy	IoT-enabled rehabilitation equipment for physical therapy.	Improved therapy outcomes, remote tracking of progress, and customised treatment plans.

(Continued)

TABLE 7.1 (*Continued*)
Examples of IoT Applications in Smart Healthcare

Case Study	Location	IoT Application	Key Outcomes
IoT for Mental Health Monitoring [54]	Spain	Wearable and mobile sensors to monitor stress levels, sleep patterns, and mood.	Early intervention for mental health issues, improved counselling services, and personalised treatments.
Smart Beds for Patient Comfort and Monitoring [55]	Netherlands	IoT-equipped smart beds for posture monitoring and pressure relief.	Reduced bedsores, improved patient comfort, and enhanced nurse efficiency.
IoT Solutions for Prenatal Monitoring [56]	Brazil	Wearable devices and sensors to monitor fetal health and maternal well-being.	Improved prenatal care reduced pregnancy-related complications and better health outcomes for mothers and babies.
Smart Contact Lenses for Eye Health [57]	Israel	IoT-enabled contact lenses for monitoring intraocular pressure and hydration levels.	Early detection of eye diseases, proactive treatment plans, and improved vision health.
IoT-Based Air Quality Monitoring in Hospitals [58]	UAE	Smart sensors to monitor air quality and environmental conditions.	Improved infection control, better ventilation management, and healthier indoor environments for patients and staff.
IoT-Enabled Home Health Monitoring [59]	Mexico	Remote monitoring systems for chronic illnesses and post-surgery care.	Better recovery outcomes, fewer hospital visits, and enhanced patient satisfaction.
Smart Wearables for Paediatric Monitoring [60]	South Africa	IoT wearables for monitoring children's vital signs and activity levels.	Improved child health tracking, early detection of illnesses, and proactive parental support.

7.7 CHALLENGES AND FUTURE DIRECTIONS

The adoption of IoT technologies in healthcare is rapidly evolving, offering transformative opportunities for improving patient care, operational efficiency, and access to healthcare services. However, while the potential for IoT in healthcare is vast, there are several challenges that must be addressed to fully realise its benefits. This section explores future directions for IoT in healthcare and the key challenges that will shape its development in the coming years.

Opportunities of IoT in the future of healthcare are vast, contributed by developments in AI, ML, smart hospitals, telemedicine, and interoperability. Adding AI and ML with IoT will enable the analysis of data in real time, and also predict information and treatment plans specific to the healthcare receiver, so as to shift from a curative model to a preventive one in the healthcare sector. By analysing data collected by wearables, AI systems can forecast health events like heart attacks or a diabetic crisis, which improves diagnosis and treatment. The birth of smart hospitals on wheels will take the next step in the evolution of hospitals and revolutionise the delivery of healthcare through automating patient flow, resources, and patient luxury through IoT sensors and devices. Other rising trend includes telemedicine and remote health monitoring, which will be used to provide real-time care and diagnostics for patients, particularly for chronic disease and in remote areas. These are advancements that would bring about more accessible, equitable, and convenient healthcare. Also, the ability to standardise and interconnect different IoT devices and systems in healthcare and EHRs will be the next big problem to solve. The adoption of open standards and APIs will encourage the interchange of data, and common data interchange grants collaborators a coordinated environment, making healthcare more united. Both of these innovations will transform how care is delivered, improving patient satisfaction and organisational flow.

The adoption of IoT in healthcare has also featured some limitations, such as security and data privacy, architecture compatibility, integration of IoT with legacy systems, size of the network, cost of implementation of IoT systems, and ethical and regulatory concerns. IoT devices in healthcare settings produce personal patient information and are hence prone to risks like hacking and information leakage. Thus, security protocols like encryption, use of two-factor authentication, and obeying HIPAA and GDPR must be adopted. Recent advancements in information technology have increased the susceptibility of healthcare organisations to various security threats, thus the need to implement solid data governance policies. Another concern is how to connect IoT with healthcare platforms that were not implemented with IoT platforms in mind; this has an implication that the integration methods need to be scalable and flexible, and rely on the cloud to be able to support the current generation of healthcare information systems while at the same time supporting IoT systems. The prices of implementing and expanding on IoT solutions are also high, as it may prove too expensive for the small and developing service providers and regions, but with the improvement of cloud and edge compute technologies, IoT solutions can be brought to a more attainable level. Ethical and regulatory issues add to the IoT challenges because the collection of data with great frequency raises issues as to ownership of the data and consent to its use, among other things. The identification of patients involves handling of their personal data, and therefore, patients should be informed how this information will be used, and AI algorithms in IoT applications have to be trustworthy and explainable, and should not inherit bias. Overcoming these challenges will be important for the effective and efficient implementation of IoT in healthcare.

7.8 CONCLUSION

In the aspect of applying IoT solutions to the healthcare system, mainly the continuation of the 6G network technology, medical systems will be changed and improved for real-time, mining, and patient care. Through 6G technology, IoT will enable efficient exchange of information between devices and care providers as it supports fast communication. However, achieving security for these devices and ensuring compliance with data protection laws is going to be very important for the uptake of IoT on the market. In addition, pairing IoT with SDGs can enable affordable, accessible, and sustainable healthcare. By optimising the consumption of resources, as well as increasing access to healthcare for populations hitherto not well catered to, IoT solutions can help build a more sustainable and fairer healthcare sector. IoT, when integrated with 6G, aims to better transform the sphere of healthcare, as the ultimate goal of the innovative development is to make healthcare systems smarter, more efficient, and patient-oriented.

REFERENCES

1. Kaur, A., & Tiwana, P. S. (2023, December). Analysis of IoT devices data using Bayesian learning on fog computing. In *Artificial Intelligence, Blockchain, Computing and Security Volume 2: Proceedings of the International Conference on Artificial Intelligence, Blockchain, Computing and Security (ICABCS 2023)*, Gr. Noida, UP, India, 24–25 February 2023 (p. 194). CRC Press.
2. Marques, G., Pitarma, R., Garcia, N. M., & Pombo, N. (2019). Internet of things architectures, technologies, applications, challenges, and future directions for enhanced living environments and healthcare systems: A review. *Electronics*, *8*(10), 1081.
3. Ahmadi, H., Arji, G., Shahmoradi, L., Safdari, R., Nilashi, M., & Alizadeh, M. (2019). The application of internet of things in healthcare: A systematic literature review and classification. *Universal Access in the Information Society*, *18*, 837–869.
4. Ahmed, A., Khan, M. M., Singh, P., Batth, R. S., & Masud, M. (2022). IoT-based real-time patients vital physiological parameters monitoring system using smart wearable sensors. *Neural Computing and Applications*, *34*(22), 19397–19673.
5. Taimoor, N., & Rehman, S. (2021). Reliable and resilient AI and IoT-based personalised healthcare services: A survey. *IEEE Access*, *10*, 535–563.
6. Karale, A. (2021). The challenges of IoT addressing security, ethics, privacy, and laws. *Internet of Things*, *15*, 100420.
7. Rajak, S., Summaq, A., Kumar, M. P., Ghosh, A., Elumalai, K., & Chinnadurai, S. (2024). Revolutionizing healthcare with 6G: A deep dive into smart, connected systems. *IEEE Access*, 12, 194150–194170.
8. Salh, A., Audah, L., Shah, N. S. M., Alhammadi, A., Abdullah, Q., Kim, Y. H., ... & Almohammedi, A. A. (2021). A survey on deep learning for ultra-reliable and low-latency communications challenges on 6G wireless systems. *IEEE Access*, *9*, 55098–55131.
9. Pattnaik, S. K., Samal, S. R., Bandopadhaya, S., Swain, K., Choudhury, S., Das, J. K., ... & Poulkov, V. (2022). Future wireless communication technology towards 6G IoT: An application-based analysis of IoT in real-time location monitoring of employees inside underground mines by using BLE. *Sensors*, *22*(9), 3438.
10. Martinelli, M., & Furtado, A. T. (2024). Remote surgery with 5G or 6G: Knowledge production and diffusion globally and in the German case. *arXiv preprint arXiv:2411.17899*.
11. Kasula, B. Y., & Whig, P. (2023). AI-driven machine learning solutions for sustainable development in healthcare—pioneering efficient, equitable, and innovative health service. *International Journal of Sustainable Development through AI, ML and IoT*, *2*(2), 1–7.
12. AbuKhousa, E., Mohamed, N., & Al-Jaroodi, J. (2012). e-Health cloud: Opportunities and challenges. *Future Internet*, *4*(3), 621–645.
13. Dwivedi, R., Mehrotra, D., & Chandra, S. (2022). Potential of Internet of Medical Things (IoMT) applications in building a smart healthcare system: A systematic review. *Journal of Oral Biology and Craniofacial Research*, *12*(2), 302–318.
14. Ullah, S., Li, J., Chen, J., ALI, I., Khan, S., Ahad, A., ... & Leung, V. (2024). A survey on emerging trends and applications of 5G and 6G to healthcare environments. *ACM Computing Surveys*, 57(4), 1–36.

15. Daah, C., Qureshi, A., Awan, I., & Konur, S. (2024). Enhancing zero trust models in the financial industry through blockchain integration: A proposed framework. *Electronics, 13*(5), 865.
16. El Zouka, H. A., & Hosni, M. M. (2021). Secure IoT communications for smart healthcare monitoring system. *Internet of Things, 13*, 100036.
17. Omolara, A. E., Alabdulatif, A., Abiodun, O. I., Alawida, M., Alabdulatif, A., & Arshad, H. (2022). The internet of things security: A survey encompassing unexplored areas and new insights. *Computers & Security, 112*, 102494.
18. Yaqoob, T., Abbas, H., & Atiquzzaman, M. (2019). Security vulnerabilities, attacks, countermeasures, and regulations of networked medical devices—A review. *IEEE Communications Surveys & Tutorials, 21*(4), 3723–3768.
19. Smid, M. E. (2021). Development of the advanced encryption standard. *Journal of Research of the National Institute of Standards and Technology, 126*, 126024, 18.
20. Ahmed, I., & Asghar, A. (2023). Evaluating the efficacy of biometric authentication techniques in healthcare. *International Journal of Responsible Artificial Intelligence, 13*(7), 1–12.
21. Vo, H. V., Du, H. P., & Nguyen, H. N. (2023). AI-powered intrusion detection in large-scale traffic networks based on flow sensing strategy and parallel deep analysis. *Journal of Network and Computer Applications, 220*, 103735.
22. Xu, K., Zhang, Z. L., & Bhattacharyya, S. (2008). Internet traffic behavior profiling for network security monitoring. *IEEE/ACM Transactions on Networking, 16*(6), 1241–1252.
23. Salima, O., Asri, N., & Hamid, H. J. (2013). Machine learning techniques for anomaly detection: An overview. *International Journal of Computer Applications, 79*(2), 33–41.
24. Tariq, U., Ibrahim, A., Ahmad, T., Bouteraa, Y., & Elmogy, A. (2019). Blockchain in internet-of-things: A necessity framework for security, reliability, transparency, immutability and liability. *IET Communications, 13*(19), 3187–3192.
25. Folorunso, A., Wada, I., Samuel, B., & Mohammed, V. (2024). Security compliance and its implication for cybersecurity. *World Journal of Advanced Research and Reviews, 24*(1), 2105–2121.
26. Kallatsa, M. (2024). Strategies for network segmentation: A systematic literature review. (Master's thesis). University of Jyväskylä.
27. Bonkra, A., & Dhiman, P. (2021, December). IoT security challenges in cloud environment. In *2021 2nd International Conference on Computational Methods in Science & Technology (ICCMST)* (pp. 30–34). IEEE.
28. Singh, S. P., Nayyar, A., Kumar, R., & Sharma, A. (2019). Fog computing: From architecture to edge computing and big data processing. *The Journal of Supercomputing, 75*, 2070–2105.
29. Khan, S., Parkinson, S., & Qin, Y. (2017). Fog computing security: A review of current applications and security solutions. *Journal of Cloud Computing, 6*, 1–22.
30. Bellavista, P., Corradi, A., & Zanni, A. (2017). Integrating mobile internet of things and cloud computing towards scalability: Lessons learned from existing fog computing architectures and solutions. *International Journal of Cloud Computing, 6*(4), 393–406.
31. Kitsiou, A., Sideri, M., Pantelelis, M., Simou, S., Mavroeidi, A. G., Vgena, K., ... & Kalloniatis, C. (2024). Specification of self-adaptive privacy-related requirements within cloud computing environments (CCE). *Sensors, 24*(10), 3227.
32. Mishra, A., Tripathi, A., Khazanchi, D., Hiran, K. K., Vyas, A. K., & Padmanaban, S. (2021). A framework for applying artificial intelligence (AI) with Internet of NanoThings (IoNT). In K. K. Hiran, D. Khazanchi, A. K. Vyas, & S. Padmanaban (Eds.), *Machine learning for sustainable development* (pp. 1–16). Berlin, Germany: Walter de Gruyter GmbH & Co. KG.

33. López-Vargas, A., Fuentes, M., & Vivar, M. (2020). Challenges and opportunities of the internet of things for global development to achieve the United Nations sustainable development goals. *IEEE Access*, *8*, 37202–37213.
34. Guégan, J. F., Suzán, G., Kati-Coulibaly, S., Bonpamgue, D. N., & Moatti, J. P. (2018). Sustainable Development Goal# 3, "health and well-being", and the need for more integrative thinking. *Veterinaria México OA*, *5*(2), 1–18.
35. Fernandez, R. M. (2020). SDG3 good health and well-being: Integration and connection with other SDGs. In well-being: Integration and connection with other SDGs. In W. Leal Filho, T. Wall, U. Azeiteiro, A. M. Azul, L. Brandli, & P. Gökcin Özuyar (Eds.), *Good Health and Well-Being* (pp. 629–636). Cham: Springer.
36. Tomaselli, M. F., Timko, J., Kozak, R., Bull, J., Kearney, S., Saddler, J., & Zhu, X. (2019). SDG 9: Industry, innovation and infrastructure–anticipating the potential impacts on forests and forest-based livelihoods. In P. Katila, C. J. P. Colfer, W. de Jong, G. Galloway, P. Pacheco, & G. Winkel (Eds.), *Sustainable Development Goals* (pp. 279–314). Cambridge, UK: Cambridge University Press.
37. Manavalan, E., & Jayakrishna, K. (2019). A review of Internet of Things (IoT) embedded sustainable supply chain for industry 4.0 requirements. *Computers & Industrial Engineering*, *127*, 925–953.
38. Fink, M. A., Höltl, A. N., & Brandtweiner, R. O. (2020). Potential impact of information systems tackling SDG10 on the dimensions of health care and gender. *WIT Transactions on Ecology and the Environment*, *241*, 221–234.
39. Lamichhane, B., & Neupane, N. (2022). Improved healthcare access in low-resource regions: A review of technological solutions. *arXiv preprint arXiv:2205.10913*.
40. Rojahn, K., Laplante, S., Sloand, J., Main, C., Ibrahim, A., Wild, J., ... & Johnson, K. I. (2016). Remote monitoring of chronic diseases: A landscape assessment of policies in four European countries. *PLoS One*, *11*(5), e0155738.
41. Quazi, F. eHealth Services in Comprehensive Smart Environments: Enhancing Healthcare through Integrated Technologies. *International Journal of New Media Studies*, *11*(2), 74–85.
42. Trmčić, B. R., Labus, A., Mitrović, S., Buha, V., & Stanojević, G. (2018). Internet of things in E-health: An application of wearables in prevention and well-being. In *Wearable Technologies: Concepts, Methodologies, Tools, and Applications* (pp. 880–885). IGI Global.
43. Mishra, S. K., Kapoor, L., & Singh, I. P. (2009). Telemedicine in India: Current scenario and the future. *Telemedicine and e-Health*, *15*(6), 568–575.
44. Chen, J. K., & Ho, H. H. (2023). Asset resource optimization solution for smart hospital facilities and energy management through an interpretive structural model. *Buildings*, *13*(12), 3064.
45. Fujiwara, T., Sheppard, J. P., Hoshide, S., Kario, K., & McManus, R. J. (2023). Medical telemonitoring for the management of hypertension in older patients in Japan. *International Journal of Environmental Research and Public Health*, *20*(3), 2227.
46. Faisal, S., Ivo, J., Lee, C., Carter, C., & Patel, T. (2022). The usability, acceptability, and functionality of smart oral multidose dispensing systems for medication adherence: A scoping review. *Journal of Pharmacy Practice*, *35*(3), 455–468.
47. Fry, M. M. (2011). A systematic review of the impact of afterhours care models on emergency departments, ambulance and general practice services. *Australasian Emergency Nursing Journal*, *14*(4), 217–225.
48. Wisotzky, E. L., Rosenthal, J. C., Meij, S., van den Dobblesteen, J., Arens, P., Hilsmann, A., ... & Schneider, A. (2025). Telepresence for surgical assistance and training using eXtended reality during and after pandemic periods. *Journal of Telemedicine and Telecare*, *31*(1), 14–28.

49. Kwon, H., An, S., Lee, H. Y., Cha, W. C., Kim, S., Cho, M., & Kong, H. J. (2022). Review of smart hospital services in real healthcare environments. *Healthcare Informatics Research, 28*(1), 3–15.
50. Mouazzen, A. K., Blomberg, K., & Jaensson, M. (2024). Experiences of psychological, social and organisational work environments in occupational health service in Sweden: A cross-sectional survey. *BMC Health Services Research, 24*(1), 1300.
51. Dong, Y., & Yao, Y. D. (2021). IoT platform for COVID-19 prevention and control: A survey. *IEEE Access, 9*, 49929–49941.
52. Morbidi, F., Devigne, L., Teodorescu, C. S., Fraudet, B., Leblong, E., Carlson, T., … & Ragot, N. (2022). Assistive robotic technologies for next-generation smart wheelchairs: Codesign and modularity to improve users' quality of life. *IEEE Robotics & Automation Magazine, 30*(1), 24–35.
53. Quarto, A., Soldo, D., Gemmano, S., Dario, R., Di Lecce, V., Guaragnella, C., … & Lombardi, A. (2017, July). IoT and CPS applications based on wearable devices. A case study: Monitoring of elderly and infirm patients. In *2017 IEEE Workshop on Environmental, Energy, and Structural Monitoring Systems (EESMS)* (pp. 1–6). IEEE.
54. Gutierrez, L. J., Rabbani, K., Ajayi, O. J., Gebresilassie, S. K., Rafferty, J., Castro, L. A., & Banos, O. (2021). Internet of things for mental health: Open issues in data acquisition, self-organization, service level agreement, and identity management. *International Journal of Environmental Research and Public Health, 18*(3), 1327.
55. Ghersi, I., Mariño, M., & Miralles, M. T. (2018). Smart medical beds in patient-care environments of the twenty-first century: A state-of-art survey. *BMC Medical Informatics and Decision Making, 18*, 1–12.
56. Marques, J. A. L., Han, T., Wu, W., do Vale Madeiro, J. P., Neto, A. V. L., Gravina, R., … & de Albuquerque, V. H. C. (2020). IoT-based smart health system for ambulatory maternal and fetal monitoring. *IEEE Internet of Things Journal, 8*(23), 16814–16824.
57. Armarnik, S., Lavid, M., Blum, S., Wygnanski-Jaffe, T., Granet, D. B., & Kinori, M. (2021). The relationship between education levels, lifestyle, and religion regarding the prevalence of myopia in Israel. *BMC Ophthalmology, 21*, 1–6.
58. Rana, A., Chauhan, R., Sajwan, V. S., Devliyal, S., & Gupta, S. (2024, January). Implementation of Internet of things in building smart cities. In *2024 International Conference on Intelligent and Innovative Technologies in Computing, Electrical and Electronics (IITCEE)* (pp. 1–8). IEEE.
59. Alam, F., Almaghthawi, A., Katib, I., Albeshri, A., & Mehmood, R. (2021). IResponse: An AI and IoT-enabled framework for autonomous COVID-19 pandemic management. *Sustainability, 13*(7), 3797.
60. Mvelase, P., Dlamini, Z., Dludla, A., & Sithole, H. (2015, November). Integration of smart wearable mobile devices and cloud computing in South African healthcare. In *eChallenges e-2015 Conference* (pp. 1–10). IEEE.

8 Implementation of IoT Solutions in Healthcare

Anjuli Goel and Chander Prabha

8.1 INTRODUCTION

"Internet of Things (IoT)" encompasses numerous technologies, protocols, applications, and initiatives. Essentially, it is an Internet-connected network of objects. These items consist of both physical objects equipped with IoT and IoT gadgets. Electronic components and software are installed in the IoT equipment to facilitate data gathering, organisation, and data exchange [1].

Recently, there has been significant growth in the healthcare industry, which may have resulted in more employment and income. A physical examination was necessary in hospitals a few years ago for the identification of illnesses and other abnormalities in the human body. Most of the patients were required to stay in the hospital for their therapy sessions. This increased healthcare costs and placed pressure on outlying and rural medical facilities. Because of technological improvements over time, small gadgets like smartwatches may now be used to monitor health and detect a range of conditions [2].

The healthcare system is becoming more patient-focused than hospital-focused thanks to technology. Motivated by the necessity for personalized treatment, enhanced healthcare systems, and a growing desire for superior healthcare results, the IoT in healthcare, commonly known as H-IoT, has evolved the broader IoT ecosystem. The ageing of the world's population, the rise in chronic illness incidence, and the pursuit of digital healthcare are significant factors propelling the spread of IoT within the healthcare sector [3].

The application of IoT in healthcare and medicine is among its most enticing uses. The IoT may be useful for a variety of medical applications, such as geriatric care, deathbed care, fitness routines, and remote health monitoring. As a result, many types of sensors, imaging, diagnostic, and medical devices can be considered smart devices that are a component of the IoT internally. IoT-driven healthcare solutions are expected to save costs, increase life quality, and improve user experience [4]. Within an IoT healthcare system, patients are connected to various wireless systems, apps, and sensors that gather data. An expert system then forwards this data to a physician or other healthcare professional. Sensors, actuators, and other communication devices can be used to remotely access medical devices for the IoT, which are devices that are linked to the Internet and that monitor patient health [5]. The expert system uses these devices to send patient data and information over a gateway to a safe platform in the cloud, where it is stored and prepared for scanning [6]. The IoT enables networks of smart devices, cloud applications, and services that enable data storage

DOI: 10.1201/9781003586456-8

and transportation. Given the prevalence of remote wireless healthcare platforms and the growing interest in electronic patient care datasets, the IoT communication architecture may serve as the main facilitator for decentralized ubiquitous healthcare applications [7].

Further, the chapter is divided into various sections. Section 8.2 provides a detailed explanation of IoT in healthcare systems, problems that persist in the conventional healthcare system, and the components included in the IoT healthcare system. Section 8.3 presents an IoT healthcare framework and layered architecture of healthcare systems and illustrates the IoT protocols used in smart healthcare systems. Section 8.4 defines the various technologies used in the healthcare system. Section 8.5 provides information about the IoT applications of healthcare systems. Section 8.6 gives a brief idea of the benefits of IoT in the healthcare system. Sections 8.7 and 8.8 illustrate the various challenges and the future of IoT in the healthcare system. Finally, the conclusion of the chapter is provided.

8.2 IOT IN THE HEALTHCARE SYSTEM

IoT technologies are revolutionizing patient care and healthcare administration, with substantial potential in the healthcare sector. Fitness trackers and smartwatches are wearable health monitors that have become essential tools for those who want to monitor their physical activity, sleep habits, and vital signs. These gadgets encourage a proactive approach to well-being by enabling patients to participate in their health management actively [3]. In the healthcare industry, intelligent healthcare may facilitate communication between all parties involved, guarantee that participants accept the assistance and resources they need, and support groups in making wise asset allocation choices [8].

Technology is required to build an IoT system for healthcare. This is so that a variety of technologies may enhance its functionality. Numerous cutting-edge technologies have been used to connect diverse healthcare applications to an IoT system. This technology may be roughly categorized into three areas, as seen in Figure 8.1: identification technology, communication technology, and location technology [2].

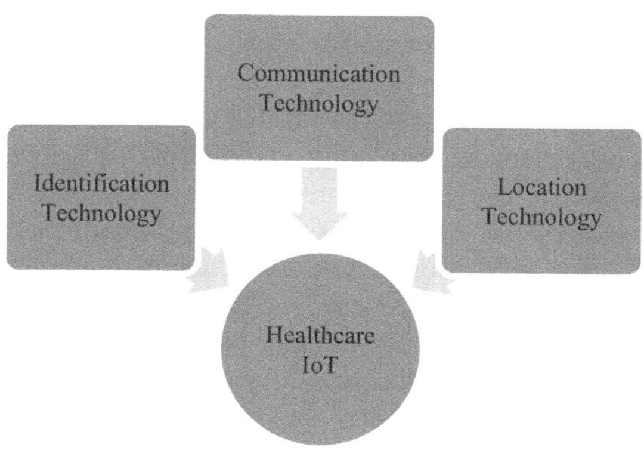

FIGURE 8.1 Technology classification of IoT healthcare.

Identification Technologies: To track and identify patients, medical staff, and medical devices in a unique way, identification technologies are utilized. This ensures accurate patient data management, personalized care, and efficient resource utilization.

- **Radio Frequency Identification (RFID)**: RFID readers and tags are frequently employed for inventory management, medical equipment tracking, and patient identification. For instance, wristbands with RFID tags can be used to track patient movements and match them with their medical records, reducing the risk of errors.
- **Barcode Systems**: Barcodes are often used for medication management and patient identification. Barcoded wristbands and medication labels ensure that the right medication is administered to the right patient.
- **Biometric Systems**: Healthcare institutions employ technologies such as computerized facial recognition and scanners for fingers to provide secure access to patient information and restricted locations. They offer the convenience of access together with a high degree of security.

Communication Technologies: Data transmission between devices, healthcare practitioners, and patients is made possible via communication technology. They are necessary for data exchange, remote consultations, and real-time monitoring.

- **Bluetooth Low Energy (BLE)**: BLE is frequently found in wearable medical equipment, such as continuous glucose monitoring and fitness trackers. It enables these gadgets to connect with cellphones and other devices to send health information effectively and with little battery usage.
- **Wi-Fi**: Wi-Fi networks are crucial for connecting various IoT devices within healthcare settings. They enable the transmission of data from remote patient monitoring devices to healthcare providers and support telemedicine consultations.
- **Cellular Networks (3G/4G/Fifth-Generation, 5G)**: Cellular networks provide connectivity for mobile health applications and remote patient monitoring devices, especially in areas with limited Wi-Fi access. They are crucial for telemedicine and emergency response systems.

Location Technologies: Location technologies help track the physical location of patients, medical staff, and equipment within healthcare facilities. They enhance operational efficiency, safety, and asset management.

- **Global Positioning System (GPS)**: While more common in outdoor settings, GPS can be used for tracking patients or medical staff in large healthcare campuses or for managing emergency medical services.
- **Indoor Positioning Systems (IPS)**: Ultra-wideband (UWB) and beacon technologies are employed in hospitals for asset monitoring and indoor navigation. They aid with the location of medical supplies and guarantee that personnel can easily locate vital resources.
- **Real-Time Location Systems (RTLS)**: These systems track the whereabouts of personnel, equipment, and patients in real time within healthcare facilities by combining technologies such as RFID, infrared, and ultrasound. This improves patient safety, streamlines processes, and manages inventories.

8.2.1 Problems in a Conventional Healthcare System

The seamless connectedness that an ideal healthcare system offers should allow patients and healthcare professionals to work together across organizational and geographic barriers. To improve diagnosis and healthcare services, it should also promote the creation of safe platforms and apps for exchanging health data.

Conventional healthcare systems face several obstacles to personalized healthcare, including network latency, imprecise and incomplete health data, workflow gaps brought on by incompatible and vendor-specific solutions for healthcare, an inadequate degree of privacy and data integrity, imprecise and comprehensive patient health histories, and a lack of a dependable, secure platform to handle data gathered from multiple healthcare systems. These types of problems are summarized in Figure 8.2. Within an established healthcare system, after diagnosing the patient, the doctor manually enters their medical records into the hospital's central database or their local system. Because there is no communication between the user's smartphone and the local system, there is no option for real-time diagnostics. The central database, which houses all patient health records, is accessible to the hospital. But consumers don't own their health data, and there are issues with data quality, privacy, and openness [9].

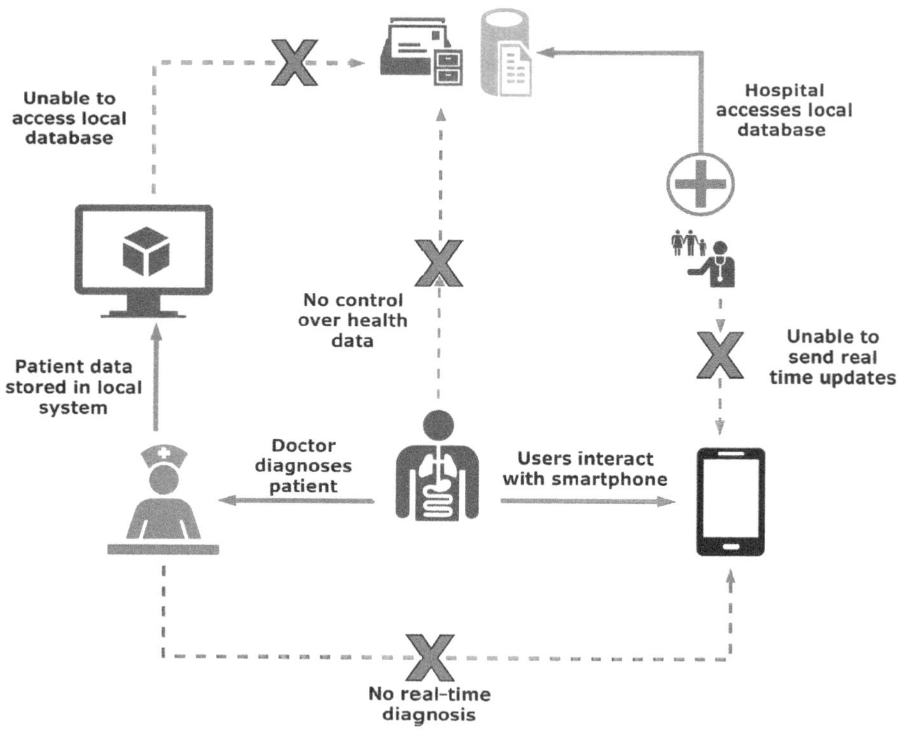

FIGURE 8.2 Issues in the existing healthcare system.

8.2.2 H-IoT Components

IoT healthcare comprises five components to support an intelligent healthcare system. These five components are shown in a circle, as the outcome of one part is the input of another part. Figure 8.3 depicts the five components of a smart healthcare system. Sensors are the first component to be discussed. Utilizing sensor technologies, patient data is gathered. After that, the intelligent network receives the gathered data and uses it to communicate with the entire system. So, sending data is stored on cloud computing for further use. Cloud computing provides security and privacy for the stored data. After storing the data on the cloud, the data is analysed in big data analysis and helps in making proper decisions. Once the data has been filtered, it is sent to the hospital for taking the appropriate action, i.e., monitoring and diagnosis. The system again collects the data to know whether the right medication is provided to the patient or not through the sensors. In this way, this circle is going on and continuing [10].

These five components have five characteristics [10]:

Stability: Sensor technology must be stable for reliable data collection for ongoing real-time monitoring.

Continuity: Support for interoperability in intelligent networks is needed to ensure that users can continue to communicate with one another and the Internet.

Confidentiality: To store rapidly changing information, a powerful cloud computing resource demands confidentiality.

Reliability: Reliability is the need for big data analysis to convert rapidly changing data into valuable information.

Efficiency: For accurate diagnosis and treatment, a smart hospital demands efficiency.

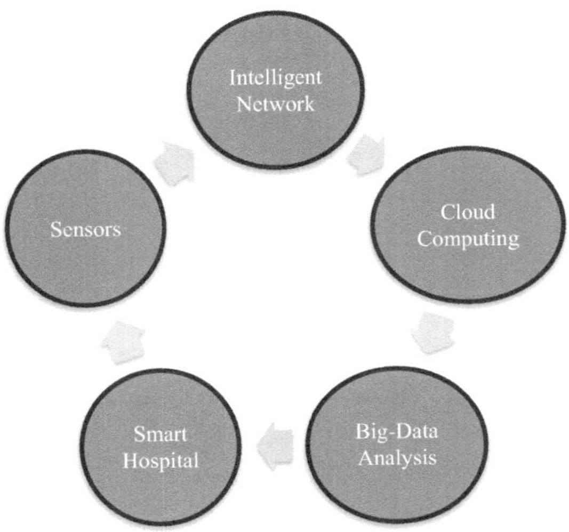

FIGURE 8.3 Five components of a smart healthcare system.

8.3 IoT HEALTHCARE FRAMEWORK

For information processing, IoT apps heavily rely on the middleware layer. Intelligent transportation, smart agriculture, smart homes, smart cities, smart grids, and smart health are a few examples. The basic architecture of a smart healthcare system consists of two functions: the first function is related to the data captured by sensor devices, and the second function is to process and analyse the captured data and generate the resulting information regarding patient monitoring.

Figure 8.4 shows the intelligent healthcare system architecture, which includes devices to monitor the patients, such as patch sensors, wearable devices, and medical homes, and capture the related data. This data is captured with the help of Internet services such as Wi-Fi, Bluetooth, etc. Now the captured data has been processed and analysed to convert it into meaningful information. This information is helpful for scientists, medical researchers, and health insurance companies for monitoring patients. Medical officers can provide medical aid to patients through the network and vice versa. Medical officers use the Cloud to store data and information for future use [4,11].

8.3.1 Layered Architecture of Smart Healthcare Systems

Smart cities, smart industries, and smart healthcare are just a few of the sophisticated applications that have emerged from the IoT' rapid advancement and expansion. By 2020, it is projected that there will be over 30 million smart devices on the network, improving user experience through the combination of billions of connected devices with sensors and actuators [12]. Figure 8.5 shows the typical IoT layered architecture in the healthcare system. It is a six-layer architecture, where middleware layers are important for IoT applications to process all the information.

The application layer, which links the items to the IoT network, is the topmost layer in the architecture. It comprises several healthcare apps that provide the user with e-health features and services [13]. Connecting to servers, network devices, and

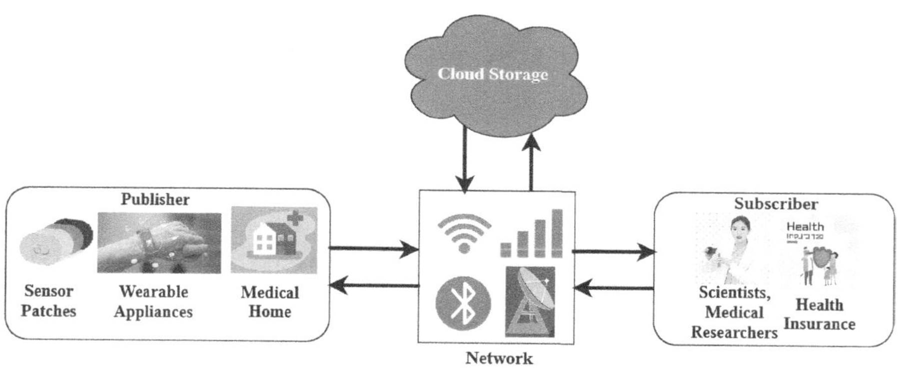

FIGURE 8.4 IoT healthcare framework.

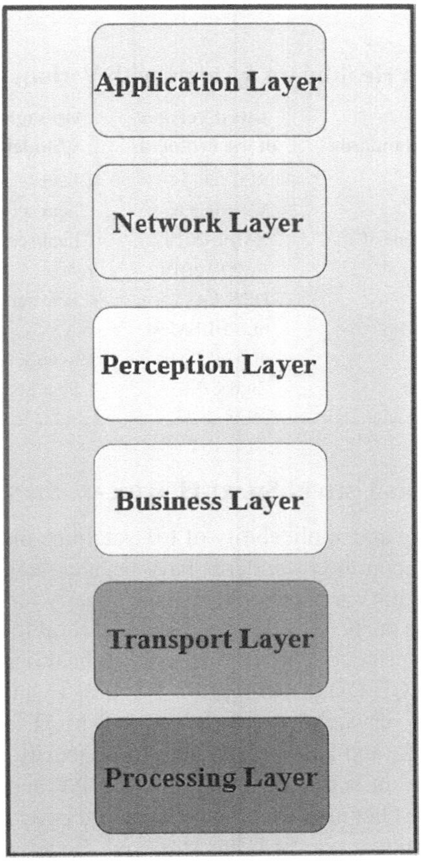

FIGURE 8.5 IoT layered architecture in a healthcare system.

other smart objects is the responsibility of the network layer. Typically, sensor data processing and transmission are its uses. A perception layer identifies physical items and sensor devices to identify other smart things in the surrounding environment. It also senses the object's physical characteristics, including information on temperature, motion, orientation, humidity, location, acceleration, etc. This layer is the initial step towards more intelligent and individualized healthcare; it consists of a generation and actuation subsystem. The business layer manages all IoT systems, including apps, businesses, privacy, and profit models. Sensor data is moved from the perception layer to the processing layer and back using networks such as infrared, wireless, LAN, RFID, Bluetooth, Zigbee, Wi-Fi, UMTS, and 3G by the transport layer, which is the general function of other layers. The vast quantity of information received may be stored and analysed by the middleware or processing layer. To handle the enormous quantity of data, it can interface with big data, cloud computing, and databases. It can also oversee and provide the lower levels with a variety of services [8].

TABLE 8.1
IoT Protocols Used in Healthcare Systems with Various Aspects

Protocols	Network Standards	Latest Version of the Protocols	Messaging Model	Application Levels
CoAP	IETF	RFC 8323	Client/server	Utility fields
MQTT	OASIS	Version 5.0	Client/server	Messaging
XMPP	Various versions of RFC	XMPP 2019	Client/server	Remote management
AMQP	ISO/IEC	AMQP 2019	NA	Enterprise integration
DDS	OMG	DDS 1.4	Client/server	Military
LoWPAN	IEEE 802.15.4	6Lo-BLEMesh	NA	Structural monitoring
BLE	802.15.1	6Lo-BLEMesh	Client/Server	Wearable devices
Zigbee	IEEE 802.15.1	Zigbee 3.0	Client/Server	Consumer electronics

8.3.2 IoT Protocols Used in Smart Healthcare Systems

To evaluate the services and applicability of IoT solutions that connect many objects to the Internet, several common standards have been created. Even if several protocols have been created, not every one of them is necessary for every IoT application at once. Depending on the purpose of the application, several IoT protocols are selected. The most popular IoT protocols, which are used in healthcare system applications, are CoAP, MQTT, AMQP, DDS, XMPP, LoWPAN, BLE, and Zigbee. Regarding the security and quality of service, the protocols CoAP, MQTT, and AMQP have both, XMPP, LoWPAN, BLE, and Zigbee have only the security parameter, whereas the rest include the quality of service. Additionally, CoAP and MATT are tree-based architectures, XMPP is client/server, AMQP is peer to peer, DDS has bus, LoWPAN has star and mesh architectures, BLE has star, and Zigbee includes star, mesh, and hybrid architectures. CoAP and Zigbee use the UDP transport protocol, TCP protocol is used by MQTT, XMPP, AMQP, and LoWPAN, whereas DDS and BLE use both protocols for transporting the data and are explained in Table 8.1, with various related aspects mentioned below [7].

8.4 IOT TECHNOLOGIES IN THE HEALTHCARE SYSTEM

Various IoT technologies used in healthcare systems are shown in Figure 8.6 and explained further.

Sensors: Sensors are widely utilized in the medical field, capable of communicating with digital networks and increasing measurement accuracy. These sensors can now keep track of all health-related parameters. Patients' blood pressure, temperature, oxygen concentrator, sugar levels, and infusion and fluid management systems are easily monitored. The information collected can assist in evaluating the health status, defective bones, and the surrounding biological tissue. Besides sensors, these devices can be integrated with wearables, medical equipment, and even implants in the human body. Some of the common sensors and their measurement are illustrated in Table 8.2.

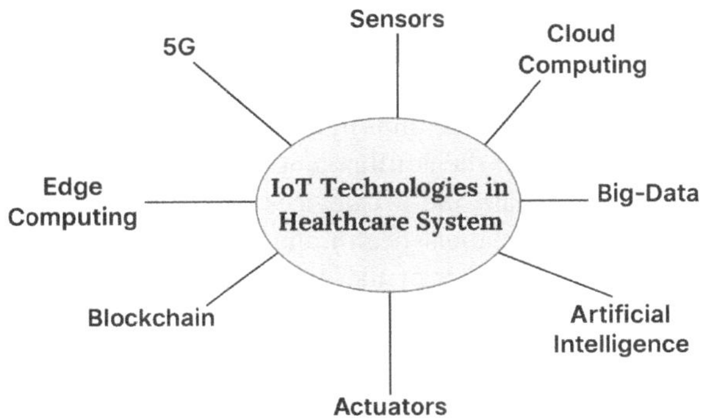

FIGURE 8.6 IoT technologies in healthcare systems.

TABLE 8.2
Common Sensors and Their Measurement

Sensors	Measurement
Electrocardiogram	To record the heart's activity
Pulse oximeter	To record the oxygen level and heart rate
Blood pressure	To display the blood pressure
Glucose	To figure out the level of glucose in the blood
Temperature	To know about the body temperature
Respiratory	To measure the rate of respiration in the body

Cloud Computing: Despite being a cloud computing service, cloud is a software platform that offers a way to store, manage, and analyse enormous volumes of data from IoT devices in the healthcare industry. Because it is now more flexible and less expensive to store and analyse data, it has changed how individuals and businesses utilize and acquire computer resources.

Within the field of medical IoT, cloud computing for the storage and processing of medical data allows healthcare providers to save expenses, improve patient outcomes, and raise the overall quality of care. For example, cloud computing may be used to securely store electronic health records, which will make it easier for medical personnel to access and exchange patient data across various systems and organizations. It can also be analysed and stored at will for huge data sets from wearables or even smart IoT devices, thus enabling medical professionals to be more precise in their treatment of diseases.

Big Data: Indeed, big data has become the most fantastic instrument for the capture, storage, and analysis of patient data. Traditionally, medical institutions stored data on paper copies, and this added an extra cost. One of the biggest advantages of Big Data in healthcare is the ability to manage all patient data, billing information, and clinical records in digital form. Naturally, the efficiency of storing the data systematically is the major contributor to the enhanced quality of healthcare services.

Artificial Intelligence: AI algorithms have newly developed into very dynamic devices in medicine, but their full potential is only realized when AI algorithms combine with data and IoT devices. AI, on the other hand, can gain real-time insights into patient health and allow for personalised interventions through the utilization of the extensive data that these devices can produce. Machine learning models can also be used to analyse and identify the disease pattern and advise treatment accordingly. It learns all this from historical data and allows the changes in medication according to the patient's health [14,15]. Wearable devices are also used to record functions such as sleep patterns, heart rate, and other relevant factors [16,17].

Actuators: Actuators are machines that provide a system's motion and regulate how the system behaves in a certain environment. The primary functions of actuators in medical applications are parameter control and accuracy maintenance. It offers support in creating a hospital bed whose height may be changed to accommodate the patient's demands.

Blockchain: Regardless of organizational or geographic barriers, healthcare practitioners may safely access and share pertinent patient data thanks to blockchain. Because it gives medical professionals access to a patient's full health record and empowers them to make well-informed decisions, this simplified data exchange facilitates improved care coordination. By giving some healthcare providers permission to view their data, patients may have greater control over their data. By automating data-sharing agreements and guaranteeing compliance with privacy standards, smart contracts—self-executing code on the blockchain—give patients more transparency and control over their data.

Edge Computing: One of the major advantages of edge computing in healthcare is the decrease in network latency. Edge computing for the IoT uses data locally to minimize the trips the information authenticates to and from cloud servers. The cloud services that are available to patients can make real-time analysis of health-related data which is necessary for very time-sensitive applications. This includes remote patient monitoring, an emergency response system, and a continuous health monitoring system.

5G: In the context of the IoT in healthcare, 5G networks provide real-time connectivity, allowing medical professionals to continually interact with patients and monitor remotely. 5G networks' increased speed and reduced latency make it possible for data to be sent and received practically instantly, allowing for prompt reactions to urgent medical emergencies. The capacity of 5G networks to accommodate many connected devices is one of its main benefits. In healthcare settings, where many IoT devices are networked, this scalability is essential. Thanks to 5G networks' high capacity, a lot of data can be sent at once, making it possible for healthcare professionals to efficiently gather and process patient data in real time.

8.5 IOT APPLICATIONS IN HEALTHCARE SYSTEMS

The growth of IoT is exciting for everyone because of its diverse applications across industries.

The healthcare industry contains several applications. The IoT has brought about several wearables and gadgets that have enhanced patient comfort [1,14,18,19].

Glucose Monitoring System: Devices with sensors embedded in them are available for patients with diabetes. The gadgets' sensors will contact patients on their phones to alert them when their blood sugar drops dangerously low and store previous data for them. Patients will be able to ascertain when they are currently and maybe in the future most prone to hypoglycaemia.

Moodables: A gadget that elevates our mood all day long is called a moodable. Although science fiction may seem improbable, it is quite close to reality. Halo Neurosciences and Thync are developing it, and it is coming along rather well. Moodables are wearable devices that are worn on the head and work by sending low-intensity currents to the brain to improve mood.

Cancer Treatment Activity Tracker: The ideal course of therapy for a cancer patient is usually determined by criteria other than just their age and weight. The optimum course of treatment for them also heavily depends on their habits and degree of fitness. Activity trackers keep tabs on a patient's dietary habits, sleep schedule, and level of fatigue. Moreover, the data collected from the tracker prior to and during the start of therapy will assist medical professionals in deciding if adjustments to the recommended treatment plan are necessary.

Smart Video or Ingestible Pills: With the help of a smart pill that passes through the patient's digestive system, an accurate image of their intestinal tract is captured. After that, such images may be accessed via a wearable gadget that is linked to a medical app. The gastrointestinal system and colon can be remotely represented with the use of smart drugs.

Hearables: Hearables, the newest gadget on the market, are devices that help people with hearing loss. These days, Bluetooth can be used to sync hearing aids with smartphones and listen to music through headphones. Sounds from the actual world can be processed with equalizers, filters, and layering effects. Doppler Laboratories is a prime example.

Depression Wearables: Those who suffer from mental illness may now utilize an Apple Watch programme to treat their depression. The application tracks a patient's episodes that happen outside of regularly planned sessions and keeps an eye on their mood and cognitive abilities.

Location Services: Wheelchairs, scales, defibrillators, nebulizers, pumps, and monitoring devices are among the things that medical professionals may come across and tag with IoT sensors. Staff workers never have to worry about losing or finding physical equipment because of the IoT.

Connected Contact Lenses: With connected contact lenses, patients with diabetes may now monitor their blood sugar levels. Eventually, though, they will be able to help with improving eyesight and getting the eye's focus back.

8.6 BENEFITS OF THE HEALTHCARE SYSTEM

There are several benefits to using IoT for medical equipment for both patients and doctors. The implementation of IoT in healthcare is revolutionizing the way hospitals take care of patients. Despite the challenges posed by rapid population growth, which directly or indirectly contributes to increased illness and ageing populations, IoT healthcare solutions undeniably have the potential to effectively address these issues. It is a fact that IoT and healthcare are closely interconnected [1,18,20–22]. Figure 8.7 depicts some IoT healthcare benefits.

Real-Time Monitoring: Certain emergency conditions require quick medical attention, such as heart attacks or asthma attacks. There were several instances where a person died because of not receiving the right care promptly. However, real-time monitoring can assist in entirely preventing this problem.

There are several varieties of wearables on the market. By wearing them constantly, it could aid in the creation of real-time data that a smartphone app would deliver to the physician. To be informed of any type of health scenario, real-time health monitoring is essential.

Data Analysis and Assortment: In the healthcare sector, one benefit of IoT connections is the ability to store enormous amounts of patient medical history data. Even while analysing such huge statistics can seem overwhelming, IoT devices can complete the task in a couple of minutes. Devices connected to the IoT exchange data. Data may be gathered, processed, monitored, sent, and received via the Internet of Medical Things through the use of a cloud database.

Timely Notifications, Alerts, and Tracking: Some circumstances result in the patient not receiving assistance because of poor communication, and life-threatening incidents are very unexpected. However, these unexpected periods are now manageable with the aid of IoT. Using real-time tracking

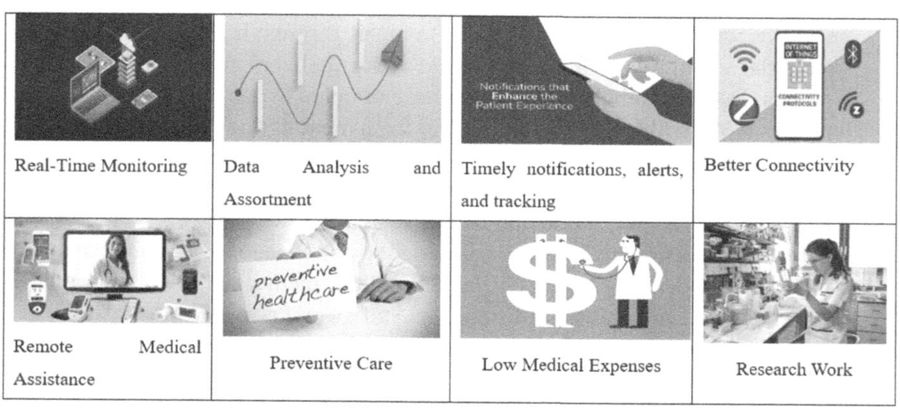

FIGURE 8.7 IoT smart healthcare system benefits.

devices, medical IoT devices will collect the necessary data and send it to the physicians. Every one of your emergency contacts receives a notification too. Tracking, monitoring, and hands-on therapy with appropriate intervention and improved accuracy are all made possible by IoT in the healthcare industry. Both the necessary outcomes and overall patient care are enhanced.

Better Connectivity: Automation is the main benefit of IoT, and the healthcare sector will benefit from these same aspects as well. The automation of all the information about a specific patient would be made possible by the healthcare mobility solution. As previously said, communication is essential to the healthcare industry, and IoT improves communication.

Remote Medical Assistance: Patients will have phone access to the doctor, who is just a few miles away, in case of an emergency. The doctor will be able to aid the patient even while they are away, thanks to the real-time tracking and alerting function, which will notify them of problems immediately.

Preventive Care: As preventive treatment helps avoid serious repercussions and new diseases, it should be the primary emphasis for physicians. The healthcare industry may be rejuvenated by utilizing IoT in medical equipment, which offers extremely precise data about every patient and helps patients live healthy lives and prevent illnesses.

Low Medical Expenses: With the rising expense of healthcare, insurance companies are moving more and more towards paying healthcare providers based on performance. For physicians to get paid for their services, they must be able to demonstrate that they benefit their patients. As a result, hospitals may cut costs by providing wearables and releasing patients from the hospital far earlier. This facilitates remote patient health monitoring for clinicians. Further cutting costs is the fact that patients feel more at ease at home and that physicians don't have to force them to stay in the hospital.

Research Work: Every other patient you may across may have a lengthy medical history. It is hard to know about a patient's childhood vaccinations, but with a little bit of investigation, you can find out about them in a matter of seconds. Physicians can swiftly gather and store all patient data in the cloud. This information will be useful for patient analysis and statistical calculations. Typically, gathering this data takes a lot of time for the researchers. However, with the use of IoT, they can save time and money on their study, which makes it simpler for them to get all the data quickly.

8.7 CHALLENGES OF THE HEALTHCARE SYSTEM

Numerous academics, researchers, engineers, and healthcare professionals are devoting a great deal of effort to developing and executing diverse IoT-based healthcare services, as well as resolving a range of technological and architectural issues related to such services. Several other issues in this setting require serious thought [23]. As seen in Figure 8.8, three categories may be used to categorize the challenges that IoT in healthcare presents: technical, financial, and ethical.

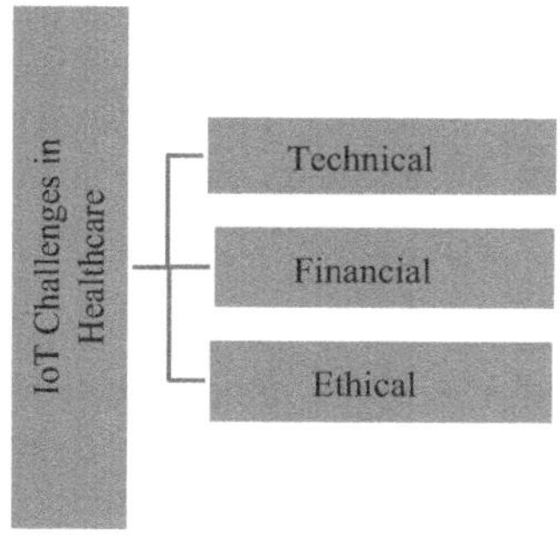

FIGURE 8.8 IoT challenges in healthcare systems.

Technical Challenges: Technical difficulties arise from the fact that IoT is not yet a widespread occurrence in daily life. Most nations lack access to IoT services and 5G wireless technologies. The great majority of healthcare professionals, academics, and patients alike are ill-informed about IoT's capabilities and applications outside of the healthcare industry. The initial technological problems with IoT applications in healthcare can be attributed to 5G. The integration of data presents the next technological difficulty, as various devices imply various data sources. For technological and budgetary reasons, the multitude of wearables and data collection devices in the healthcare industry cannot be readily converted to a single pattern of data gathering. Manufacturers are still unable to agree on standards and communication methods [1,22].

Financial Challenges: The number of medical patients worldwide is rising daily, which has led to a sharp growth in the usage of IoT-enabled devices over the past several years. In addition to rising in price, IoT device demand is also evolving. As far as the author is aware, there is no comparison research on IoT healthcare devices. Cost analysis is a contentious topic in both developed and developing nations because of this. For the average person, the expense of adopting IoT-based healthcare facilities is currently prohibitive [23].

Ethical Challenges: In terms of privacy and data security, IoT in healthcare might result in major problems. Real-time data is recorded and sent by IoT security devices. Nevertheless, most IoT-capable devices disregard security and data regulations. Furthermore, the laws governing data ownership with electronic devices are rife with uncertainty. As a result, hackers might simply get access to the system and take advantage of patients' and doctors' Personal Health Information (PHI). A patient's electronic health information may be used by cybercriminals to create fictitious identification documents so they may buy prescription medications and medical equipment to resell later. Hackers may potentially use a patient's name to file false insurance claims [18].

8.8 FUTURE TRENDS IN IOT HEALTHCARE SYSTEMS

The potential and reach of IoT in healthcare are enormous. There will probably be a large expansion in the uses and effects of IoT in healthcare as technology develops and becomes more pervasive. The fascinating future prospects and developments in the area of IoT healthcare systems are highlighted in the following [24,25].

Advanced Health Monitoring and Diagnostics: Real-time, continuous health monitoring for a range of ailments will be available with IoT devices, such as for diabetes and heart disease. This will enhance early intervention and detection. To provide more thorough health insights, wearables of the future generation may track a greater range of indicators, including blood oxygen levels, metabolic rates, and even hormonal fluctuations [26,27].

Integrated Health Ecosystems: Better interoperability between various health devices will be a feature of future IoT systems, forming a unified health ecosystem. This linked strategy will make an accurate and comprehensive approach to health management possible. Health monitoring systems and the IoT-enabled smart home appliances will collaborate to handle anything from prescription reminders to optimizing house conditions for health [28,29].

Personalized Medicine: IoT devices will support precision medicine by offering comprehensive data that can be utilized to customize therapies and treatments based on patient-specific information obtained from real-time health data. Combining IoT data and genetic information will allow for highly customized treatment regimens, risk prediction, and optimizing therapeutic approaches [30,31].

Enhanced Patient Engagement and Empowerment: In the future, IoT devices will provide patients with additional resources to autonomously manage their health, such as interactive features for tracking progress and applications that deliver actionable insights. Intelligent patient portals will include IoT data, providing patients with a thorough overview of their health indicators and encouraging increased patient involvement in their care [32,33].

AI and Data Analytics: Proactive monitoring and preventive treatment will be possible thanks to AI-driven analytics that evaluate IoT data to anticipate possible health concerns before they become serious. Large-scale IoT data analysis by AI will help medical practitioners make better treatment decisions and improve clinical judgement [34,35].

Telehealth and Remote Care Expansion: The integration of remote diagnostics, virtual consultations, and real-time health monitoring into more complex telehealth platforms will be made possible by IoT. IoT advancements may make it possible for professionals to conduct complicated surgeries from a distance using robotic assistance and remote surgery [36].

Improved Healthcare Operations: IoT will improve hospital operations' efficiency in areas including building maintenance, equipment tracking, and patient flow management.

IoT sensors will track medical equipment and anticipate maintenance requirements, saving downtime and guaranteeing the dependability of important equipment.

Data Security and Privacy Innovations: It will be necessary to develop new techniques for data encryption and security to guard against breaches and unwanted access to private health information. Transparency and trust might be increased by using blockchain to generate safe, unchangeable recordings of health data [37].

Global Health Monitoring and Management: IoT devices will be essential for tracking the spread of illnesses and informing public health actions in the event of a global health emergency. IoT will facilitate the effective distribution of medical supplies and resources, especially in underprivileged areas.

Integration with Emerging Technologies: Smart Technology in conjunction with AR and VR can help with difficult surgical operations and offer immersive training environments for medical personnel. Through quicker and more dependable data transmission—a critical component of telehealth and real-time health monitoring—the 5G network's deployment will expand the possibilities of IoT devices.

8.9 CONCLUSION

Due to the Internet's ability to facilitate virtual interactions in a range of contexts, including social and professional ones, the Internet has had a profound impact on how people live. The goal of "any media, anytime, anything, anywhere" connectivity might be realized with the help of the IoT, which facilitates communication between linked objects. IoT in healthcare, often known as Health IoT, has grown out of the broader IoT ecosystem and is driven by the need for more individualised treatment, more effective healthcare systems, and rising demand for better healthcare results. The implementation of IoT in healthcare enhances and resolves the shortcomings of the conventional healthcare system. The IoT in healthcare systems improves high pharmaceutical costs and precise treatment diagnosis. The elements of smart healthcare are as follows: sensors, intelligent networks, cloud computing, big data analysis, and smart hospitals. This chapter also illustrates the basic workings of the IoT framework for smart healthcare, in which devices are connected to communication services and the cloud, from where practitioners can collect the data and use it in the future. Also, the layered architecture of smart healthcare is mentioned. Perception, network, and application layers play an important role in a layered architecture. Furthermore, this study briefly incorporates the IoT communications protocols along with their unique properties. There is a full discussion of the IoT's core technologies in smart healthcare. IoT integration in healthcare systems has further advantages. The last few decades have seen the development of numerous IoT applications for smart healthcare, which ensure specific facilities for medical professionals. These include tracking and diagnosing various health issues, measuring various health variables, and offering diagnosis and treatment at remote locations. In summary, this article has looked at the current roadblocks and shortcomings in the development of

IoT-based healthcare applications as well as the projected course of development and research in the years to come. It is planned to provide academics and scientists who are eager to study more about IoT technologies with a strong foundation so they can fully comprehend the overall architecture and functionalities of the numerous parts and processes that make up the idea of the IoT.

REFERENCES

1. R. A. Rayan, C. Tsagkaris, and R. B. Iryna, "The internet of things for healthcare: Applications, selected cases and challenges", in *IoT in Healthcare and Ambient Assisted Living*, Springer, pp. 1–15, 2021.
2. H. H. Alshammari, "The internet of things healthcare monitoring system based on MQTT protocol", *Alexandria Engineering Journal*, vol. 69, pp. 275–287, 2023.
3. A. Atadoga, T. T. Omaghomi, O. A. Elufioye et al., "Internet of Things (IoT) in healthcare: A systematic review of use cases and benefits", *International Journal of Science and Research Archive*, vol. 11, no. 1, pp. 1511–1517, 2024.
4. D. A. M. Budida and R. S. Mangrulkar, "Design and implementation of smart healthcare system using IoT", in *International Conference on Innovations in Information, Embedded and Communication Systems (ICIIECS)*, 2017.
5. D. Tiwari, D. Prasad, K. Guleria et al., "IoT based smart healthcare monitoring systems: A review", in *6th IEEE International Conference on Signal Processing, Computing and Control (ISPCC 2k21)*, pp. 465–469, 2021.
6. K. Hameed, I. S. Bajwa, S. Ramzan et al., "An intelligent IoT based healthcare system using fuzzy neural networks", *Scientific Programming*, vol. 2020, p. 8836927, 2020.
7. M. M. Islam, S. Nooruddin, and F. Karray, "Internet of things device capabilities, architectures, protocols, and smart applications in healthcare domain: A review", *IEEE Sensors Council*, vol. 10, no. 4, pp. 3611–3641, 2022.
8. M. N. Mohammed, S. F. Desayansah, S. A. Zubaidi et al., "An internet of things-based smart homes and healthcare monitoring and management system: Review", *Journal of Physics: Conference Series*, vol. 1450, pp. 1–15, 2020.
9. S. Krishnamoorthy, A. Dua, and S. Gupta, "Role of emerging technologies in future IoT-driven Healthcare 4.0 technologies: A survey, current challenges and future directions", *Journal of Ambient Intelligence and Humanized Computing*, vol. 14, pp. 361-407, 2021.
10. J. S. Jeong, O. Han, and Y. Y. You, "A design characteristics of smart healthcare system as the IoT application", *Indian Journal of Science and Technology*, vol. 9, no. 37, pp. 1–8, 2016.
11. S. Agnihotri and K. R. Ramkumar, "IoT and healthcare: A review", in *International Conference on Emerging* Technologies*: AI, IoT, and CPS for Science & Technology Applications*, 2021.
12. A. O. Almagrabi, R. Ali, D. Alghazzawi et al., "Blockchain-as-a-utility for next-generation healthcare internet of things", *Computers, Materials & Continua*, vol. 68, no. 1, pp. 359–376, 2021.
13. V. Aivaliotis, K. Tsantikidou, and N. Sklavos, "IoT based multi-sensor healthcare architectures and a lightweight-based privacy scheme", *Sensors*, vol. 2022, p. 4269, 2022.
14. C. Li, J. Wang, S. Wang et al., "A review of IoT applications in healthcare", *Neurocomputing,* vol. 565, pp. 1–12, 2024.
15. Z. Chang, F. Catani, F. Huang et al., "Landslide susceptibility prediction using slope unit-based machine learning models considering the heterogeneity of conditioning factors", *Journal of Rock Mechanics and Geotechnical Engineering*, vol. 15, no. 3, pp. 1127–1143, 2023.

16. F. S. R. Pavier, S. Kaluvan, S. J. Fearn et al., “A highly scalable and autonomous spectroscopic radiation mapping system with resilient IoT detector units for dosimetry, safety and security”, *Journal of Radiological Protection*, vol. 43, pp. 1–13, 2023.
17. B. J. Dhakale, M. Sharma, M. Arif et al., “An automatic sleep-scoring system in elderly women with osteoporosis fractures using frequency localized finite orthogonal quadrature Fejer Korovkin kernels”, *Medical Engineering & Physics*, vol. 112, no. 2023, p. 103956, 2023.
18. N. Venu, A. ArunKumar, K. K. Vaigandla, “Investigation on Internet of Things (IoT): Technologies, challenges and applications in healthcare”, *International Journal of Research*, vol. 11, no. 2, pp. 143–153, 2022.
19. Wireless Watchdogs, “IoT Applications in Healthcare”, https://www.wirelesswatchdogs.com/blog/iot-applications-in-healthcare (Accessed 10 Sep 2024).
20. Technostacks, “Impact of the Internet of Things (IoT) in the Healthcare Industry”, https://technostacks.com/blog/iot-in-healthcare/ (Accessed 10 Sep 2024).
21. S. Lypchenko, “The Internet of Medical Things: Benefits of IoT in Healthcare”, https://gowombat.team/blog/posts/the-internet-of-medical-things-benefits-of-iot-in-healthcare (Accessed 11 Sep 2024).
22. S. Upadhyay, M. Kumar, S. Verma et al., “Challenges and limitation analysis of an IoT-dependent system for deployment in smart healthcare using communication standards features”, *Sensors*, vol. 23, p. 5155, 2023.
23. M. N. Bhuiyan, Md. M. Rahman, Md. M. Billah et al., “Internet of Things (IoT): A review of its enabling technologies in healthcare applications, standards, protocols, security, and market opportunities”, *IEEE Internet of Things Journal*, vol. 8, no. 13, pp. 10474–10498, 2021.
24. K. H. Almotairi, “Application of internet of things in healthcare domain”, *Journal of Umm Al-Qura University for Engineering and Architecture*, vol. 14, pp. 1–12, 2023.
25. Vance Duong, “IoT Healthcare: The Future Landscape of Healthcare Industry”, https://savvycomsoftware.com/blog/iot-healthcare-industry/#ftoc-heading-3 (Accessed 15 Sep 2024).
26. C. Prabha, M. Gupta, and S. Ahmed, “Introduction to AIoT for smart environments,” in *AIoT Technologies and Applications for Smart Environments*, Institution of Engineering and Technology, 2022, pp. 1–19.
27. C. Prabha and A. Garg, “IoMT healthcare emerging technologies and implementation challenges: A brief review,” in *International Conference on Intelligent and Smart Computation (ICIASC-2023)*, 2024.
28. C. Prabha, P. Mittal, K. Gahlot, and V. Phul, “Smart healthcare system based on AIoT emerging technologies: A brief review,” in *Lecture Notes in Electrical Engineering*, Singapore: Springer Nature Singapore, 2023, pp. 299–311.
29. S. Agarwal and C. Prabha, “Analysis of lung cancer prediction at an early stage: A systematic review,” in *Lecture Notes on Data Engineering and Communications Technologies*, Singapore: Springer Nature Singapore, 2022, pp. 701–711.
30. G. Sharma and C. Prabha, “A systematic review for detecting cancer using machine learning techniques,” in *Innovations in Computational and Computer Techniques: ICACCT-2021*, 2022.
31. D. Sharma and C. Prabha, “Security and privacy aspects of electronic health records: A review,” in *2023 International Conference on Advancement in Computation & Computer Technologies (InCACCT)*, 2023.
32. J. Singh, C. Prabha, G. Singh, Muskan, and A. Verma, “Addressing role of ambient intelligence and pervasive computing in today’s world,” in *Lecture Notes in Electrical Engineering*, Singapore: Springer Nature Singapore, 2023, pp. 257–269.

33. C. Prabha, S. Kaur, J. Singh, and M. Malik, "An overview of IoT and smart application environments: Research and challenges," *Communication and Intelligent Systems*, pp. 111–124, 2023.
34. N. Sharma and C. Prabha, "Computing paradigms: An overview," in *2021 Asian Conference on Innovation in Technology (ASIANCON)*, 2021, pp. 1–6
35. C. Prabha, J. Singh, S. Agarwal, A. Verma, and N. Sharma, "Introduction to computational intelligence in healthcare," in *Computational Intelligence in Healthcare*, Boca Raton: CRC Press, 2022, pp. 1–15.
36. P. Abdul Hafeez, G. Singh, J. Singh, C. Prabha, and A. Verma, "IoT in agriculture and healthcare: Applications and challenges," in *2022 3rd International Conference on Smart Electronics and Communication (ICOSEC)*, 2022.
37. D. Sharma and C. Prabha, "Electronic medical images security and privacy techniques," in *2024 International Conference on Expert Clouds and Applications (ICOECA)*, Bengaluru, India, 2024, pp. 102–106, doi: 10.1109/ICOECA62351.2024.00031

9 Securing 6G-Enabled IoT Using Fog Computing

Shalini Kumari and Chander Prabha

9.1 INTRODUCTION

The International Telecommunication Union (ITU) makes standards for every new generation of wireless technology once a decade. The next generation of advanced wireless technology, 6G, will dominate the Internet of Things (IoT) technologies. In the IoT age, thousands of sensors, cars, automobiles, drones, and robots could be networked, making the world more intense and less expansive. 6G is the successor to fifth-generation (5G) technology. 6G networks will use higher frequencies than 5G, resulting in significantly improved capacity and reduced latency. One of the goals of the 6G internet is to allow communication with a latency of one microsecond. The ability of 5G to support a significant shift in the IoT has been an important selling point. However, many applications are deemed to be decades away, as they rely on 5G specifications that have yet to be set. Future 6G communication systems will be far more advanced than the existing 5G communication technology. IoT can play an important role in 5G and beyond communications. The applications of IoT and Multi-access Edge Computing (MEC) in everyday life will help individuals track and monitor objects, which makes interaction with the physical world easier. Due to the decentralized nature of blockchain, it has become an excellent tool to address IoT problems. This blockchain technology is key in fulfilling the potential of 6G networks and addressing the present and future requirements. 6G enables wireless transmission of data that IoT devices collect from their environment to enable a host of new applications. A 6G network rectifies some deficiencies of the 5G network, hence augmenting non-public communications and enabling smart platforms in 5G networks such as industrial Internet, remote education, healthcare, and smart homes and cities. The IoT is vital in accomplishing these new platforms across this paradigm shift [1].

The mass usage of wireless communication devices has profoundly impacted peoples lives and fast-tracked the growth of the manufacturing sector. Online communication and entertainment companies have been the most obvious beneficiaries of the current fourth-generation (4G) infrastructure. The future fifth-generation communication networks will govern the IoT, driverless vehicles, and virtual and augmented reality (AR/VR) applications. Lacking a human component in communications for 5G, it is no surprise researchers and businesses alike are geared to the future for the next stage of wireless communications. Having innumerable advantages, technology has been deployed devoid of human substance [2]. The connectivity structure of the 6G generation turns "connected things" into "associated intelligence." The significant investments in research and development for the 6G communication system

DOI: 10.1201/9781003586456-9

primarily focus on the media, the product, and the essential components. The advent of enhanced Intelligent Computing, the ability to collect, transmit, and evaluate data, and the provision of a wide range of applications and intelligent services are expected developments with the rollout of 6G technology. Human development will be framed in the 6G framework, above information, technology, and services. The 6G communication network will likely enhance secrecy and security by eliminating vulnerabilities, making the network more robust. However, conventional machine learning techniques may expose the central server architecture to privacy and security attacks from multiple internal nodes and create a single failure point [3].

Deep Learning (DL) is a form of AI that can learn patterns and characteristics with the help of vast datasets created by lens detectors, ultrasonic detectors, and many more devices. Autonomous vehicles, which facilitate self-driving cars, make it possible to make real-time alterations of a vehicles acceleration and braking functions under road conditions. The further accuracy of the assessments is contingent on the quality of input data and the proficiency of DL models. Recently, the emphasis on networked vehicles under the internet-of-vehicle (IoV) programme has shifted focus to integrated cognition. The subsequent data interchange and communication among the mobile nodes will sooner or later employ a hybrid method, incorporating vehicle-to-anything (V2X) and a specified limited-range gearbox. Wireless access in-vehicle environments (WAVE) relies on the 802.11p IEEE protocol, which must be adapted to support large data transmissions over the internet. Future advances in 6G technology are expected to enhance V2X capabilities, focusing on privacy-preserving properties in these applications [4]. Hence, in the future, 5G can support a limited data volume or a vast customer base. According to the current 5G standard, managers can manually adjust and optimize their networks. Therefore, the manual management of a large, dynamic, and highly distributed network is a great challenge. In a B5G or after the advent of 6G, introducing AI technology will likely remove these shortcomings. The IoT is one of the industries expected to benefit the most from introducing 6G networks. Many applications of IoT remain unrealized as the current limitations of mobile networks in terms of latency, throughput, and density of devices make it impossible for them to be realized. IoT is a technology that connects physical things with the Internet to collect and share data. Some examples of these devices include washing machines, cell phones, refrigerators, wearable devices, headphones, self-driving cars, televisions, and other things. 5G is expected to serve only a section of the future demand of the next ten years and then onwards, as 6G is expected to offer global connectivity, improved spectrum and energy efficiency, cost-efficiency, more intelligence, superior security, etc.

9.1.1 Structure of the Chapter

Further, the rest of the chapter is organised as follows Section 9.2 examines 6G technology, 6G networks in conjunction with the IoT, and 6G applications in Fog Computing. Section 9.3 examines the research about 6G in conjunction with IoT. Section 9.4 delineates the QoS-aware approaches for energy-efficient IoT Fog Computing. Section 9.5 discusses potential solutions for implementing Fog

Computing in 6G while considering security aspects. Section 9.6 addresses the security improvements in 6G through Fog Computing and includes relevant case studies. Finally, the chapter is concluded at the end.

9.2 6G NETWORKS

Although 5G networks are being implemented now, academics have conceptualised the next generation of mobile networks, known as 6G, that can accommodate much larger data rates and several potential applications. Table 9.1 shows the features of 5G and 6G communications. 5G communication emphasizes three main application services: improved mobile broadband, ultrareliable low-latency communications, and massive machine-type communications. 6G networks, on the other hand, enable applications including agumented reality and virtual reality, autonomous and networked drones and vehicles, and intelligent and automated machines. These applications require up to terabits per second in data throughput, a value far from what 5G networks will offer. A fundamental characteristic of 6G networks is the utilization of the Terahertz band. 5G networks currently use sub-6GHz and millimetre wave frequencies. 6G networks will have a latency of under 1 ms and a dependability of 99.99999%, allowing ubiquitous connectivity and many future applications.

6G will use AI to make autonomous systems possible, apply distributed learning models to improve model accuracy, employ big data analytics for predictive analysis and informed decision-making, and actualize intelligent edge, fog, and cloud nodes. Energy efficiency will be a major challenge for 6G networks [5]. The main reason for the increased energy requirements is the support for massive IoT applications [6]. Future network nodes, including IoT devices and fog nodes, will need to handle multiple tasks, thus increasing energy consumption. Another energy-related challenge for 6G networks is using Terahertz communication, a short-range communication technology. Tera Hz communication, although it does provide increased data rates, is highly attenuating due to the spreading loss and molecular absorption loss [7]. Figure 9.1 shows the improvements to communication systems.

1G was an analogue technology introduced in the 1980s. The original wireless cellular technology, 1G, only supported voice conversations and lacked SMS and data transmission capabilities. The biggest drawback of this service was

TABLE 9.1
Applications and Features of 5G and 6G Communications

5G	6G	Features
Enhanced mobile broadband massive machine-type communication	Augmented Reality virtual reality	Applications
Up to Gigabits per second	Up to terabits per second	Data Rate
Up to 1 ms	<1 ms	Latency
Sub-6 GHz	Tera Hz	Spectrum

FIGURE 9.1 Improvements to communication systems.

noise from its surroundings, leading to poor data transfer quality and speed. The maximum speed of 1G technology was 2.4 kbps.

2G was the digital technology that replaced the analogue in 2G. Multiplexing enabled 2G capabilities by allowing users to channel into a channel. 2G offers benefits such as secure digital phone calls, improved long-distance efficiency for probe-level mobile phones, SMS, video messaging, and MMS. The 2G technology has a maximum speed of 64 kbps [8].

3G was introduced in response to increasing demands for multimedia transmission, video conferencing, and online gaming; 3g wireless mobile communications technology was developed with a multimedia approach. Compared to the previous generation, it offers faster Internet connections, video calls, video downloads, TV image reception on mobile phones, and additional digital content on the mobile network platform. The top speed of 3G technology was only 2 Mbps [8].

4G offers enhanced bandwidth for Internet access and typical 3G functions. Laptops with wireless modems can connect to smartphones and other mobile devices. The 4G generation offers expanded services such as mobile browsers, IP technologies, high-quality TV, video conferencing, online gaming, 3D TVs, and cloud computing. The 4G launch aimed to enhance service capacity, speed, quality, coverage, and cost reduction. 4G is maximum speed was 1 Gbps [8].

5G is the wireless communication concept. It can link humans, robots, cars, and ordinary objects. 5G technology guarantees high speed, low latency, enhanced security, increased access, and enhanced performance, resulting in an exceptional user experience [9]. The benefits of 5G technology include faster download and upload speeds, improved file stability, self-driving cars, remote surgery, IoT, network slicing, and reduced energy use. 5G networks offer download speeds up to 20 Gbps and upload speeds up to 10 Gbps.

6G is the advanced version of wireless communication technology for mobile data networks. This technology replaces 5G. 6G networks function as broadband cellular networks with limited geographic coverage, similar to previous generations. By 2030, 5G is expected to be the most extensively used mobile communication technology deployed in 2019. Despite this projection, the 6G rollout may commence between 2030 and 2035 [10].

9.2.1 6G and IoT

IoT is a vast connectivity involving thousands of devices, sensors, and machines. This type of IoT classification is based on the number of connected devices and the volume of traffic generated. During the last decade, the number of devices connected to the Internet has increased drastically. Shortly, it will increase at a high scale, which demands a strong network to connect all these devices. In large-scale IoT, the network density is as high as about 1 million devices per square kilometre. This will generate vast volumes of data to be disseminated between different devices, an enormous amount of tasks to be performed, and broad information that would have to be stored and analyzed as big data [11]. As such, 6G and fog computing will be crucial enabling factors for vast-scale IoT applications.

9.2.2 6G and Fog Computing

Fog computing is a technique that allows edge devices to do computational and storage tasks nearer to the edge. By deploying many fog nodes at diverse locations, devices can offload duties and cache data onto these nodes. One of the most significant advantages of fog computing is that it provides decentralized computing services instead of the centralised computing provided by the cloud. Further, reduced latency is achieved because data and workloads are fetched and processed close to the end devices. Fog computing optimizes frequency spectrum usage and increases network capacity. Placing fog nodes improves compute and storage capability, thus improving application reliability. Fog computing will be a key component of 6G large IoT. The sheer volume of the devices that exchange data and the numerous applications have made the assistance that fog nodes provide to end devices regarding computing and storage critical. Moreover, communications for 6G will almost entirely use these fog nodes to provide ubiquitous connectivity and achieve low latency. Many issues regarding resource allocation, job offloading, energy efficiency, latency reduction, equity, and security will arise in the new paradigm of fog computing-based 6G-enabled huge IoT.

Figure 9.2 illustrates the compatibility of 5G and 6G with IoT. The development of 6G networks, like 5G networks, is expected to revolutionise Internet connection and the IoT platform. Their key purposes and functions remain unknown until 6G networks are fully built and commercialised. As 6G is more efficient than 5G, low-power IoT devices can be charged through the network, a well-established economy. As 4G and 5G share a central network, 6G and 5G may coexist for some time [12].

FIGURE 9.2 5G to 6G Interaction.

9.3 STUDY RELATED TO 6G WITH IOT

Deebak et al. [4] analyzed integrating IoT wireless technologies, satellites, and UAVs to represent the IoT application paradigm. Future 6G technologies were extensively discussed to assess cellular network capabilities. This paper introduces 6G communication networks by showing a possible application scenario and the underlying technology. Chowdhury et al. [2] conducted an in-depth study on deploying future 6G and IoT networks using optical wireless communication technology. It also described 5G, 6G, and IoT systems. Sekaran et al. [13] addressed the challenges of integrating blockchain and IoT technologies with a 6G network, which reached high-level solutions by transcending their limitations. The report mentions that blockchain technology for IoT networks suffers due to increased communication overhead, energy consumption, and execution time. In some research methods, data confidentiality and integrity are inadequate. Future research recommendations for blockchain-enabled IoT with a 6G connection were also given. Barakat et al. [14] presented a new approach to providing industry IoT application scenarios with differing performance demonstrations.

Cases from popular research fields are supported through 6G and its technologies. The domains include healthcare, smart grid, transportation, and industrial. The scenarios principal criteria and suitability to the future generation of wireless communications were analyzed. This evaluation raises open issues and concerns about 6G and IoT. The 6G concept by Zhang et al. [15] has three components: AI, super IoT, and mobile ultra-broadband. They examined essential technologies for each section. Basic principles, major difficulties, and solutions for each technology were discussed. Gupta et al. [16] studied analysing dependability and estimating availability in developed modelling 6G. Wireless systems create cloud, edge, and fog computing-based wireless communication applications. In recent years, IoT technology has been used to build consumer and provider networks.

9.3.1 IoT Energy-Efficient Fog Computing Techniques

Task Offloading with Energy Consideration: Most IoT devices are battery-powered and have limited computing and communication capabilities. Such devices are also limited in speed and task capabilities. As applications become more computationally intensive, they require real-time answers and reduced latencies. Such applications can easily drain the power resources of a device, leaving it unavailable for network use. Conserving/boosting energy to maximise the lifespan of IoT devices is a big challenge due to limited energy resources. Local devices can rely on task helper nodes, like fog nodes, to do tasks to preserve power.

Device-to-fog offloading: Local devices transfer calculation duties to remote computing platforms like fog or cloud to compensate for limited computational and power resources. This job offloading conserves energy resources and extends network availability. Remote cloud servers make offloading work directly from local devices impractical due to high transmission energy consumption and processing delays. However, fog nodes, positioned near end devices, can be used to offload jobs, saving energy and improving latency. For local devices, task offloading to fog nodes is preferable. Gai et al. [17] utilised a device-to-fog offloading method to reduce energy resources.

Offloading fog to fog: Fog nodes can offload jobs to the cloud or neighbouring nodes to save energy or for peak workloads. Vertical offloading may not be possible for time-sensitive applications due to high transmission energy consumption and latency due to the vast distance between fog nodes and the cloud [18]. On the other hand, horizontal task offloading results in numerous task helper fog nodes. However, their capabilities are limited compared to the cloud.

Deploying from fog to cloud: Cloud servers provide powerful computing, storage, and energy capacities. Common IoT networks apply cloud servers to process data from different devices. Intermediate processing nodes like fog nodes and edge devices act as bridges that bridge the significant distance between the cloud servers and local devices. Fog nodes leverage the clouds computing, storage, and energy capabilities. The authors assessed task popularity to determine offloading locations. The task popularity is determined using a probabilistic model based on task launch frequency. Popular and low-latency operations are offloaded to the fog, while the remainder are offloaded to the cloud. The energy efficiency and load balancing solutions in data centres. The authors explored fog node utilisation options that relieve data centre load. By assisting data centres in task computing, fog nodes can improve energy efficiency and reduce load.

9.3.2 Strategic Location of Fog Nodes for Energy Efficiency

The Fog-based IoT network does not have a common architecture and is best represented as a layered representation. Researchers recommend three, four, or six layers for fog architecture [19]. Most people use the three-layer design (cloud, fog, and edge). The resources are distributed hierarchically. Cloud servers have the most, followed by fog nodes with medium resources that extend the cloud layer, and edge devices have the least. There are three node placement models for energy conservation in fog-based IoT networks: (1) clustered, (2) centralised, and (3) distributed.

Clustered Architecture: Clustering is a common approach in Wide Sensor Networks (WSNs) for energy-efficient communication. Sensor nodes are clustered by size in a clustered WSN. Size depends on the position of sensor nodes, density, and distance from the base station/fog node. In each cluster, a resourceful sensor node is elected as the head, managing the flow of communication. The sensor nodes within the cluster forward data to the cluster head, aggregate it, and then send it to the base station/fog node.

Centralized Architecture: A clustered approach for job offloading among fog nodes, utilising centralised and distributed architecture to optimise energy and time. The proposed architectures are assessed for their pros and cons in various scenarios. All fog nodes are divided into two tiers, dependent on their power supply.

Distributed Architecture: No clusters are generated in this architecture, and low-power fog nodes (LPFN) can choose among high-power fog nodes (HPFN) and fog access point (F-AP) fog nodes in their coverage region. These two policies are employed.

- **Policy 1**: Offloading uses only the fog node layer, which means that LPFNs can offload partially to HPFNs within their coverage region.
- **Policy 2**: Employs fog nodes and F-AP layers for job offloading. LPFNs can partially delegate work to HPFNs and F-APs.

9.3.3 Energy-Aware Device Control

Task offloading saves energy. Using task offloading with device control-based energy conservation approaches can increase energy savings. These strategies govern device functions for improved performance and energy conservation. Device control allows local devices and task helper nodes to manage characteristics, including transmission power, switching time, battery voltage, frequency, and modulation method.

- **Optimisation of Transmission Power:** Although it uses more transmission energy to transfer part of the work to the task helper node, offloading the work to them saves energy in the local computation. Factors such as air conditions, path losses, collision, interference, and the distance between the local and task assistance devices determine the transmission energy. Several methods exist for controlling these aspects to conserve or enhance gearbox energy.

- **Powering Up and Off Electronic Devices:** To conserve energy, turn off gadgets when not in use and only needed for specified tasks.
- **Voltage and frequency scaling dynamically:** Dynamic voltage and frequency scaling (DVFS) is hardware and software-based to save energy. DVFS is used in most modern architectures to save power consumption by dynamically altering the frequency and supply voltage of processors, controller chips, and components according to the workload. As the frequency and voltage directly impact the energy consumption, an increase in workload increases consumption, whereas a decrease in workload decreases consumption. Adjusting frequency and supply voltage based on workload optimises resource utilisation and energy conservation. Most modern computer hardware utilises the DVFS scheme to optimise power savings, battery life, and device durability while retaining computing availability.
- **Varying Voltage:** Dynamic voltage scaling (DVS), unlike DVFS, controls device energy usage solely through dynamic voltage. The fog node job offloading choice was optimised using Lagrangian dual theory and the DVS method to enhance energy efficiency by optimising offloading ratio, transmission power, central processing unit (CPU) computation speed, and transmission time.
- **Adaptive Modulation Diagnosis:** To maximise transmission efficiency while minimising transmission energy consumption, dynamic modulation scaling (DMS) continuously adjusts the modulation level. The transmission speed can be fine-tuned by varying the bit rate with the packet rate required at any given time interval. The DMS method typically settles on a compromise between transmission energy and transmission latency.

9.3.4 Conserving Energy Through Harvesting

On-device power supplies power to small, wireless IoT devices and uses different battery capacities based on application needs. Due to their limited energy storage capacity, rechargeable batteries are necessary for large-scale IoT deployments, sensors inside the body, and devices in toxic environments. The loss of battery power might cause device nonavailability, harming the network. Developing alternative charging methods for gadgets, particularly sensors, is a major problem of the IoT revolution.

9.4 QOS-AWARE METHODS FOR ENERGY-EFFICIENT IoT FOG COMPUTING

Improving network energy efficiency reduces operational costs, extends network life, and makes computation-critical devices available. Numerous ways prioritise QoS features beyond energy efficiency. We divide energy-efficient QoS approaches into seven categories: latency reduction, fair offloading, fog node cooperation, load balancing, fault tolerance, accuracy and privacy assurance, and resource allocation, as shown in Figure 9.3.

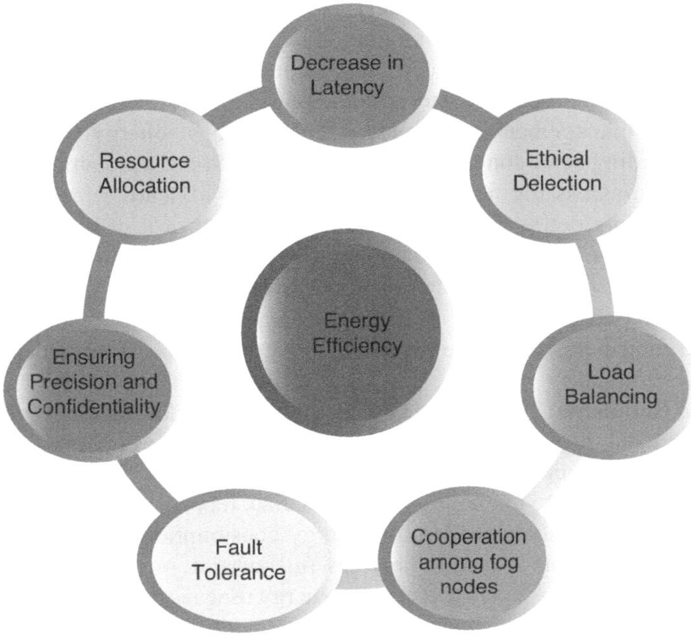

FIGURE 9.3 QoS-aware methods for energy-efficient IoT Fog Computing.

- **Decrease in Latency:** The maximum delay is the primary limitation when deciding how to offload computation workloads. Data lag and application failure are the outcomes of not meeting the latency criteria. The user experience is critical to the survival of an application in a competitive context. Hence, latency improvements that reduce the execution time of tasks have the opposite effect. Overall service time/latency of often requested jobs is improved because the fog node processes several of these tasks simultaneously, which is necessary because the number of frequently requested tasks is always huge. The waiting time for the execution of an offloaded task is very minimal, as it is forwarded to a device that has a very short queue for tasks. This ensures that the network is not overburdened and reduces the total task waiting time.
- **Ethical Deletion:** The fog IoT network consists of heterogeneous devices with different computing, storage, and power capacities owned by different parties. The energy efficiency and fair offloading of the devices in the network are a must for sustainability and resource sharing among the owners of the fog nodes. Although fairness is desirable, it introduces delay. The trade-off between fairness and delay is needed to accomplish tasks within time constraints. The fog node with the lowest remaining energy will receive fewer jobs, while the one with the highest will receive more. To meet time limits, fog nodes are selected based on their remaining energy and processing delay to determine the optimum option.

- **Cooperation among fog nodes:** Fog nodes are expected to process offloaded jobs using their computation skills and accept tasks from neighbouring devices. However, certain fog nodes may not share resources and prefer distinct processing, storage, and power allocation methods. Reasons for this management may include cost savings, energy conservation, and avoiding road congestion and long lines. Some academics use algorithms to assess fog node participation before allocating resources. Fog node owners define resource allocation for each task offloading assignment. The goal is to foster a cooperative environment among fog nodes by ensuring fair task distribution and implementing a reward/punishment system based on resource sharing percentage. The more resources a fog node contributes, the more assistance it receives from other nodes during high workloads.
- **Load Balancing:** This crucial QoS feature prevents devices from overworking and stabilises network operations. Load balancing commonly reduces device task queues, resulting in faster task completion. Devices broadcast task information to fog nodes, which calculate reports based on past energy usage and current task offloading energy consumption. This report determines whether the fog node provides full, partial, or no resources to the device. A fog nodes historical average energy usage is compared to its queue length to achieve load balancing. A fog node with a high historical rank will be assigned tasks last. A higher probability may be assigned to options that have shorter wait durations.
- **Fault Tolerance:** It is the capability of maintaining desired service despite system failures. Li et al. [20] used load balancing and fault tolerance for energy-efficient job offloading. A work can be delegated to three layers: edge, fog, and cloud. To balance the load, jobs are offloaded to the next higher tier based on the minimum/optimal workload for the edge and fog layers. Fog devices will send tasks to be processed in the same layer if their arrival rate is smaller than the service time. Fog nodes implement fault tolerance to ensure that all assigned tasks are completed. The system monitors offloaded jobs and transfers them to another fog node if one node disappears due to a power outage and has unfinished tasks.
- **Ensuring Precision and Confidentiality:** Researchers have combined data accuracy and privacy with energy economy. Data accuracy and timelines are crucial for apps like Amazon Echo Speaker and Apple Watch, as delayed information may lose relevance. The authors employed queuing theory to divide the workload among edge, fog, and cloud layers, assuming timely task completion and accurate data. Sensor nodes may acquire inaccurate data due to noise and malicious attacks, increasing energy consumption and shorter lifespans. Fog nodes employ two strategies to assess sensor node trust.
    - The direct evaluation method calculates trust for each direct neighbour based on historical communication interactions, node residual energy, and packet loss rate.

  - The indirect assessment method computes indirect trust for nodes not directly associated with fog nodes through the direct trust rating of the neighbour. Nodes are clustered, and cluster heads are selected based on the trust value.
- **Resource Allocation:** Fog computing uses existing network devices, like cell phones, for computation without dedicated resources. Resource allocation and scheduling algorithms in fog IoT networks are challenging regarding availability and efficiency because of resource flexibility. Moreover, fog devices focus on their calculation jobs and do extra computations later. Fully using the fog device for its tasks prevents it from processing other devices. Running IoT applications in fog computing presents resource allocation and job scheduling challenges due to constraints [21]. Resource schedulers must evaluate all computing devices/nodes, their resources, and ongoing computation jobs for effective resource allocation. To achieve energy efficiency, they must balance computations, connectivity, and latency limitations to disperse workload without overburdening computer devices. This facilitates resource allocation in fog computing-based IoT networks.

9.5 THE POTENTIAL SOLUTIONS

Designing, modelling, and implementing fog-based energy-aware 6G-enabled huge IoT architectures, policies, and applications presents various problems. Further research is needed to enable energy-efficient solutions for huge IoTfog environments using 6G technology. Some of the points taken into consideration are as follows:

Managing Resources: Resource management in distributed systems involves managing compute and storage resources. Locating suitable fog nodes near IoT devices is a tough task. Resource allocation and job offloading are crucial for energy-efficient resource management in an IoT fog network. Fog resources must be used efficiently to save energy. In these networks, resource utilisation varies, and IoT traffic patterns often vary. Thus, resource management is more difficult in this changing setting. Thus, smart techniques are needed to balance resource utilisation, energy usage, and QoS requirements for applications and services. In this context, load-balancing strategies can reduce delays, optimise bandwidth, and enhance resource utilisation. The techniques should also assure energy efficiency in the network. Lower network traffic conditions can make load-balancing solutions less effective, as neighbouring fog devices can process IoT operations more efficiently. Redirecting jobs to other fog devices can increase communication and administration overheads. Additionally, balancing the load across multiple fog devices can be inefficient if a few can perform the job. A scalable, cost-effective resource management and load-balancing technique is necessary to manage fog resources efficiently.

Differences in quality: Heterogeneity is software-hardware differences. This concept applies to fog nodes processing various job types. An IoT-enabled fog network includes diverse IoT and fog devices with distinct hardware and software. In addition, the fog network supports numerous IoT applications. Various applications need various storage and computing resources. Energy-aware techniques are difficult to use in such a varied setting. Different operating systems have varying power consumptions that can be optimised for reduced usage. This issue is important because different versions of the same operating system may require varying amounts of electricity. Energy management is possible for an operating system. To store and compute various IoT requests on fog devices with varying hardware and software characteristics, execution time and synchronisation difficulties must be handled. The heterogeneity of the IoT-enabled fog environment makes it hard to apply traditional energy-efficient rules. The mechanisms of smooth operations need to be compatible with the 1G to 6G generations of wireless technologies. To ensure good QoS, efficient handover techniques are required. Semantic-based techniques [22] prove to be helpful in combating heterogeneity in fog computing. The application developers of IoT can also deal with heterogeneity by incorporating energy-aware techniques into their work. Policies must address energy and thermal limits, matching resource capabilities with IoT work requirements.

Optimising Energy Efficiency with QoS Awareness: Most energy-aware strategies in distributed computing significantly affect Quality of Service. Some of fog networks most popular energy-saving strategies include consolidating devices and turning off or putting idle components into sleep mode to save energy consumption. This method effectively reduces both dynamic and static energy usage in fog-based networks. However, this power-saving strategy may impact QoS in several ways. IoT communication must be quickly redirected to neighbouring fog devices when a device is shut off or put into sleep mode. This approach can increase latency in IoT-based fog environments since requests are relayed to neighbouring fog nodes. Serving the request from a fog device far from the IoT device that generated it will increase the delay. This will also cause some of the IoT requests to be dropped, affecting network performance. Task migration between fog devices must be effective and fluid before being applied to underutilised devices. Increased IoT traffic necessitates dynamically activating fog devices, which can become time-consuming and hinder request processing.

Smart Surface with Reconfigurability: Reconfigurable Intelligent Surfaces (RIS) technology in 6G increases throughput and decreases latency. An RIS has numerous passive parts with controllable and reconfigurable phases. Altering the phase of these surfaces changes their reflection characteristics and allows for signal propagation between wireless nodes. Using RIS could improve network throughput and communication dependability. RIS reduces outage risks, end-to-end time, and transmit volume. Tasks at IoT nodes will be efficiently transmitted in fog nodes due to reduced energy consumption and transmissions. After completing its task, fog nodes can send their task to the IoT nodes via reduced transmission powers. In data caching applications employing RIS, content can be accessed from fog nodes with less energy [23]. RIS-assisted fog computing faces the difficulty of efficiently allocating RIS to various IoT nodes. Matching theory-based algorithms can produce a stable and optimal solution.

Federated Learning: The amount of data generated by 6G-enabled devices is so enormous that massive processing is required. The effective usage of computing resources requires intelligent learning algorithms. To reduce the complexity of energy-aware task offloading methods, fog nodes can create models based on local data, including task number, size, arrival rate, and energy usage. Local learning models from the fog nodes can be transmitted to a centralised controller to improve the accuracy of global learning. Federated learning provides security, collaborative data sharing, easier computation, and more accurate global models. Fog nodes can save energy by using fewer complex algorithms and making more accurate job offloading decisions. A significant challenge in federated learning for energy-aware task offloading is determining a global model from local models trained on various data sources [24].

Energy Collaboration: 6G-based fog computing will be based on energy collaboration for energy conservation in fog and IoT nodes. Future 6G devices will harness energy from several sources, such as solar and Radio Frequency (RF). Energy collaboration between IoT and fog nodes is essential for large-scale computation. Higher energy nodes can support lower energy nodes in the execution of tasks. Smart spectrum utilization solutions will be designed for 6G communication and computing. A major challenge for energy collaboration is the incentives to encourage truthful and fair collaboration among devices [25].

Energy-Efficient Method Selection: In an IoT-enabled 6G fog environment, energy-aware offloading, fog node placement, and device control are employed to optimise resource utilisation and conserve energy. Choosing the right strategy is critical for maximising energy savings without impacting fog network functionality. Two basic categories of energy usage exist in fog computing environments: static and dynamic. Some approaches focus on static energy usage, while others tackle dynamic energy use. The static energy consumption of a gadget is defined as the energy consumed when the gadget is on but not in use. Dynamic energy is the energy consumed by fog devices when processing IoT requests. Static energy consumption is more frequently higher than dynamic energy usage. The energy-efficiency approach and targeting energy usage in fog networks is difficult because of the diverse environment and service-level agreements of IoT applications. Further exploration of methodologies focusing on dynamic energy consumption, a significant component of fog node energy consumption, is necessary.

9.6 SECURITY ENHANCEMENT IN 6G USING FOG COMPUTING

Various measures are available to enhance the security in 6G using fog computing. Some of these are discussed as follows:

- **Distributed Security Processing:** Fog nodes optimise security performance through proactive threat monitoring near endpoint systems. Security workloads distributed among fog nodes enhance system resistance against attacks. Executing security functions at local processing nodes ensures network performance enhancement and failure resilience.
- **Context-Aware Security:** The monitoring action of fog nodes consist of inspecting local network operations alongside device performance. This improves usage-based threat detection. Localised analysis of behaviour allows systems to detect anomalies more efficiently.
- **Privacy Enhancement:** Fog nodes perform sensitive data operations locally without requiring sending information to cloud systems. All personal and geographical information stays within the original data storage area. Reducing data movement between endpoints decreases the opportunity for interception [26].
- **Dynamic Security Policies:** Each network section operates with distinct security rules. The security capabilities of fog nodes enable them to adjust protection measures according to regional requirements. Exact network addresses can obtain specific policy updates while they are happening.
- **Enhanced Authentication:** Local authentication procedures at Fog nodes provide faster access control methods. End-device multi-factor authentication is possible. Distributed authentication relieves central servers.
- **Improved Distributed Denial of Service (DDoS) Protection:** Core network communication threats can be blocked by fog layers. The location of threat identification and mitigation occurs nearer to where attacks originate. Multiple defences hinder attackers.
- **AI-Powered Security:** The AI security models that Fog nodes execute operate from a local node architecture. The adoption of distributed AI processing methods accelerates of security threat detection. Localised data patterns are input for teaching machine learning models [27].

9.6.1 Case Study 1

A Secure Fog-Assisted Architecture for IoT Networks in 6G Healthcare Systems introduced a widespread security framework for fog computing at Shanghai General Hospital through research. The research deployed a three-tiered fog infrastructure which connected more than 200 medical IoT devices throughout multiple hospital sectors. Overall security performance during implementation achieved two major accomplishments: reduced data vulnerability by 65% and faster detection times, reaching up to 40 milliseconds below typical cloud security standards. Patient data protection at real-time intervals used AI-powered fog nodes, which delivered 99.99% continuous uptime for vital medical systems through distributed security operations [28].

9.6.2 Case Study 2

Another relevant case study, "Fog-Enhanced Security for Industrial IoT in 6G Networks," studied industrial security operations at German manufacturing facilities.

With zero-trust security methodology, this research deployed decentralized authentication programme through 1,500 industrial sensors and actuators connected by fog computing nodes. Results showed that threat detection accuracy improved by 70%, while security events could be addressed within seconds instead of minutes. Research demonstrated that joint operations between edge and fog infrastructure deliver effective real-time anomaly detection to minimize cyber-physical security risks [29].

9.7 CONCLUSION

The amalgamation of fog computing with 6G-enabled IoT offers considerable prospects and obstacles for forthcoming wireless networks. As networks advance to accommodate extensive IoT implementations with millions of devices per square kilometre, fog computing becomes essential for managing computational duties and storage requirements near the edge. This distributed methodology has multiple benefits, such as diminished latency, increased spectrum efficiency, and augmented network capacity. Nonetheless, numerous critical hurdles must be confronted to actualise the promise of this integration properly. Energy efficiency is a critical issue necessitating advanced management strategies in device control, task offloading, and resource allocation. The diverse characteristics of IoT devices and fog nodes complicate this environment, requiring adaptive systems to accommodate differing hardware capabilities and software specifications. Quality of Service considerations must be meticulously balanced with energy conservation initiatives, as conventional power-saving techniques such as device consolidation and sleep modes may adversely affect network performance and reliability. Anticipating future developments, emerging technologies like Reconfigurable Intelligent Surfaces and federated learning exhibit potential in improving network efficiency and optimising resource utilisation. Collaboration in energy across nodes and efficient spectrum use will be paramount in 6G networks despite ongoing issues in creating equitable and incentivised collaboration mechanisms. Choosing suitable energy-efficient techniques must account for static and dynamic energy consumption trends while upholding service-level agreements for various IoT applications. Ongoing research in this domain should prioritise creating scalable solutions capable of adeptly managing the intricate interactions among energy efficiency, resource allocation, and quality of service in fog-based 6G IoT environments.

REFERENCES

1. F. Guo, F. R. Yu, H. Zhang, X. Li, H. Ji, and V. C. M. Leung, "Enabling massive IoT toward 6G: A comprehensive survey," *IEEE Internet Things J.*, vol. 8, no. 15, pp. 11891–11915, 2021.
2. R. Chataut and R. Akl, "Massive MIMO systems for 5G and beyond networks—Overview, recent trends, challenges, and future research direction," *Sensors*, vol. 20, no. 10, p. 2753, 2020.
3. M. Z. Chowdhury, M. Shahjalal, M. K. Hasan, and Y. M. Jang, "The role of optical wireless communication technologies in 5G/6G and IoT solutions: Prospects, directions, and challenges," *Appl. Sci.*, vol. 9, no. 20, p. 4367, 2019.

4. B. D. Deebak and F. Al-Turjman, "Drone of IoT in 6G wireless communications: Technology, challenges, and future aspects," in *Unmanned Aerial Vehicles in Smart Cities*, Cham: Springer International Publishing, 2020, pp. 153–165.
5. T. Huang, W. Yang, J. Wu, J. Ma, X. Zhang, and D. Zhang, "A survey on green 6G network: Architecture and technologies," *IEEE Access*, vol. 7, pp. 175758–175768, 2019.
6. U. Gustavsson et al., "Implementation challenges and opportunities in beyond-5G and 6G communication," *IEEE J. Microw.*, vol. 1, no. 1, pp. 86–100, 2021.
7. M. Polese, J. M. Jornet, T. Melodia, and M. Zorzi, "Toward end-to-end, full-stack 6G terahertz networks," *IEEE Commun. Mag.*, vol. 58, no. 11, pp. 48–54, 2020.
8. S. Abdelwahab, B. Hamdaoui, M. Guizani, and T. Znati, "Replisom: Disciplined tiny memory replication for massive IoT devices in LTE edge cloud," *IEEE Internet Things J.*, vol. 3, no. 3, pp. 327–338, 2016.
9. J. Liu, M. Agiwal, M. Qu, and H. Jin, "Online control of preamble groups with priority in massive IoT networks," *IEEE J. Sel. Areas Commun.*, vol. 39, no. 3, pp. 700–713, 2021.
10. A. Husen, M. H. Chaudary, and F. Ahmad, "A survey on requirements of Future Intelligent Networks: Solutions and future research directions," *ACM Comput. Surv.*, vol. 55, no. 4, pp. 1–61, 2023.
11. U. M. Malik, M. A. Javed, S. Zeadally, and S. ul Islam, "Energy-efficient fog computing for 6G-enabled massive IoT: Recent trends and future opportunities," *IEEE Internet Things J.*, vol. 9, no. 16, pp. 14572–14594, 2022.
12. T. Sharma, A. Chehri, and P. Fortier, "Review of optical and wireless backhaul networks and emerging trends of next generation 5G and 6G technologies," *Trans. Emerg. Telecommun. Technol.*, vol. 32, no. 3, p. e4155, 2021.
13. R. Sekaran, R. Patan, A. Raveendran, F. Al-Turjman, M. Ramachandran, and L. Mostarda, "Survival study on blockchain based 6G-enabled mobile edge computation for IoT automation," *IEEE Access*, vol. 8, pp. 143453–143463, 2020.
14. B. Barakat et al., "6G opportunities arising from Internet of Things use cases: A review paper," *Future Internet*, vol. 13, no. 6, p. 159, 2021.
15. L. Zhang, Y.-C. Liang, and D. Niyato, "6G Visions: Mobile ultra-broadband, super internet-of-things, and artificial intelligence," *China Commun.*, vol. 16, no. 8, pp. 1–14, 2019.
16. X. Fernando, A. Gupta, and O. Das, "Reliability and availability modeling techniques in 6G IoT networks: A taxonomy and survey," in *2021 International Wireless Communications and Mobile Computing (IWCMC)*, 2024.
17. K. Gai, X. Qin, and L. Zhu, "An energy-aware high performance task allocation strategy in heterogeneous fog computing environments," *IEEE Trans. Comput.*, vol. 70, no. 4, pp. 626–639, 2021.
18. Mostafa, M. A. A. A., & Khater, A. M. (2019, April). Horizontal offloading mechanism for IoT application in fog computing using microservices: Case study: Traffic management system. In 2019 IEEE Jordan International Joint Conference on Electrical Engineering and Information Technology (JEEIT) (pp. 640–647), Jordan. IEEE. https://doi.org/10.1109/JEEIT.2019.8717462
19. R. K. Naha et al., "Fog computing: Survey of trends, architectures, requirements, and research directions," *IEEE Access*, vol. 6, pp. 47980–48009, 2018.
20. G. Li, J. Yan, L. Chen, J. Wu, Q. Lin, and Y. Zhang, "Energy consumption optimization with a delay threshold in cloud-fog cooperation computing," *IEEE Access*, vol. 7, pp. 159688–159697, 2019.
21. P. Habibi, M. Farhoudi, S. Kazemian, S. Khorsandi, and A. Leon-Garcia, "Fog computing: A comprehensive architectural survey," *IEEE Access*, vol. 8, pp. 69105–69133, 2020.
22. H. Rahman and M. I. Hussain, "Fog-based semantic model for supporting interoperability in IoT," *IET Commun.*, vol. 13, no. 11, pp. 1651–1661, 2019.

23. S. Jia, X. Yuan, and Y.-C. Liang, "Reconfigurable intelligent surfaces for energy efficiency in D2D communication network," *IEEE Wirel. Commun. Le*tt., vol. 10, no. 3, pp. 683–687, 2021.
24. O. A. Wahab, A. Mourad, H. Otrok, and T. Taleb, "Federated machine learning: Survey, multi-level classification, desirable criteria and future directions in communication and networking systems," *IEEE Commun. Surv. Tutor.*, vol. 23, no. 2, pp. 1342–1397, 2021.
25. Y. Zhan, P. Li, Z. Qu, D. Zeng, and S. Guo, "A learning-based incentive mechanism for federated learning," *IEEE Internet Things J.*, vol. 7, no. 7, pp. 6360–6368, 2020.
26. N. Sharma and C. Prabha, "Computing paradigms: An overview," in *2021 Asian Conference on Innovation in Technology (ASIANCON)*, 2021.
27. A. Goel and C. Prabha, "A detailed review of data deduplication approaches in the cloud and key challenges," in *2023 4th International Conference on Smart Electronics and Communication (ICOSEC)*, 2023.
28. R. Gupta and D. Yadav, "6G-enabled IoT for e-healthcare systems," in *6G-Enabled IoT and AI for Smart Healthcare*, Boca Raton: CRC Press, 2023, pp. 121–141.
29. S. A. Ansar, J. K. Samriya, M. Kumar, S. S. Gill, and R. A. Khan, "Intelligent fog-IoT networks with 6G endorsement: Foundations, applications, trends and challenges," in *6G Enabled Fog Computing in IoT*, Cham: Springer Nature Switzerland, 2023, pp. 287–307.

10 Cyber Threat Hunting in 6G-Enabled IoT Devices
Future Scope and Challenges

Aditya Pathania, Raghav Goel, Chetna Sharma, and Vansh Kwatra

10.1 INTRODUCTION

Global communications are expected to enter a new era with the emergence of 6G networks, which might have revolutionary implications on sectors globally and individual countries. Internet of Things (IoT) is being continually connected at speeds over 100 Gbps thanks to the use of AI, terahertz communication, and next-generation edge computing. This redefines what IoT can mean; this is not the IoT of your parents. In the 6G situation, this leads to an enormously connected system that includes smart homes, driverless cars, and critical infrastructure. The various devices are connectable to each other through a variety of high-speed communication modalities, such as Ultra-Reliable Low Latency Communication (URLLC), Multi-Type Traffic Communication (MTTC), and significantly improved Machine-to-Machine (M2M). The ease with which the entire lives of people can be powered with a single button push, thanks to this hyper-connectivity, however, also opens the door for cyber enemies to penetrate and creates an omni-channel attack surface that goes beyond what was previously recognised. The fundamental design of 6G networks presents opportunities as well as challenges. By putting sophisticated AI and Machine Learning (ML) models at the core of 6G, this enables IoT to make decisions in real time, perform forecasting analysis, and perform self-efficient or automated system management functions [1]. However, increased reliance on AI creates new attack surfaces as well. Cybercriminals may use any bias, inaccuracy, or potential vulnerabilities in the algorithms themselves to corrupt or infiltrate IoT devices, or they may target the model behaviour that the algorithms are designed to recognise and respond to. Second, data integrity and confidentiality issues will become a major focus of security research in this highly distributed scenario as 6G accelerates the trend towards decentralised networks driven by blockchain and distributed ledger systems. One particularly important component of an active offensive security posture is cyber threat hunting. Cyber threat hunting differs fundamentally from the traditional cybersecurity measures which the current generation has grown accustomed to because the former are primarily reactive (use of firewalls, antivirus software, and signature-based intrusion detection systems that claim to protect the systems but frequently end up failing). But in this instance, it's the bulk of Advanced Persistent

DOI: 10.1201/9781003586456-10

Threats (APTs), which are hackers that have gained access to a system and spend weeks or months disguising themselves as authorised users while secretly lingering somewhere within your network and not setting off any alerts. Cyber threat hunting is particularly important since next-generation 6G-IoT settings are now vulnerable to the same advanced tactics, methods, and procedures (TTPs) used by adversaries. IoT devices on a 6G network present security issues. Specifically, there are a tonne of networked devices that are open to attack [2]. Even while the IoT is a more specialised and diverse field than it ever was, with each device having its own operating system, processing power, and communication protocol, it becomes harder to maintain the overall scope of a uniform security architecture throughout the entire interconnected network. Every gadget, from health monitoring in wearables to smart sensors in industrial systems using operational technology, is a possible point of entry for cybercriminals [3]. The fact that 6G networks are more dynamic and enable devices to connect and disconnect from the network instantly only serves to exacerbate the issue. This presents a terrifying possibility for security supervision.

The enhancement of data confidentiality, integrity, and availability, also known as the CIA trinity, by 6G has further implications for IoT systems. Apart from offering novel prospects, the instantaneous capabilities of 6G and its long-range connectivity provide easier access to data interception, manipulation, and theft. Preventing illegal access to highly sensitive personal health, financial, and government communications data is a priority, as it is likely that 6G networks will be utilised for their transmission. Additionally, 6G has such low latency that over time, the machine will be able to understand and respond without the need for human involvement. It is now much more difficult to maintain confidence; thus, being aware of cyber threats is essential. Even in 6G-enabled IoT contexts, there are new threat landscapes that demand a paradigm shift in the perception of cybersecurity. The majority of conventional defensive systems in operation today identify and neutralise known threats using historical data. However, a faster and more active threat detection method might be needed to counteract attacks on a 6G-enabled IoT system, which could contain zero-day threats or AI threats unique to a single homeowner [4]. The increasing popularity of cyber threat hunting, which entails identifying unusual patterns, unusual behaviours, and possible indicators of compromise before a significant attack, is indicative of this emphasis. Cyber threat hunters utilise a combination of advanced analytics, threat intelligence, and skilled human judgement to find sophisticated attack routes that conventional systems would not be able to detect in situations where standard defences may fail. This is made more difficult by the fact that 6G networks rely on edge computing. The key here is to process data at the point of creation, which is in the IoT devices themselves, rather than having to route all of it through centralised cloud-enabled infrastructure. While this lowers latency and boosts performance, it also significantly increases the security risk surface because every one of these edge devices is a target for easy exploitation [5]. This means that threat hunters must use a more thorough security approach, searching all endpoint devices within the 6G-IoT ecosystem rather than simply the primary network infrastructure [6].

In this book chapter, Section 10.2 discusses 6G networks. In Section 10.3, the depth about various threats and attack factors in 6G IoT networks, which pose a great deal of threat to the goal of a secure 6G IoT network model, has been discussed.

Section 10.4 lists the major cybersecurity challenges in 6G IoT models to give a more detailed description of threats in the network. Section 10.5 focuses entirely on the causes of vulnerabilities in 6G IoT models, which will help us build more robust network architectures.

To mitigate all the cyber threats in the network, classification of threats has been made based on different parameters, which is done in Section 10.6. After successful classification, the next and last step is to explore various ways to mitigate these threats. Section 10.7 discusses the countermeasures for threat mitigation in 6G IoT networks. Section 10.8 describes one of the major technologies used for mitigating risks in the 6G IoT networks, which is Distributed Ledger Technology (DLT). Section 10.9 has discussed the most effective way to protect the network from all types of attacks, which is authentication.

10.2 6G NETWORKS

6G networks are the successors to 5G wireless technology—a bold redesign of high-speed connections developing into reality, representing oriented levels of connectivity and reliability with data transfer content as high as 1 terabit per second. Seamless connections of the aforementioned digital and physical worlds will be harnessed through new standards based on these networks integrated by technologies such as AI, edge computing, terahertz frequency bands, and quantum communication. 6G's objective is not only to accommodate incumbent services—high-definition streaming, autonomous systems—but also to enable leading-edge breakthroughs—holographic communications, tactile internet, and connected intelligence. 6G will provide ultra-low latency, improved energy efficiency, and massive device connectivity to power smart cities, Industry 4.0, and next-gen healthcare. The network will additionally be green as a part of the sustainability effort by lessening organisations' carbon footprint through green technologies. Nonetheless, the making and deploying of 6G will meet hurdles such as spectrum dispersion, infrastructure preparedness, and policy systems [2]. Already beset by global research and standardisation activities, 6G is expected to repeat the cycle of transformation only a decade or so later, to open new capabilities for wireless communications and for the way humankind interacts with technology in the following decade. The expectation from 6G networks is to bring fundamental changes in technology through unprecedented levels of integration between connectivity, computation, and intelligence. The promising feature of 6G technology will be the deployment of THz frequency bands, which provide ultra-wideband channels with extremely high data rates and very low interference. From within the frequency range, networks will also have the capacity to support connectivity in high-density areas like smart factories and urban environments that are notoriously hard for existing systems to accommodate [7]. Additionally, 6G will utilise more AI and machine learning algorithms designed to improve network optimisation by adjusting the operation for real-time circumstances and offer predictions of user needs with unprecedented reliability. Integration of this will allow dynamic resource allocation, enabling to make use of the bandwidth and energy more efficiently, even in extreme conditions. The other primary function of 6G will be to enable immersive and experiential technologies. Holographic telepresence, AR/VR applications, and brain computer interfaces will soon be possible, transforming sectors such as entertainment, education, and remote work. As an example, surgeons will be able to operate remotely with the tactile internet providing instantaneous haptic feedback via 6G networks [1]. The new inter-architecture will support autonomous vehicles, drones, and other IoT devices in smart cities, creating a holistic urban ecosystem that can respond to and control its infrastructure seamlessly. Furthermore, through improved satellite integration, 6G will further connect unconnected and remote areas that lack coverage by bringing together the benefits of improved global inclusiveness while bridging the digital divide. Inversion of the thinking paradigm will also lead to a revolution in security and privacy with 6G, as blockchain and quantum cryptography will be employed to secure networks for unprecedented levels of protection against cyberattacks, so that even as the amount of sensitive data is increasing exponentially, reliability can be ensured in terms of integrity, confidentiality, and availability. Edge computing will improve data privacy by processing data closer to its source, where it is less vulnerable [3]. Still, the journey towards 6G will require massive amounts of hardware and infrastructure innovation, such as new antennas, low-power devices, and a flexible base station (BS). To overcome these hurdles, organisations across the world—governments, industries, and research institutions—are working together to establish concepts around unified standards and frameworks that will shape how 6G technologies will be defined. While creating 6G unfolds, it will not merely boost the capabilities of already embedded applications but also initiate some innovations that are going to be indispensable. Figure 10.1 shows the 6G security architecture along with the threats to the data and privacy in the network.

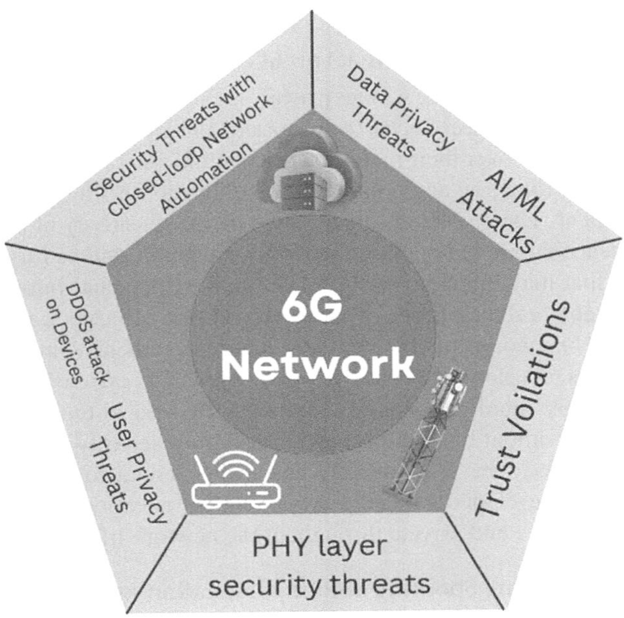

FIGURE 10.1 6G security architecture.

10.3 TYPES OF THREATS AND ATTACKS TO 6G IoT NETWORKS

In this section of the chapter, various threats to 6G IoT networks resulting in vulnerable networks have been listed. The types of threats will help us find ways to hunt the threats to ensure the robust nature of networks

10.3.1 Confidentiality Threats

10.3.1.1 Breach of Confidentiality in 6G IoT Networks

As 6G IoT networks involve sensitive data, e.g., real-time industrial process information or personal health information, privacy leakage is the most significant threat to security. On the one hand, the decentralised architecture of 6G networks will make a large number of edge computing-powered entry points accessible to hackers seeking to intercept and steal confidential information. The kinds of cyberattacks that will be feasible in a 6G future might be beyond the scope of the encryption approaches you can use to safeguard data, especially when hackers begin to employ quantum computing or AI-driven ways for fingerprinting and cracking your codes. Once more, the prevalence of wide-ranging encrypted communication in 6G networks poses a significant risk for man-in-the-middle (MITM) attacks, which enable rogue actors to intercept communications between IoT devices and their control systems [7]. 6G IoT network security requires Singular Value Decomposition (SVD) identity verification, AI anomaly detection (AIDS), and cryptography to take confidentiality to the next level. The stakes are high, though, as failing to address these concerns about violating confidentiality might result in unauthorised data access, intellectual property theft, and damage to the reputation of companies working on 6G IoT technologies. It's risky.

10.3.1.2 Associated Privacy Threats on 6G IoT Networks

The anxiety stems from the more serious privacy issue caused by the 6G IoT networks' unparalleled volume of data collected, transported, and processed in real time. People eventually lose control over their privacy when routine and critical data is collected, ranging from GPS routes in autonomous automobiles to health conditions in smart medical equipment. The problem is made worse by the sensitivity of the data that AI and machine learning are fed on these networks, since this data is further processed by algorithms that may unintentionally release private personal information due to network exploitable vulnerabilities or an increased risk of large-scale breaches [5]. Also, because 6G networks are, by their very nature, fragmented and involve a large number of devices, developing a cohesive privacy policy is extremely difficult, if not impossible [8]. This could occasionally inadvertently lead to the disclosure of personal information without the required authorisation or security [9]. Strong encryption techniques, privacy-aware AI models, and stringent data governance procedures will all need to be developed to secure privacy in 6G networks and guarantee that user data is kept secure and private throughout the network life cycle.

10.3.1.3 Attacks Using Spoofing and Impersonation in 6G IoT

With 6G, IoT devices have greater autonomy and the ability to make decisions in real time. This introduces significant risk in compromised devices and effective impersonation or spoofing attacks. In such an instance, the attacker creates a mask to masquerade as an IoT device that the targeted edge device trusts, and gains unauthorised access to the network or alters data. A security attack can gain extremely high efficiency, which can't be achieved at this stage of 5G technology, or intercept data without being discovered or tamper with data (or "data manipulation") without being discovered, due to the corresponding high speed and low latency of 6G networks [1]. One of the best examples out there would be if an attacker could impersonate a medical device within a healthcare environment and provide test results or treatments that were incorrect. In general, malware-like authentication protocols (e.g., biometric locks or multi-factor authentication, MFA) need to be adopted to mitigate the risk of spoofing and impersonation attacks in 6G IoT systems [7]. Moreover, anomaly detection through AI-based identity verification techniques should be applied to detect unconventional device communication patterns or user behaviour.

10.3.2 Integrity Threats

10.3.2.1 Security Risks: IoT Devices with 6G Devices That Compromise Integrity

The integrity of the smart devices that power an IoT 6G network has a major impact on how reliably information is sent. In a 6G context, IoT devices will need to manage autonomous operations, and compromised data integrity could have disastrous consequences. For example, even if an attacker can do things like modify data in a health care system, patient data may not be compromised, but the data can be used to generate false medical decisions. However, malware injections, fake firmware manipulations, and other potentially dangerous firmware updates represent further incoming elements on the data integrity from these devices, and in consequence, the entire system. Moreover, with the integrated IoT devices increasingly relying on either processing data in real time or reacting to data changes, the integrity of the data transmission and reception becomes a growing necessity [2]. To protect the weaknesses in the 6G network against AI poisoning, false data injection attacks, and more sophisticated attacks, machine learning algorithms are built that can defend themselves against such attacks. The complexity of the problem domain, shaped by scalability needs, secure software development practices, and effective authentication mechanisms to prevent tampering with data, is an issue crucial for the operation of IoT devices in a 6G network.

10.3.2.2 Issues with Physical Layer Security (PLS) in 6G Networks

This requires shoring up a distinct class of security challenges related to the physical layer of 6G networks to help secure IoT devices. Due to multi-terahertz frequencies and heavy beamforming technology in support of high-speed communication in 6G, adversaries may exploit the physical layer and disrupt communication subjects or silently listen into the encryption of key information. Furthermore, there is an extra challenge of protecting data processing on physical edge devices such as IoT sensors and gateways over 6G networks and decentralised edge computing. Attackers may find themselves physically close to the devices, which raises the odds that they will interfere with or jam their signals. Under these circumstances, 6G networks must adopt techniques such as secure beamforming, jamming-resistant communication protocols, and physical layer encryption to ensure that network transmissions are not intercepted or penetrated by an adversary's broadcast.

10.3.2.3 Remote Exploitation and Vulnerabilities of 6G IoT Devices

These massive volumes of IoT devices show that an unknown number of internet-enabled devices can likely be remotely hacked, especially if they can be seen and/or controlled from anywhere in the world, instead of from a closed network. As 6G is an advanced technology compared to 5G, it also demands more resources in terms of hardware and software [2]. Software bugs and weak authentication protocols are ways for hackers to access IoT devices, compromising the entire security of the network [10]. This way they can attack the devices for changing the default settings, gather data, and create a backdoor for plans. There's a greater chance of remote weaponisation in 6G networks, given that there are more IoT devices and less protection [5].

10.3.3 Availability Threats

Availability is a red-hot topic for a network in general, but there is a risk in the case of IoT systems driven by 6G, for multiple reasons that 6G entails continuous connection and real-time operations of multiple devices in a large network. The performance characteristics of available 6G networks, such as ultra-low latency and vast bandwidth, bring about a new set of challenges in provisioning the constant availability of services, particularly in large-scale IoT applications. The impact of network unavailability may be rather dramatic, for instance, the stopping of industrial activities, plunging smart-city infrastructures into disarray, and prohibiting autonomous cars from running. Network resources can be exhausted by intentional attacks like Distributed Denial of Service (DDoS) assaults as well as technical issues like hardware malfunctions and software faults. Besides, because 6G networks are widely decentralised, it is more difficult to provide instantaneous service changes in remote locations like mountains or deserts, where edge computing nodes are common [4]. In these environments, edge computing nodes tend to have limited access to electric power sources, so they may occasionally turn off to recharge their batteries or become inaccessible due to communication outages. As a result of the adoption of sophisticated fault-tolerant systems, appropriate radio resource allocation algorithms, and predictive maintenance devices that should be able to foresee future failures, the 6G network will need to have high availability and be available at all times.

Figure 10.2 is a visual depiction of the 6G network overview. This figure also lists major fields under 6G IoT networks and qualities in terms of security aspects.

Figure 10.3 shows the Cyber Threat Problem with 6G Network in terms of attackers and transfer of data through different components of the network.

10.3.4 APT in 6G

APTs can be a key concern as more 6 G-enabled IoT devices come to the market, potentially compromising the security and integrity of systems [11].

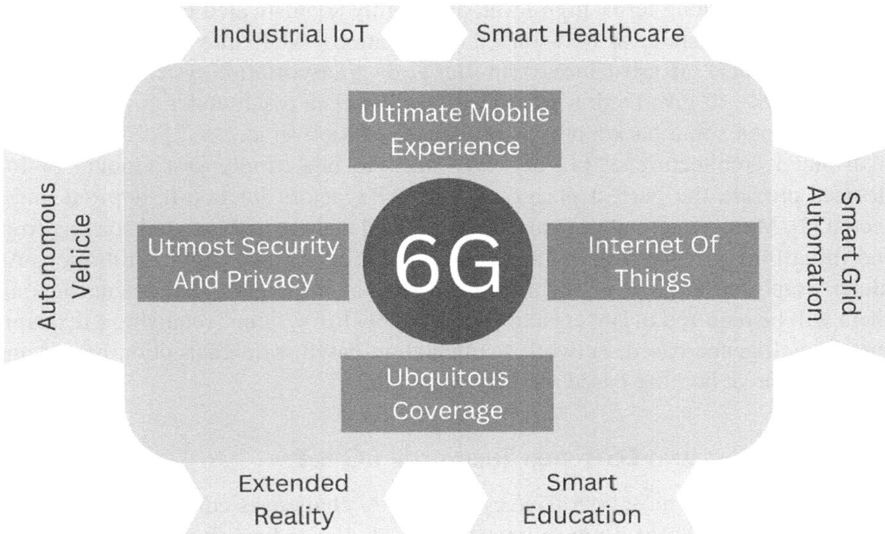

FIGURE 10.2 6G network overview.

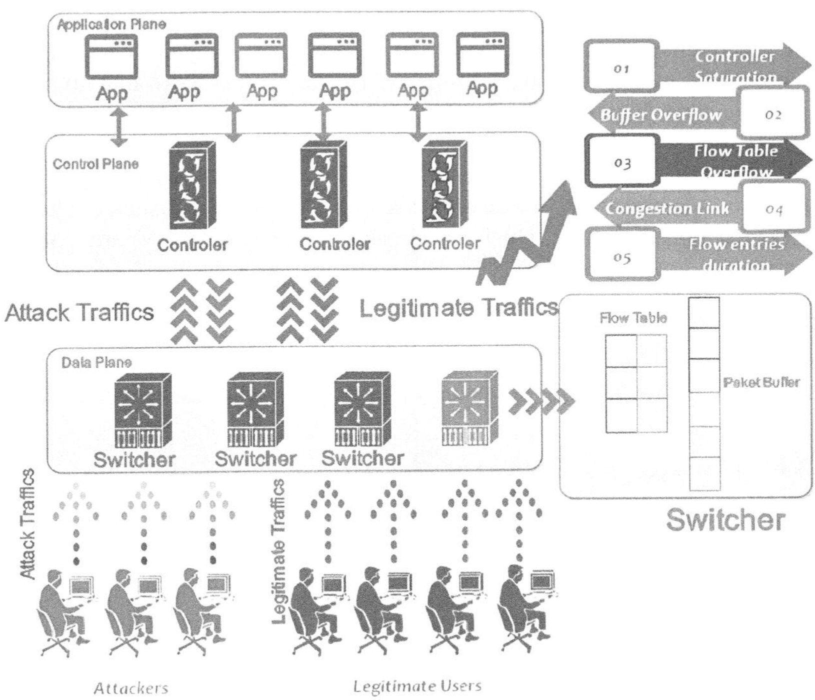

FIGURE 10.3 Cyber threat problem with the 6G network.

An APT is a long-term, high-level, and highly sophisticated cyberattack implemented by nation states or well-funded organisations. APTs may focus on 6G USN, which are critical infrastructure like power grids, transportation systems, and healthcare networks, to give them a significant head-start to reach under-cover, break in, and then open some backdoors for prolonged disruptive access [2]. The extremely distributed architecture of 6G networks and the blisteringly vast amount of IoT devices provide the perfect playground for APT actors for hop-in-hop-out undetected [2]. Most of these attacks are multi-stepped and are orchestrated starting from the initialisation to breach, to lateral movement, and finally data exfiltration. An attack graph for APTs at a very new type of system highlights that defence against them will be required in current and upcoming 6G IoT systems, requiring a real-time threat intelligence-based, network segmentation-based, safe encryption-based, and proactive threat hunting-based security approach.

10.3.5 Insider and Ecosystem Threat for 6G IoT

It is also one of the most pervasive cybersecurity challenges confronting the world today, and 6G networks seem to intensify the challenge for IoT systems. Using the unprecedented permission that 6G-enabled IoT environments provide them (as workers, contractors, or compromised individuals), insiders can mount attacks that challenge a network's availability, confidentiality, and integrity. Because 6G networks are decentralised with many internal workings, insider threats can fester for a time without being noticed, especially if they are using stolen credentials to commandeer IoT assets or manipulate AI-driven decision processes from within [1]. This introduces significant risk in compromised devices and effective impersonation or spoofing attacks. In such an instance, the attacker creates a mask to masquerade as an IoT device that the targeted edge device trusts, and gains unauthorised access to the network or alters data. An unauthorised person may, for instance, alter sensor data in vital infrastructure, which could cause catastrophes like power outages or turmoil in the transportation system. By implementing a strong access control configuration, real-time monitoring, and behavioural analytics to identify anomalous activity that may point to an insider operating behind the scenes, organisations can guard against insider threats [12]. Maintaining ongoing company training is also essential to lowering the risk of insider threats. This will assist in their decline.

10.3.6 Malware and Cyberattack Threats

10.3.6.1 Cyberattacks and Malware on 6G Networks

The open floodgates of IoT devices operating on 6G networks will not only be the solution desired to achieve from leveraging these next-generation networks, but will also expose people and organisations to a fresh wave of malware, cyberattacks, and ransomware. Malware aimed at the IoT might exploit a 6G network to swiftly move from one susceptible device to another, infecting millions of people in an hour. Fresh methods of attack: The combined decentralised architecture of 6G, edge computing, and AI-driven deterministic decision-making creates new avenues for exploiting security flaws, misconfiguring solutions, or even tampering with AI models. IoT devices have the potential to turn into zombie botnets, arbitrarily release large amounts of data online, or even be used maliciously by hackers who infect them with malware designed to target 6G, affecting things like medical equipment and industrial machinery [4]. To mitigate these risks, 6G network operators need to offer robust cybersecurity frameworks with features such as threat intelligence, automatic malware detection, and patch management that will defend IoT devices against modern-day attacks.

10.3.6.2 IoT Devices Powered by 6G and Supply Chain Attacks

When 6G rolls around, the supply chain for IoT devices will be much more exposed because these devices will be indispensable to entire businesses. 6G-enabled IoT systems, where devices can be hacked by hostile actors at any stage during the lifecycle (the manufacture, distribution, or deployment), are particularly vulnerable to supply chain attacks. An attacker can introduce malware, backdoors, or hardware flaws that can be used to gain access to a network in the future via forking cryptocurrency projects. With the billions of connected IoT devices under 6G's purview, even a single compromised device might provide an entry point for an exploit to go worldwide [3]. Hence, for this reason, in scenarios based on 6G IoT, supply chain security ought to be a top priority, from stringent vendor assessments to safe hardware manufacturing procedures and field monitoring for indications of IoT compromise or tampering.

10.3.7 6G Denial of Service (DoS) Attacks on IoT Devices

Despite DoS assaults being among the most traditional forms of cybercrime, they can become some of the most effective and complex when launched against IoT devices equipped with 6G. While the potential for DoS attacks is increased by terahertz-com and massive bandwidths in 6G networks, an attacker may inundate IoT devices with so much fraudulent traffic that it renders them completely inoperable. 6G could enable federated and edge computing paradigms that allow such coordinated assaults to take place under at least some legal cover, making it harder, if not impossible, to detect and eliminate them in real time. This is in contrast to previous network generations, where DoS attacks were always associated with a single party or group seeking to disable the services of another remote one [5]. Most significantly, a short-term outage can have a direct impact on sectors like smart manufacturing, transportation, and health care because of how heavily IoT devices depend on continuous network connectivity. 6G-enabled IoT systems must effectively manage DoS attacks, necessitating the installation of sophisticated traffic filtering techniques that detect attacks at an early stage and notify or isolate specific compromised devices before they propagate throughout the network. A DoS attack has been visualised in Figure 10.4, showing various components of the attack, like the attacker, users, IoT ecosystem, and the basic flow of data in the attack.

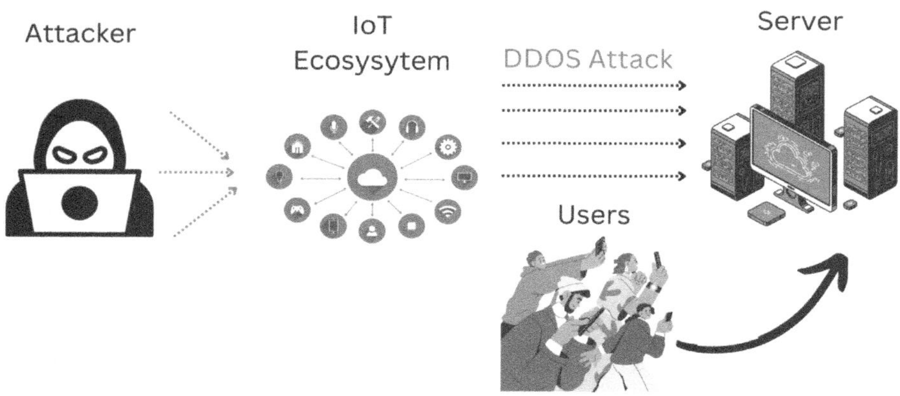

FIGURE 10.4 6G on DoS attacks in IoT.

10.4 CYBERSECURITY CHALLENGES IN 6G IOT MODELS

With the emergence of the sixth generation of mobile networks (6G), a large number of devices, thanks to the IoT, would be able to exhibit new functionalities and be present in new channels. Great expectations lie in this next generation. Nevertheless, with the promise of enhanced performance, there arises an issue of enhanced cyber threats endangering the integrity of the connected systems. Due to the fusion and complexity of 6G IoT networks, several vulnerabilities exist, making it a good target for adversaries. Data tampering, DoS, and MITM attacks are some of the threats that are increasingly becoming common. As cities and Electric Vehicle Chargers (EVCs) increasingly rely on IoT-enabled devices, the impact of successful events could be significant and negative on the health and safety of the public. On the other hand, the combination of new capabilities such as AI and blockchain will create both opportunities and challenges. All these kinds of cybersecurity threats need to be identified and worked on. This chapter is to explain the goals of organisations that defend and undertake deterrent activities in order to make sure these threats do not happen. In this way, a bright future lies ahead with this new class of technology.

10.4.1 Importance of Data Confidentiality

6G IoT technology and systems have in place some level of confidentiality, but attacks such as MITM are a concern regarding intercepted communications directed between two parties, especially in the context of false base station attacks (FBSAs) within mobile environments. So, employing solutions such as AI-based intelligent routing and public key cryptography can mitigate the threat posed by attackers. Besides, eavesdropping and snooping risk compromise the data confidentiality of any information transmitted by intercepting such information. As a solution to this problem, authors in 6 also proposed blockchain-based confidentiality matrices to protect the information as well as strengthen the data security. Targeted data attacks also represent threats, but exposure-based overlay strategies using various metrics of statistics can be useful in counteracting these risks. Misrepresentation and stalking are other threats as well, as those who have no secured channels can entice users to disclose information seen as confidential. However, as the 6G world is growing larger and larger day by day, loss of important information has to be prevented with strong regulatory policies and measures. Network networks disrespect client data in case of each lateral attack, so a lot of research regarding blockchain technology has been made to find suitable approaches for data storage and data security.

10.4.2 Risks to Data Integrity

In fact, it is conceivable that integrity threats in 6G networks can cause communication disruption with disastrous effects. The primary risk is when a message is altered, that is, when someone modifies or changes a message without authorisation, which can have a detrimental effect on data accuracy. While methods like blockchain and cryptography can guarantee data integrity, they potentially introduce processing overheads in densely populated networks. Potential attack vectors include session hijacking, an attack where an attacker takes control of an authenticated session, resulting in multiple security incidents. This is especially an issue in fog computing contexts, where the ability to maintain the integrity of the session is needed. Data diddling—the manipulation of data before it enters a system—is also a serious concern. One method for resolving this is abstracting QR code-based secret sharing, especially in industrial IoT applications.

10.4.3 Threats to System Availability

Availability threats pose serious concerns for 6G systems and include a range of risks. DDoS attacks flood networks and interrupt services. Innovative solutions are emerging to improve the identification techniques and virtual shadow networks that enhance resilience to these kinds of attacks. Threats of physical harm are also extremely dangerous, in particular for infrastructures, such as smart grids, where intrusions can have devastating consequences. Therefore, in other words, you need to implement solid physical security measures. In addition, FIFO and free-riding attacks use timing to compromise data flow and system availability. Developing one way to detect and defend against these vulnerabilities flawlessly is through machine learning-based systems.

10.4.4 Vulnerabilities in Authentication

There are different and significant authentication threats in the 6G networks. Common methods include brute force, where attackers try to guess passwords, and replay attacks, where attackers reuse the username and password combination from previous breaches. These approaches have been countered effectively by AI-driven security measures. Another weakness is forged signals, such as collision attacks, where the attacker must defeat not only the controller but also any authentication applied by the control system. Lite protocols are being investigated to bolster defences against those threats. Furthermore, it is important to realise that there are potential vulnerabilities within current authentication protocols that adversaries can leverage (e.g., brute force), and as the advancement toward 6G networks, re-evaluation of techniques being employed has to be done, considering their relevance and context for usage.

10.4.5 Risks to Access Control

Systems dealing with sensitive information should be aware of access control threats. Social segmentation and phishing are common techniques that leverage human psychology to obtain unauthorised access to data. How to avoid these threats? The organisations must put in place sustainable access control measures. Moreover, a data mining attack that collects malicious digital data to conduct large-scale attacks has encouraged exploration in blockchain-supported solutions to improve security across various networks. As quantum computing hums along, there is a necessity to design more resilient algorithms against birthday and cloning attacks and implement certain access control mechanisms for 6G networks.

10.5 CAUSES OF VULNERABILITIES IN 6G IOT MODELS

10.5.1 Insecure Devices

The security vulnerabilities in IoT devices are a significant concern, especially as the proliferation of these devices increases with advancements like 6G technology. The issues outlined are critical to understanding the landscape of IoT security.

10.5.1.1 Weak Default Passwords

Many IoT devices are shipped with default usernames and passwords that are often easy to guess. If users fail to change these credentials during setup, devices become accessible to attackers. This situation is compounded by the fact that default passwords are frequently shared online, making them targets for automated attacks.

10.5.1.2 Limited Encryption

A lack of robust encryption is another major vulnerability. Devices that do not encrypt data during transmission or storage expose sensitive information to interception. This lack of protection can lead to unauthorised access to personal data, compromising user privacy and security.

10.5.1.3 Neglected Firmware Updates

Manufacturers often neglect the importance of regular firmware updates. Many devices are not designed to receive automatic updates, leaving them exposed to known vulnerabilities. Users may be unaware of the need to perform these updates, and without them, devices remain susceptible to exploits that could have been mitigated.

As 6G technology enables a higher density of connected devices, the risks associated with poorly secured IoT devices grow substantially. The potential for mass exploitation increases, which could lead to widespread security breaches, affecting not only individual users but also larger networks and systems. Addressing these vulnerabilities is essential to ensure the safety and security of IoT ecosystems as they expand.

10.5.2 Complex Network Architecture

As such, there are various communication protocols, network types, and device types that 6G networks will inevitably be built on top of, leading to the inherent complexities of such networks. This complexity can lead to:

- **Configuration Errors:** The most common vulnerability is misconfigurations of networking settings, particularly in large-scale deployments. These mistakes can effectively leave security backdoors for an attacker to use [13].
- **Challenges in Monitoring:** As more devices get connected, effectively monitoring all network activity can become complex. It becomes challenging to identify suspicious behaviour or possible intrusions.
- **Interconnected Insecurities:** While devices and protocols may each have their own security, the interactions between these different layers can yield unexpected vulnerabilities. By leveraging the fact that in such protocols, the weakest link can drive the reliability of the whole network.

10.5.3 Interoperability Challenges

Due to the large number of different devices and manufacturers in the IoT ecosystem, many interoperability issues arise, including:

- **Inconsistent Security Standards**: Different manufacturers often have varying approaches to security. This inconsistency can lead to devices that are incompatible with robust security measures, creating vulnerabilities when these devices communicate with one another [13].
- **Protocol Disparities:** Security efforts can be further complicated because multiple communication protocols are used for each system. This, in turn, means that if one protocol's security features are inadequate, that protocol is a weak link in the overall network security.
- **Vendor Lock-In:** Companies may be locked into certain vendors, which may prevent the integration of devices satisfying the higher levels of secure standards. It also allows them to continue using less secure devices inside their infrastructure.

10.5.4 Data Transmission Risks

The high-speed capabilities of 6G, enabling rapid data transfer, come at the cost of several risks:

- **Interception of Data:** Different types of traffic can be more vulnerable to capture (assuming traffic is not encrypted). This is a problem for sensitive data—personal data, controlling critical infrastructure, and the like.

- **Weak Encryption Implementation:** When devices fail to implement strong cryptography protocols such that their data is continuously encrypted while being transmitted, it raises a significant risk of eavesdropping and data manipulation.
- **Weakness in Hand-Offs:** Data can be vulnerable as devices transition from one type of network to another or between incompatible protocols. Without sufficient protections, data can be intercepted or modified as it's passed from one machine to another.

10.6 CLASSIFICATION OF THE 6G IoT THREAT MODEL

The growth of IoT systems with quicker, more dependable, and more effective connections is made possible by the advent of 6G networks. However, these successes have also led to a rise in sophisticated cyberthreats, which has forced us to reassess our conventional security paradigms. Since security is a multi-layered issue, a comprehensive threat model for 6G-enabled IoT should address every aspect of the problem, from device-level physical security to network-wide worries about data breaches and system-wide attacks. Since the 6G IoT threat model encompasses multiple dimensions, classification of these risks as attacks on Privacy as well as Confidentiality, Integrity, and Availability has to be done. In addition, new concerns arising from contemporary technologies include machine learning vulnerabilities and cyberattacks powered by AI. In addition to edge computing risks, 6G also presents its own security and distributed ledger management challenges due to its decentralised architecture that is unlike its predecessors [7]. The 6G IoT threat model classification should cover the following threat scenarios, including insider threat, spoofing, malware attack scenarios, APT TTPs, and supply chain compromise assaults. Experts from this space will clearly identify these forms of malicious attack vectors and provide strong security solutions for the evolving threat vectors in 6G IoT systems. The 6G threat model is shown in Figure 10.5, also describing various elements in the model and their subparts.

10.6.1 Classification of Cyber Threats

In order to mitigate the potential risks, it is crucial to understand and assess the nature of cyber threats; especially with the development of more exciting and complicated technologies such as 6G IoT networks, the need for cyber threat classification is apparent. Classifying by the goal or damage done is one of the most common techniques used [14]. For example, malware is a blanket term for any and all malicious software, such as viruses, worms, and even ransomware, all designed to take advantage of the weaknesses within computer systems. Phishing is another, often insidious tactic involving exploiting the social aspects and persuading users to provide sensitive data via fraudulent emails or fake websites. DoS is an attack whereby systems are bombarded with unnecessary requests so that legitimate users cannot use the systems.

Furthermore, another beneficial strategy is to classify the threats in terms of their targets. Threats that individuals face are often less complex than those that an

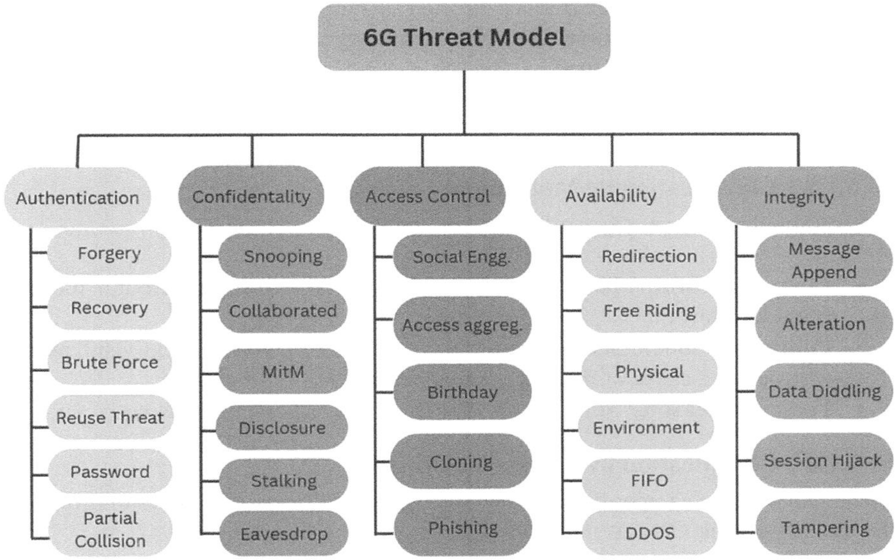

FIGURE 10.5 Classification of the 6G threat model.

organisation or critical infrastructure faces. For example, in cases where people are targeted, the restriction may be only on theft of credentials or funds, but for business targets, sensitive information or the ability to maintain operations may be the target. Certain types of threats are only directed to critical infrastructural systems like the health care systems and energy plants, which are publicly harmful, and therefore, protective mechanisms are needed.

10.6.1.1 Classification by Nature of Attack

Classification of cyberattacks is one of the major aspects of mitigating cyber threats. Here, classification of the attacks by their nature of attack is made.

- **Malware:** Malicious software, such as viruses, worms, ransomware, and spyware, designed to damage or disrupt systems.
- **Phishing:** Social engineering attacks that trick users into revealing sensitive information through deceptive emails or websites.
- **DoS:** Attacks that aim to make a system or service unavailable by overwhelming it with traffic.

10.6.1.2 Classification by Target

Another effective classification method is based on the target of the attack. The threats classified by target are:

- **Individuals:** Threats aimed at personal devices, often focusing on identity theft or financial fraud.

- **Organisations:** Cyber threats directed at businesses, typically aiming to steal data, disrupt operations, or extort money.
- **Critical Infrastructure:** Attacks targeting essential services, such as healthcare systems or power grids, which can have severe public safety implications.

10.6.1.3 Classification by Motivation

Motivation also plays a major role in cyberattacks; some of them are:

- **Financial Gain**: Many attackers seek monetary rewards through theft, fraud, or ransom.
- **Political or Ideological Reasons**: Some attacks, known as hacktivism, aim to promote a cause or make a political statement.
- **Personal Reasons**: Some attacks are done due to personal reasons.

10.6.1.4 Classification by Sophistication Level

Complexity is also a major factor in classifying cyber threats. The types of threats on the basis of complexity are:

- **Basic Threats**: Simple attacks, such as basic phishing scams, that can be executed with minimal skill.
- **APTs**: Attacks organised by a large, well-organised group, which involve a long-term engagement and advanced skill set.

10.6.1.5 Classification by Impact

Below are given the threats on the basis of their potential impact on the system:

- **Low Impact**: Threats that may cause minor disruptions or data loss, often easily recoverable.
- **Moderate Impact**: Attacks that could lead to significant data breaches or system downtime, requiring substantial recovery efforts.
- **High Impact**: Severe threats that could cripple critical systems, leading to extensive data loss and substantial financial or reputational damage.

These classification methods play a vital role in understanding cyberattacks and contribute to effective measures to mitigate them. Some mitigation methods are discussed in the later section of this chapter.

10.7 COUNTERMEASURES FOR THREAT MITIGATION IN THE 6G IoT NETWORK

10.7.1 Cryptographic-Based Countermeasures

The cryptographic-based techniques provide a strong and dependable security architecture by incorporating fundamental ideas of symmetric and asymmetric cryptography [15]. In 6G mobile networks, a variety of techniques use Public Key

Infrastructure (PKI) to authenticate BSs and Access Points (AP) [16]. The Paillier cryptosystem, which consists of three primary algorithms—key generation, encryption, and decryption—was proposed by the authors in [17]. Two huge, independent, and randomly selected prime numbers are necessary for this strategy to work. Even though this method was created with 6G communication in mind, more research is required to assess its processing power requirements and any susceptibilities to MITM attacks, particularly in light of factorisation concerns based on quantum computing. Putting in place a group signature system works well for enabling conditional. Likewise, powerful Rivest–Shamir–Adleman (RSA)-based group signatures are ideally suited for a compact group signature scheme. As a result, group-based signature architectures' potential is constantly being investigated for 6G IoT network applications, especially in terms of evaluating how well they operate against quantum computing difficulties [18].

6G networks use a variety of symmetric key-based techniques to protect privacy. Cybertwin is an example of a novel, unconventional 6G network architecture. In the context of the 6G communication system, this special framework serves as a network helper, records communication habits, serves as a caching mechanism, organises resources, facilitates localisation, and incorporates embedded security. For the Cybertwin-driven 6G environment, a cryptography-based security architecture is considered appropriate [19,20].

10.7.2 The PLS

The communication features of context-aware security frameworks in 6G wireless networks were examined and assessed by the researchers in [21]. However, backward compatibility problems, processing constraints in devices with limited resources, and susceptibility to interference or jamming are some of the difficulties that context-aware security faces. Several examples of context-aware security using northbound Application Programming Interfaces (APIs) are offered by Software-Defined Networking (SDN) technology [22].

Distributed SDN research has investigated its application in clever techniques for identifying DDoS attacks and reducing interference. Nevertheless, maintaining consistency and controlling scalability in distributed architectures continue to be difficult tasks [23]. One of the main areas of research for 6G-enabled heterogeneous Unmanned Aerial Vehicle (UAV) networks is the management and optimisation of resources based on three-dimensional locations. When creating security measures, considerations including 3D location, resource consumption, and system behaviour are essential. Location-based strategies, however, are vulnerable to risks including GPS spoofing and legal issues. In both industry and academia, privacy and trust management in 6G-enabled vehicle networks are significant subjects of debate. A key component of safely controlling and exchanging important vehicle parameters is trust management [24,25].

6G-driven huge heterogeneous connection enables dynamic vehicle-to-infrastructure (V2I) interactions via mm-Wave 3D beam tracking. This approach uses symmetry-based environmental encoding, which can introduce a variety of risks [26]. In the 6G era, establishing an ideal balance between privacy protection and authentication via conditional features is regarded as a promising option. Physical layer features provide a variety of options for creating privacy and authentication solutions using PLS. Multi-antenna beamforming systems using reconfigurable surfaces can efficiently protect against jamming and interference in the crowded environment of 6G communications. Human-centric communication will be critical to the seamless adoption and extensive commercialisation of 6G, since it assures high levels of privacy and tremendous throughput. To address propagation and secrecy issues in long-distance THz communication, researchers suggested an antenna selection strategy that integrates optical and Radio Frequency (RF) links, with a focus on improving secrecy performance against eavesdropping assaults in 6G networks [27].

10.7.3 Countermeasures Based on Intrusion Detection

The authors presented a reduced danger matrix for identifying and assessing vulnerabilities in the emerging 6G scenario. This matrix acts as a threat library, allowing for rapid detection and classification. However, in order for the suggested system to function properly, there must be mutual agreement on how to maintain the threat library over a distributed heterogeneous network. Without this, compatibility concerns may occur, reducing the scheme's effectiveness [28].

Likewise, the authors in [29] have proposed a Non-Orthogonal Multiple Access (NOMA)-based sparse signature matrix for 6G networks, which can function as a threat detection system. However, this method is limited to the wireless domain; hence, it is more applicable to the upper layers, an extremely challenging task in multi-tier heterogeneous networks. There is also the danger of compatibility. Then, reviewing some of the most crucial challenges current on IoT-specific NOMA systems has been done, illustrated through various examples of IoT systems that would not use NOMA [29].

For the detection and recognition of malicious activity, a new security process was formed in [30]. This method makes use of a routing system that was designed to enhance the efficiency of the Low Energy Adaptive Clustering Hierarchy (LEACH) protocol in addition to improving resource utilisation. To identify the flood attacker, the minimum number of times the data is sent. But while the study provides the information about flooding attacks, further work is necessary to generalise this approach to serve a wider spectrum of possible threats.

An opportunity is provided for authentication in the protection of 6G networks based on Channel State Information (CSI) for intrusion identification in a cyber-secure environment [31]. Nonetheless, this approach is confined to the wireless domain; therefore, it may introduce backward compatibility issues and increased processing overhead. Moreover, depending on interference or jamming, it can weaken the performance of such solutions, thus lowering their overall effectiveness.

10.8 DLT

It represents a paradigm shift in the way digital systems log and validate transactions by DLT [32]. DLT shuffles data across a distributed set of nodes and provides a decentralised and cooperative environment instead of the traditional centralised systems, requiring an authoritative source [33]. To ensure transparency and minimise the risk of a single point of failure, each edge node maintains a synchronised copy of the ledger [34]. This peer-to-peer technology allows any user to participate in and validate transactions, creating a trust less landscape in which the consensus processes substitute the middlemen. The primary benefit of DLT is its distributed nature.

Each node in the network independently verifies and confirms the transaction data in order to provide data integrity and counter data fraud or manipulation verifies changes to the ledger. Any attempt to change transaction data would require agreement across the entire network, as every node is in possession of a complete copy of the ledger, rendering illegal changes all but impossible. In addition, DLT provides transparency by allowing users to access a shared, immutable transaction history. This trait can be particularly advantageous in disciplines where traceability and accountability matter. The decentralised structure also provides robust security against outages or cyberattacks since the ledger's operation is not dependent on a single server or entity. Such consensus protocols allow nodes to agree on the state of the network and validate transactions without the need for a central authority. Consequently, DLT serves as the foundational technology for blockchain systems and other innovative solutions that value security, transparency, and distributed control. DLT is like a network architect: It provides decentralised architecture, eliminating a single point of failure by removing the need for middlemen. This distributed approach enhances the system's overall resiliency by ensuring that no single party has undue control or influence over how the network operates. DLT secures the network against malicious attacks and disruptions by distributing data and verification responsibilities across multiple nodes. Transparency is one of the key elements of DLT. Moreover, the fact that DLT is transparent increases accountability to trust. Auditability and traceability are key for use cases from supply chain management to financial services, while the transparent, unchangeable track of transactions allows for both. Each entry in the ledger is secured using cryptographic techniques, which ensures that it cannot be changed or tampered with and that the integrity of the data is maintained.

It is essential to maintain synchronisation and consistency of all versions of the ledger even in cases of network outages (or worse case, network partitions with malevolent actors). Verified data integrity throughout the network ensures that the integrity of the data is preserved throughout the network.

10.8.1 Types of DLT

Various DLT models have been presented dedicated to the integration of the IoT in 6G network architecture. Having said that, each of these types has its distinguished features and strengths, making them apt for catering to the variety of requirements and challenges presented by IoT deployments in the emerging 6G ecosystem. Thereby, as 6G-enabled IoT applications demand high-speed connectivity, real-time data processing, and secure device interoperation, these DLT variations enhance scalability, security, and efficiency.

10.8.1.1 Blockchain

Blockchain, the best-known of the DLTs, is highly valued for its immutable transaction records and distributed architecture. For example, the decentralised and secure part of blockchain gives an excellent groundwork to the totally clear and impassable documents needed to empower the extensive variety of associated IoT devices and sensors in the 6G advanced era. Smart contracts, one of its key features, enable automatic and secure execution of agreements between IoT devices. It brings a great trust less interaction to enhance the efficiency and security of the overall IoT ecosystems, and it ensures that the transactions happen without any requirement for any intermediaries.

10.8.1.2 Directed Acyclic Graphs (DAGs)

DAGs are one of the most powerful alternatives to traditional blockchain technology, with the ability to attain high scalability and execute feeless transactions. In the realm of the 6G IoT revolution, DAGs could potentially be a lightweight super architecture to enable micro transactions and high-throughput data transfers between IoT devices, node to node, without relying on any middlemen. Because DAGs are inherently asynchronous, they seamlessly blend into the bursty data transfer behaviour commonly seen in the IoT setting, making them apt for supporting distributed and dynamic IoT processes.

10.8.1.3 Hashgraph

Hashgraph is a data structure and a consensus algorithm [35] based on virtual voting and exchanging information to reach distributed consensus. It is known for efficacy, low latency, equity, and resistance to bad actors. In the context of IoT and 6G networks, Hashgraph enables rapid and efficient transaction validation, ensuring secure communication and real-time data exchange between IoT devices. It enhances interaction and increases data transfer reliability, which makes this technology well-suited to the requirements of next-generation IoT systems.

10.8.1.4 Halochain

Holochain is an advanced technology invented to allow for distributed applications (dApps) without the need for centralised management. Cryptographic signatures combined with a distributed hash table (DHT) enable every node within the network to maintain its own immutable ledger. Holochain shows great potential to develop edge computing and provide seamless peer-to-peer communication between things in the context of 6G IoT. In IoT environments, this characteristic enables the creation of fault-tolerant, distributed indicators that enhance connectivity and self-sufficiency.

10.8.1.5 Multichain

New generation technology, Multichain, allows businesses to control what parties can access their distributed ledgers and to build and manage private blockchains. For 6G IoT, Multichain guarantees the sharing of data among trusted entities in a secure and permission-based manner to validate data confidentiality and help ensure compliance with the regulation. Since it enables the establishment of presence consensus processes and data access restrictions particular to individual IoT use cases, it is the ultimate answer for secure and controlled IoT implementations.

Different types of DLTs have their own pros and cons regarding scalability, security, decentralisation, and transaction speed. Through the proper selection of DLT, 6G can gainfully integrate IoT devices and decades of innovations and operations through the DLT pervasively to enable new applications and services. This integration ensures improved connectivity and enhanced reliability, paving the way for a more connected and resilient digital ecosystem.

10.8.2 Applications of DLT in IoT

DLT has a variety of use cases and advantages in IoT environments for 6G networks. Some of the fundamental applications and advantages of DLT in relation to IoT are the following:

10.8.2.1 Secure Data Sharing

DLT creates an immutable and trusted system for the sharing of data between IoT devices. DLT keeps communication in the network secure and reliable, utilising cryptographic methods and decentralised consensus algorithms. This capability is key to such applications dealing with sensitive data, including industrial automation, smart cities, and health care.

10.8.2.2 Supply Chain Management

DLT enables fully visible and traceable supply chains by tracking the entire lifecycle of a product from field to fork. IoT devices with sensors can easily capture and code real-time data related to the flow and condition of goods, which is securely stored on the distributed ledger. This transparency benefits companies with lower chances of fraud, greater overall efficiency in supply chain operations, and enhanced accountability.

10.8.2.3 Smart Contracts

DLT platforms like blockchain support self-executing contracts with defined conditions. Smart contracts can be defined and executed in each IoT ecosystem, eliminating the need for middlemen and facilitating the enforcement of business standards. In a smart grid setup, IoT devices, for instance, can conduct energy trading contracts automatically per set parameters, allowing for managed resources and reduced operational costs.

10.8.2.4 Access Control

By decentralising identity, DLT-based solutions make it significantly harder for an attacker to impersonate a device or authenticate an attack. Each device is given a unique cryptographic identity that is securely stored on the distributed ledger for secure and immutable authentication. IoT networks are usually not secure, and a strong security procedure greatly minimises unauthorised access and identity spoofing.

10.8.2.5 Data Monetisation

DLT enables the rise of decentralised autonomous organisations (DAOs), which act independently of traditional structures, governed by smart contracts and community consensus. In IoT ecosystems, DAOs are able to govern autonomous IoT devices and, hence, decentralise decision-making and resource management. Such a pattern increases scalability, resilience, and flexibility in IoT applications.

10.8.2.6 DAOs

DLT ensures data integrity and transparency by maintaining an immutable and publicly accessible record of each transaction and data exchange. It is particularly critical in industries like health care and finance that have high standards. Maintaining data integrity to comply with audit requirements is a critical necessity in these industries.

10.8.2.7 Data Integrity and Auditability

DLT provides data integrity and transparency by maintaining an immutable and readily accessible record of all transactions and data exchanges. This skill is particularly crucial in fields such as health care and finance that have rigorous standards. Maintaining data integrity and audit compliance is critical for trust in these sectors. DLT revolutionises asset tracking and transactions by allowing data to be spread across multiple network nodes. DLT is decentralised when it comes to data management and storage than in traditional databases. DLT protocols allow updates to records on distributed ledgers through concurrent reads and validations. This decentralised method statically improves transparency and resilience, which is based on a network of connected entities, venues, or nodes. A distributed ledger is a decentralised database, where there is a version of the distributed ledger running on each node. Each node processes and validates transactions on its own. Transactions serve as a single source of truth. This method helps participants reach consensus on the accuracy and validity of the data. Static and dynamic data can be stored on distributed ledgers: device-level data, or static data, like registries, and dynamic data, like financial transactions. Although blockchain is probably the best-known implementation of a DLT, other DLTs also offer transaction security and integrity.

10.9 AUTHENTICATION AS A THREAT MITIGATION TECHNIQUE

This part explores several authentication mechanisms utilised in 6G cellular networks. These techniques are classified into eight types: handover authentication, mutual authentication, physical layer authentication, deniable authentication (DA), token-based authentication, certificate-based authentication, privacy-preserving key agreement, and MFA.

10.9.1 Handover Authentication

As the number of mobile devices grows and network technologies evolve, cellular data traffic has increased. The secure and smooth transition of devices within a cellular network is mainly reliant on authentication during the handover phase. However, building an effective handover authentication scheme is a substantial task due to the limited computing capacity of devices and the fragility of wireless networks [36,37].

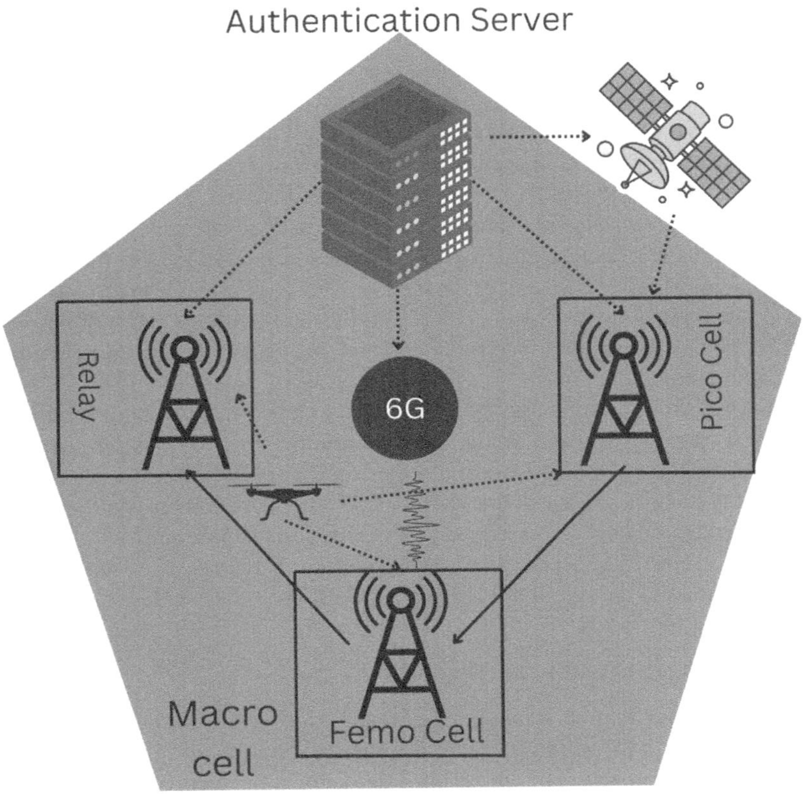

FIGURE 10.6 Handover authentication.

A strong and dependable handover authentication mechanism is required to maintain security in the highly mobile and interconnected framework of next-generation networks. Handover authentication is seen as a critical component of both security and mobility in atypical network configurations, such as terrestrial satellite systems. The 6G network is expected to serve as a comprehensive, integrated network capable of supporting applications on terrestrial satellite networks.

Handover authentication is a complicated task, particularly in multi-domain contexts, as shown in Figure 10.6.

The authors of [38] introduced a lightweight clustering and game-theory-based handover decision system, together with authenticated ground mobility management configurations, for 6G-integrated mega satellite constellations. This technique reduces handover delays, signalling overhead, and allows for faster convergence. However, it does not address issues such as handover in higher-level constellations, inter-layer management, or management structure optimisation.

In [39], the authors proposed a multi-connectivity architecture to improve handover efficiency and increase coverage in fading signal conditions. The highly dynamic and complex structure of 6G networks necessitates frequent handovers.

10.9.2 Mutual Authentication

Research into lightweight cryptographic systems for mutual authentication and key agreement in resource-constrained IoT devices is ongoing. Authors in [40] proposed a random Hash-Based Message Authentication Code (HMAC) and Elliptic Curve Cryptography (ECC)-based mutual authentication system for device to device (D2D). This technique preserves message validity with a 7.7 times delay and complexity reduction. It is also resistant to a wide range of adversarial attacks, including free-riding attacks. These techniques have attracted interest due to their efficiency in processing, as well as strong resistance to advanced attacks such as free-riding attacks and their countermeasures. This, however, facilitates the mobile devices' security policies, which continue to be complex because of their erratic actions and evolving situational demands.

Therefore, Mutual Authentication is a good practice in managing security policies. Based on fingerprints, a mutual authentication technique was proposed in [41] and a MAC Address for Safe Handshake Between Mobile Nodes.

In [42], the authors suggested a client-server key management strategy that uses bilinear pairing and ECC. This method efficiently resists most network attacks while having low overhead and computational expenses. However, it is subject to brute force attacks due to the use of a fixed curve in ECC and SHA-1 flaws.

10.9.3 Physical Layer Authentication

While most authentication procedures are applied at levels higher than the physical communication layer, physical layer authentication has developed as a distinct and distinctive approach. This method of authentication involves superimposing a properly crafted secret modulation into waveforms. Physical layer authentication has acquired prominence due to its robustness against interference and limited need for bandwidth [43].

In [44], the authors proposed an adaptive machine learning-based intelligent physical layer authentication technique for improving authentication in dynamic contexts. This method incorporates machine learning into the physical layer via a fusion model based on physical properties that uses a kernel machine. PLS solutions have evolved into an efficient framework for decreasing complexity, eliminating delays, and assuring a light footprint within context-aware security paradigms.

10.9.4 DA

DA protocols enable the sender to deny the authentication process to any third party. These methods are commonly used to avoid the "Encryption-then-MAC" strategy by preventing entities from commencing the MAC-based authentication procedure [45]. In [46], the authors presented a source-hiding approach based on projective hash functions. This approach improves Wi-Fi authentication by prohibiting third-party connections from being made. In [47], the authors presented a formal security argument for DA based on bilinear pairing and identity-based encryption. Group Key Agreement (GKA) protocols also use DA in fog computing-enabled vehicle networks, where the environment's high mobility and wireless connectivity render key parameters particularly vulnerable.

In [48], the authors presented a random oracle approach for authentication in social media network communication. This technique uses a single logical step, which considerably reduces both ciphertext length and computational cost. It is vital to note that there is little study on the implementation, solutions, and challenges of DA in 6G networks.

10.9.5 Token-Based Authentication

Multi-access Edge Computing (MEC) technology is tightly linked with token-based access control management in next-generation networks. The authors of [49] presented a token-based authentication mechanism using JavaScript Object Notation (JSON) Web Tokens.

Blockchain and token-based authentication provide a secure and transparent orchestration approach for multi-domain SDN architecture in 6G networks [50].

In [51], the authors proposed using SHA-256-based cyclic keys for narrowband IoTs. Currently, many organisations are adopting blockchain-based intelligent mechanisms to detect and monitor malicious behaviour within IoT networks.

A token-based system provides combined authentication and access control for resource-constrained, blockchain-enabled IoT devices in 6G networks. However, these token-based authentication methods are vulnerable to prediction and theft attempts. As a result, it is critical to explicitly validate the security of the token production and distribution procedures.

In [52], the authors proposed an ECC-enabled blockchain for efficient and secure data flow across businesses. However, the system is vulnerable to ECC-specific concerns, such as fixed curve flaws and potential quantum attacks.

10.9.6 Certificate-Based Authentication

A certificate-based authentication technique employs a variety of cryptographic algorithms and handshake protocols. This method is frequently used in current apps and smart infrastructure to ensure safe authentication and verification.

In [53], the authors suggested a lightweight cryptography-based certificate authentication technique that guarantees anonymity and untraceability. This technique protects against a variety of attacks, including MITM, DoS, impersonation, and replay attacks.

Certificate Authority (CA)-based schemes may become obsolete in the future as decentralised blockchain certificates (Bcert) gain popularity. In these systems, the devices produce their own blockchain certificates, which are then frequently updated with new certificates while retaining the same identification information.

10.9.7 Key Agreement with Privacy

Many authentication systems rely heavily on key agreement. Over time, many hierarchical key agreement systems have emerged to enable privacy-protected interaction among gateways, sensors, and users. It is also an essential component in numerous cryptographic algorithm-based security measures. However, the crucial agreement procedure is not without flaws. Some of the most prevalent dangers are protocol replay attacks, key reuse, deniability concerns, signature tampering, key compromise, server trust vulnerabilities, and identity manipulation. These vulnerabilities underscore the importance of strong and secure key agreement mechanisms to protect communication in 6G and beyond.

Misuse of biometric credentials, as well as the possibility of privacy abuse, can have serious, and often irrevocable, implications. To reduce these dangers, it is critical to use dynamic, privacy-protected key agreement mechanisms. This means that biometric data are stored with appropriate procedures that guarantee them has limited access and are difficult to misuse, thus securing the authentication process.

Focusing on privacy and uncloneability, a new candidate key game is proposed. The solution relies on a one-way hash function to enable a secure two-part communication scheme between the smart metres with limited resources and the grid management system. The technique enhances privacy without resorting to copying authentication credentials, making it suitable for use cases where safety and resource limitations are dominant factors.

Key management becomes more complex such that the density in interconnectivity and emergence of network expansion in modern communication systems increase. The complexities of managing multiple devices and secure communication channels necessitate more effective and scalable approaches to key management.

10.9.8 MFA

MFA is a critical component of an effective Identity and Access Management (IAM) architecture. It increases security by including numerous verification mechanisms, making it less vulnerable to brute force assaults and the theft of third-party information. MFA uses a variety of authentication factors, including one-time passwords (OTPs) provided by SMS or email, software tokens, mobile apps, biometric information, third-party keys like Gmail, USB keys, smart cards, PINs, RFID tags, physical keys, and geographically based identifiers.

[53] MFA faces numerous risks, including MITM attacks, reverse brute force, credential stuffing, keyloggers, and spear-phishing or phishing efforts. As 6G networks expand into huge and diversified communication systems, MFA is becoming more urgent to protect against new dangers such as quantum computing-based cryptographic assaults.

The diversity and interconnectedness of modern networks have created several chances for cross-verification purposes in MFA.

10.10 CONCLUSION

The integration of 6G networks with IoT devices represents a revolutionary step forward in digital connectivity, characterised by ultra-fast speeds, extremely low latency, and seamless accessibility. This technological leap offers unprecedented opportunities, enabling innovations like autonomous vehicles, immersive virtual experiences, and next-generation healthcare advancements. Despite these promising prospects, it introduces a highly intricate landscape of cybersecurity challenges that demand a significant overhaul in traditional security approaches. The decentralised nature of 6G networks, combined with the rapid growth of IoT devices, dramatically increases the potential for cyberattacks, creating a vast and complex attack surface. This expansion makes the network more susceptible to advanced threats such as AI-driven exploits, APTs, and vulnerabilities inherent in distributed systems. Tackling these challenges demands comprehensive security measures that emphasise safeguarding data confidentiality, ensuring integrity, and maintaining availability. It also calls for proactive strategies to counter risks like MITM attacks, data tampering, and DDoS assaults.

The present solutions, like cryptographic methods, intrusion detection systems, and DLTs such as blockchain, provide robust security by enabling decentralised validation and fostering transparency. AI and machine learning serve as crucial tools for detecting threats in real time, identifying anomalies, and implementing adaptive security protocols. Additionally, advanced authentication techniques, including MFA, physical layer verification, and token-based systems, are indispensable for ensuring secure communication in the highly dynamic landscape of 6G networks.

In the future, the role of AI in threat management should advance to incorporate self-learning and adaptive algorithms designed to counter increasingly complex and evolving cyber threats. With quantum computing on the horizon, implementing quantum-resistant cryptographic methods will become essential for safeguarding sensitive information within 6G networks. Strengthening the security of edge computing—a cornerstone of decentralised systems—will necessitate the adoption of efficient encryption techniques and reliable firmware update mechanisms to address potential vulnerabilities.

The establishment of comprehensive global security standards and frameworks is crucial for addressing vulnerabilities and ensuring seamless interoperability across 6G networks. Proactive threat-hunting strategies, driven by advanced analytics and real-time intelligence sharing, will play a key role in fortifying security measures. Moreover, the development of sustainable IoT systems that harmonise strong security protocols with energy-efficient practices is essential, fostering the adoption of environmentally conscious and future-ready technologies.

Effective collaboration between industries, governments, and academic institutions is vital to combine resources, exchange threat intelligence, and drive innovation in cybersecurity. The potential of 6G-enabled IoT networks can only be realised through the adoption of well-rounded, adaptive, and cooperative security frameworks and by strengthening the end users against all types of cyber threats. By leveraging advanced technologies and embracing progressive strategies, a secure, resilient, and sustainable digital ecosystem that meets the demands of the future can be built.

REFERENCES

1. Gao, P., Shao, F., Liu, X., Xiao, X., Qin, Z., Xu, F., Mittal, P., Kulkarni, S.R. and Song, D., 2021. Enabling efficient cyber threat hunting with cyber threat intelligence. In *2021 IEEE 37th International Conference on Data Engineering (ICDE)*, Chania, Greece (pp. 193–204). IEEE. doi: 10.1109/ICDE51399.2021.00024.

2. Nour, B., Pourzandi, M. and Debbabi, M., 2023. A survey on threat hunting in enterprise networks. *IEEE Communications Surveys & Tutorials*, *25*(4), pp.2299–2324. doi: 10.1109/COMST.2023.3299519.
3. Ma, Y., Chen, X., Feng, W. and Ge, N., 2022. DDoS detection for 6G Internet of Things: Spatial-temporal trust model and new architecture. *China Communications*, *19*(5), pp.141–149. doi: 10.23919/JCC.2022.05.002.
4. Sodhro, A.H., Pirbhulal, S., Luo, Z., Muhammad, K. and Zahid, N.Z., 2021. Toward 6G architecture for energy-efficient communication in IoT-enabled smart automation systems. *IEEE Internet of Things Journal*, *8*(7), pp.5141–5148. doi: 10.1109/JIOT.2020.3024715.
5. Mahmood, M.R., Matin, M.A., Sarigiannidis, P. and Goudos, S.K., 2022, A comprehensive review on artificial intelligence/machine learning algorithms for empowering the future IoT toward 6G era. *IEEE Access*, *10*, pp.87535–87562. doi: 10.1109/ACCESS.2022.3199689.
6. Ferrag, M.A., Friha, O., Kantarci, B., Tihanyi, N., Cordeiro, L., Debbah, M., Hamouda, D., Al-Hawawreh, M. and Choo, K.K.R., 2023. Edge learning for 6G-enabled Internet of Things: A comprehensive survey of vulnerabilities, datasets, and defenses. *IEEE Communications Surveys & Tutorials*, *25*(4), pp.2654–2713.
7. Al-Mohammed, H.A. and Yaacoub, E., 2021. On the use of quantum communications for securing IoT devices in the 6G era. In *2021 IEEE International Conference on Communications Workshops (ICC Workshops)*, Montreal, QC, Canada (pp. 1–6). doi: 10.1109/ICCWorkshops50388.2021.9473793.
8. Saeed, M.M., Saeed, R.A., Abdelhaq, M., Alsaqour, R., Hasan, M.K. and Mokhtar, R.A., 2023. Anomaly detection in 6G networks using machine learning methods. *Electronics*, *12*(15), p.3300.
9. Abdel Hakeem, S.A., Hussein, H.H. and Kim, H., 2022. Security requirements and challenges of 6G technologies and applications. *Sensors*, *22*(5), p.1969.
10. Ekong, M.O., George, W.K., Pandey, B.K. and Pandey, D., 2023. Enhancing the fundamentals of industrial safety management in TVET for metaverse realities. In *Applications of Neuromarketing in the Metaverse* (pp. 19–41). IGI Global.
11. Kaur, G., Bharathiraja, N., Singh, K. D., Veeramanickam, M. R. M., Rodriguez, C. R., & Pradeepa, K., 2024. Emerging trends in cybersecurity challenges with reference to pen testing tools in Society 5.0. *Artificial Intelligence and Society 5.0*, 196–212.
12. Neto, A.J.H. and dos Santos, A.F.P., 2020. Cyber threat hunting through automated hypothesis and multi-criteria decision making. In *2020 IEEE International Conference on Big Data (Big Data)*, Atlanta, GA, USA (pp. 1823–1830). doi: 10.1109/BigData50022.2020.9378213.
13. Shaukat, K., Luo, S., Chen, S. and Liu, D., 2020, October. Cyber threat detection using machine learning techniques: A performance evaluation perspective. In *2020 International Conference on Cyber Warfare and Security (ICCWS)*, Islamabad, Pakistan (pp. 1–6). IEEE.
14. Islam, C., Babar, M.A., Croft, R. and Janicke, H., 2022. SmartValidator: A framework for automatic identification and classification of cyber threat data. *Journal of Network and Computer Applications*, *202*, p.103370.
15. Jiang, H., Zhu, X. and Han, J., 2022. Instruction-fetching attack and practice in collision fault attack on AES. *Symmetry*, *14*(10), p.2201.
16. Al Mousa, A., Al Qomri, M., Al Hajri, S. and Zagrouba, R., 2020, October. Utilizing the eSIM for public key cryptography: A network security solution for 6G. In *2020 2nd International Conference on Computer and Information Sciences (ICCIS)* (pp. 1–6). IEEE.
17. Shen, X.S., Liu, D., Huang, C., Xue, L., Yin, H., Zhuang, W., Sun, R. and Ying, B., 2022. Blockchain for transparent data management toward 6G. *Engineering*, *8*, pp.74–85.

18. Partala, J., 2021. Post-quantum cryptography in 6G. *6G Mobile Wireless Networks, 1*, pp.431–448.
19. Li, G., Lai, C., Lu, R. and Zheng, D., 2021. SecCDV: A security reference architecture for cybertwin-driven 6G V2X. *IEEE Transactions on Vehicular Technology, 71*(5), pp.4535–4550.
20. Yu, Q., Ren, J., Zhou, H. and Zhang, W., 2020, March. A cybertwin based network architecture for 6G. In *2020 2nd 6G Wireless Summit (6G SUMMIT)*, Levi, Finland (pp. 1–5). IEEE.
21. Chorti, A., Barreto, A.N., Köpsell, S., Zoli, M., Chafii, M., Sehier, P., Fettweis, G. and Poor, H.V., 2022. Context-aware security for 6G wireless: The role of physical layer security. *IEEE Communications Standards Magazine, 6*(1), pp.102–108.
22. Ong, A.V. and Peradilla, M., 2021, August. An IoT framework based on SDN and NFV for context-aware security. In *2021 Twelfth International Conference on Ubiquitous and Future Networks (ICUFN)*, Jeju Island, South Korea (pp. 167–172). IEEE.
23. Kazmi, S.H.A., Qamar, F., Hassan, R. and Nisar, K., 2023. Routing-based interference mitigation in SDN enabled beyond 5G communication networks: A comprehensive survey. *IEEE Access, 11*, pp.4023–4041.
24. Qin, P., Zhu, Y., Zhao, X., Feng, X., Liu, J. and Zhou, Z., 2021. Joint 3D-location planning and resource allocation for XAPS-enabled C-NOMA in 6G heterogeneous Internet of Things. *IEEE Transactions on Vehicular Technology, 70*(10), pp.10594–10609.
25. Muniyandi, R.C., Qamar, F. and Jasim, A.N., 2020. Genetic optimized location aided routing protocol for VANET based on rectangular estimation of position. *Applied Sciences, 10*(17), p.5759.
26. Zhang, L., Zhong, W., Zhang, J., Lin, Z., Yang, Z. and Wang, J., 2022. mmWave beam tracking for V2I Communication systems based on spectrum environment awareness. *Symmetry, 14*(4), p.677.
27. Ibrahim, M., Badrudduza, A.S.M., Hossen, M.S., Kundu, M.K. and Ansari, I.S., 2021. Enhancing security of TAS/MRC-based mixed RF-UOWC system with induced underwater turbulence effect. *IEEE Systems Journal, 16*(4), pp.5584–5595.
28. Lanoue, M., Bollmann, C.A., Michael, J.B., Roth, J. and Wijesekera, D., 2021. An attack vector taxonomy for mobile telephony security vulnerabilities. *Computer, 54*(04), pp.76–84.
29. Lu, K. and Yang, H., 2021, September. Design of NOMA sparse signature matrix for 6G integrating sensing and communications networks. In *2021 IEEE 94th Vehicular Technology Conference (VTC2021-Fall)*, Helsinki, Finland (pp. 1–5). IEEE.
30. Soni, G. and Chandravanshi, K., 2021, August. Security scheme to identify malicious maneuver of flooding attack for WSN in 6G. In *2021 8th International Conference on Signal Processing and Integrated Networks (SPIN)*, Noida, India (pp. 124–129). IEEE.
31. Srinivasan, M., Skaperas, S. and Chorti, A., 2021, November. On the use of CSI for the generation of RF fingerprints and secret keys. In *WSA 2021; 25th International ITG Workshop on Smart Antennas*, France (pp. 1–5). VDE.
32. Thakar, C., Saxena, R., Himabindu, M., Rakesh, C., Dutt, A. and Dhanraj, J.A., 2024. Distributed ledger technology in 6G management strategies for threat mitigation. In Monika Gupta, Kumar Shalender, Babita Singla, Nripendra Singh (eds) *Security Issues and Solutions in 6G Communications and Beyond* (pp. 160–176). IGI Global.
33. Swapna, H.R., Bigirimana, E., Madaan, G., Hasan, A., Pandey, B.K. and Pandey, D., 2023. Impact of neuromarketing on consumer psychology in digitally connected networks. In Monika Gupta, Kumar Shalender, Babita Singla, Nripendra Singh (eds) *Applications of Neuromarketing in the Metaverse* (pp. 193–205). IGI Global.
34. George, W.K., Ekong, M.O., Pandey, D. and Pandey, B.K., 2023. Pedagogy for implementation of TVET curriculum for the digital world. In *Applications of Neuromarketing in the Metaverse* (pp. 117–136). IGI Global.

35. Bessant, Y.A., Jency, J.G., Sagayam, K.M., Jone, A.A.A., Pandey, D. and Pandey, B.K., 2023. Improved parallel matrix multiplication using Strassen and Urdhvatiryagbhyam method. *CCF Transactions on High Performance Computing*, 5(2), pp.102–115.
36. Hindia, M.N., Qamar, F., Majed, M.B., Abd Rahman, T. and Amiri, I.S., 2019. Enabling remote-control for the power sub-stations over LTE-A networks. *Telecommunication Systems*, *70*, pp.37–53.
37. Yap, K.Y., Chin, H.H. and Klemeš, J.J., 2022. Future outlook on 6G technology for renewable energy sources (RES). *Renewable and Sustainable Energy Reviews*, *167*, p.112722.
38. Ji, S., Sheng, M., Zhou, D., Bai, W., Cao, Q. and Li, J., 2021. Flexible and distributed mobility management for integrated terrestrial-satellite networks: Challenges, architectures, and approaches. *IEEE Network*, *35*(4), pp.73–81.
39. Özkoç, M.F., Koutsaftis, A., Kumar, R., Liu, P. and Panwar, S.S., 2021. The impact of multi-connectivity and handover constraints on millimeter wave and terahertz cellular networks. *IEEE Journal on Selected Areas in Communications*, *39*(6), pp.1833–1853.
40. Chow, M.C. and Ma, M., 2021. A lightweight traceable D2D authentication and key agreement scheme in 5G cellular networks. *Computers and Electrical Engineering*, *95*, p.107375.
41. Bairwa, A.K. and Joshi, S., 2021. Mutual authentication of nodes using session token with fingerprint and MAC address validation. *Egyptian Informatics Journal*, *22*(4), pp.479–491.
42. Yang, L., Chen, Y.C. and Wu, T.Y., 2021. Provably secure client-server key management scheme in 5G networks. *Wireless Communications and Mobile Computing*, *2021*(1), p.4083199.
43. Hindia, M.N., Qamar, F., Abbas, T., Dimyati, K., Abu Talip, M.S. and Amiri, I.S., 2019. Interference cancelation for high-density fifth-generation relaying network using stochastic geometrical approach. *International Journal of Distributed Sensor Networks*, *15*(7), p.1550147719855879.
44. Bahache, A.N., Chikouche, N. and Mezrag, F., 2022. Authentication schemes for healthcare applications using wireless medical sensor networks: A survey. *SN Computer Science*, *3*(5), p.382.
45. Vashistha, A., Alghamdi, T.A., Almalki, S.A. et al. 2025. AI-driven intrusion prevention system: leveraging mathematical foundations for advanced threats detection. *International Journal of Information Technology*. https://doi.org/10.1007/s41870-025-02543-8.
46. Kazmi, S.H.A., Hassan, R., Qamar, F., Nisar, K. and Ibrahim, A.A.A., 2023. Security concepts in emerging 6G communication: Threats, countermeasures, authentication techniques and research directions. *Symmetry*, *15*(6), p.1147.
47. Li, F., Hong, J. and Omala, A.A., 2018. Practical deniable authentication for pervasive computing environments. *Wireless Networks*, *24*, pp.139–149.
48. Huang, W., Liao, Y., Zhou, S. and Chen, H., 2019. An efficient deniable authenticated encryption scheme for privacy protection. *IEEE Access*, *7*, pp.43453–43461.
49. Niewolski, W., Nowak, T.W., Sepczuk, M. and Kotulski, Z., 2021. Token-based authentication framework for 5G MEC mobile networks. *Electronics*, *10*(14), p.1724.
50. Balachandran, C., Ramachandran, G. and Krishnamachari, B., 2020, November. EDISON: A blockchain-based secure and auditable orchestration framework for multi-domain software defined networks. In *2020 IEEE International Conference on Blockchain (Blockchain),* Rhodes Island, Greece (pp. 144–153). IEEE.
51. Pothumarti, R., Jain, K. and Krishnan, P., 2021. A lightweight authentication scheme for 5G mobile communications: A dynamic key approach. *Journal of Ambient Intelligence and Humanized Computing*, pp.1–19.

52. Singh, P., Masud, M., Hossain, M.S. and Kaur, A., 2021. Cross-domain secure data sharing using blockchain for industrial IoT. *Journal of Parallel and Distributed Computing, 156*, pp.176–184.
53. Nyangaresi, V.O. and Ogundoyin, S.O., 2021, October. Certificate based authentication scheme for smart homes. In *2021 3rd Global Power, Energy and Communication Conference (GPECOM),* Antalya, Turkey (pp. 202–207). IEEE.

11 Securing 6G-Enabled IoT Devices

Smart Attack Detection System

Chetna Sharma, Aryan Gupta, and Jivansh Midha

11.1 THE 6G REVOLUTION: OPPORTUNITIES AND THREATS

6G technology is changing the face of communication between devices, enabling fast, reliable, and secure connections. A technology platform handling billions of connections for wearable to critical operations will introduce revolutionary changes into the smart city, healthcare model of services, and usage patterns of transport, and much more. Security issues will have major consequences with the immense growth of devices and the massive flow of data. With the continuously rising complexity of IoT networks, the classical, centralised, and static security model would soon fail to be on par with emerging new cyber threats and fast-rising vulnerabilities [1].

The SADS provided a more forward-looking look at what could be a panacea toward the solution of security offered by a decentralised approach when unique challenges brought by the adoption of the 6G are addressed by using leading and emerging edge technologies that relate to real-time machine learning, edge, blockchain, and federated learning [2]. By having the right configuration of sensors within different sectors of security, 6G network complexity in evolving changes to a device landscape with varying forms of device interconnectivity can easily keep up. Figure 11.1 shows a schematic framework for the SADS that depicts all necessary building blocks cooperating to secure 6G-enabled IoT networks, such as IoT devices, edge computing, machine learning, blockchain, federated learning, and centralised monitoring.

11.1.1 The Need for Adaptive Security in 6G Networks

With 6G networks at the core of future digital infrastructure, it will bring interconnectivity to unprecedented heights, enabling IoT, from the smallest wearable personal devices to vital infrastructure components that can all connect seamlessly throughout the network. There is a fast growth of devices, but not without a lot of headaches, such as the huge number of connected objects leading to a dynamic and constantly changing environment, rendering most static and centralised traditional security protocols quite ineffective in their role. These gaps must be filled

DOI: 10.1201/9781003586456-11

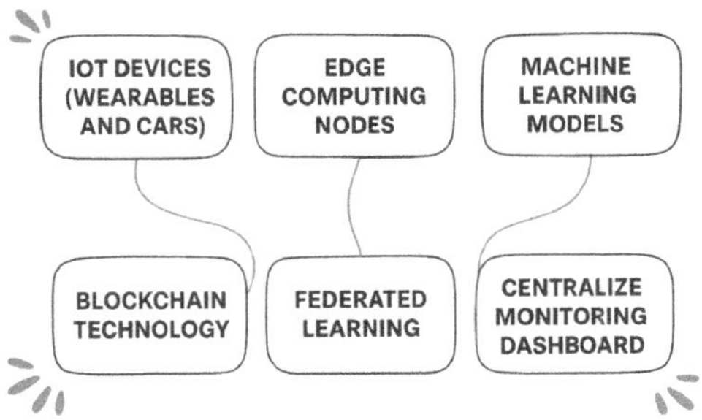

FIGURE 11.1 SADS framework.

by building security into the very fabric of the 6G network, where mechanisms are built to handle not only the volume of data but also the complexity of threats that will emerge in real time.

11.1.2 SADS: A Decentralised and Adaptive Security Solution

Thus, SADS promises a transformative security model that is adaptive and scalable. Centralised solutions of security are not working today in this world where data is spawned and processed at the edge in distributed locations. The SADS shifts from centralised, reactive detection of a threat towards a more proactive and decentralised approach [2]. The system avoids being confined by the bounds of pre-defined threat signatures, and it uses machine learning models that can identify anomalous activity based on historical and real-time data; hence, it detects zero-day attacks or novel attack methods that have not yet been encountered. It provides much-needed defence against changing threats.

11.1.3 Real-Time Threat Detection with Edge Computing

With the introduction of edge computing in SADS, therefore, it will allow processing security data immediately at or near the origin of said data at either the level of the device or the network's edge. This hence reduces the chances of loss of such data or delay of reaction time in applications when time is critical. For instance, even a fraction of a second lag in self-driving car detection of security threats could be disastrous. SADS would ensure that the threats are detected and mitigated before the full-scale attacks occur, with the real-time processing of these threats. Such localisation of processing also enhances privacy because sensitive data need not travel long distances to a central server for analysis, thus lowering the probability that it is intercepted en route to the server.

11.1.4 Blockchain for Transparency and Accountability

With blockchain technology, SADS ensures further strengthening of the security posture of the system by being transparent and accountable. The immutable ledger that blockchain creates is a verifiable record of all actions taken within the network, from data exchanges to security interventions. This is especially valuable in industries that are highly regulated and audit-intensive, such as healthcare and finance [3]. The ability of blockchain to track and verify data interactions transparently in a very meticulous manner enhances trust within the system, both among users and across devices. In a world facilitated by 6G technology, where the integrity of data transmission is important, a system that guarantees immutable records can assure both stakeholders and regulatory entities.

11.1.5 Federated Learning: Balancing Privacy and Security

Federated learning in the SADS framework addresses one of the biggest concerns of the 6G era: that is, how to balance security with privacy [2]. In the 6G era, given the exponential growth in the amount of data being generated through IoT devices, it becomes increasingly difficult to guarantee confidentiality to individual data points. The system can aggregate insights from multiple devices without even compromising the privacy of the data itself through federated learning. Devices can learn directly from their local data to improve the overall capabilities of the system in detecting threats. Only the necessary model updates are shared with the central server. This decentralised approach will allow SADS to learn, grow smarter, and get more efficient over time, without exposing sensitive user data to unnecessary risk.

11.1.6 Scalability and Flexibility in Evolving Threat Landscapes

As the IoT networks grow in size and complexity, adaptive security solutions become increasingly urgent. On the other hand, a static security system that is not able to adapt itself to the dynamically changing threat landscape gets outdated very fast. With its AI, edge computing, blockchain, and federated learning altogether, SADS provides a highly flexible and scalable solution that evolves as the IoT ecosystem evolves. This adaptability capacity ensures that measures adopted for security remain effective with the rising number of connected gadgets, as well as new approaches in cyberattacks on emerging complex networks.

11.1.7 Enhancing Resilience through Threat Containment

While being very important in offering the aspect of real-time threat detection and response, SADS can be one component in improving the IoT overall system's robustness. During a particular attack, SADS also contributes to damage limitation in isolating devices that are compromised in terms of preventing the further propagation of this contamination across other parts of the network. This containment ability is particularly critical in decentralised IoT systems in which devices are usually found dispersed and may not be seen as having direct oversight by centralised management systems. In general, SADS quickly identifies a source and the nature of the threat, thus helping limit the lateral movement of attacks around the network, further minimising the scope and consequences of security incidents.

11.1.8 Enabling Safe Adoption of 6G-Driven Innovations

As different fields consider taking up the detailed potentials provided by 6G networks, security will become an important factor in bringing all the emerging technologies together under its umbrella in a safe way. For instance, take smart cities where infrastructure is all interconnected and dependent on information transmitted instantaneously; each vulnerability in security would affect most functionalities and activities like traffic handling and power distribution. It helps create another layer of safety, ensuring that the vast volumes of data crossing urban networks will indeed be safe, and that any potential disruptions will be addressed at the earliest opportunity [1]. Health care, as always perceived, deems obligatory that information safety and functioning medical equipment be assured: SADS can help save vital information while making smooth exchange, in real time, happen.

11.1.9 Fostering Innovation by Removing Security Barriers

This is a very notable characteristic of SADS since it removes security worries that would otherwise delay the rollout of new IoT technologies. Since organisations and industries target tapping the full potential available through 6G networks, a security architecture is needed that responds to the needs of organisations and remains protective of confidential information. SADS allows the organisations to feel safe enough to experiment with novel uses such as smart agriculture, precision medicine, or a fully automated supply chain since installing the advanced security mechanisms right in the IoT infrastructure provides that type of environment where nothing frightens cyberattacks to subdue.

11.1.10 The Growing Importance of SADS in Securing 6G IoT Networks

Looking forward, the importance of SADS will increase in securing the landscape of the IoT only as 6G networks give way to more efficient interconnections. With real-time analytics, decentralised intelligence, and transparent, tamper-proof systems, SADS offers a comprehensive solution to the challenges of cybersecurity related to the next generation of IoT [2]. Thus, with the proper, reliable, proactive defence against a variety of threats, SADS ensures that every industry can have all the benefits of 6G connectivity while minimising all the risks of an increasingly complex, connected digital world. The IoT ecosystem will continually grow.

11.2 DARK SIDE OF CONNECTIVITY: EMERGING RISKS

The advent of 6G networks promises constant connection and instant data sharing. This, therefore, will transform industries and everyday life. The smart city concept could change with improvements in healthcare, transportation, and manufacturing, but those improvements bring along crucial cybersecurity risks. The same features that make 6G networks so transformative—billions of connected devices, decentralised systems, and hyper-fast communication—also make them highly vulnerable to cyberattacks. The rapid expansion of the IoT, coupled with the complexity of 6G networks, introduces new threats that could have far-reaching, catastrophic consequences [4]. This article considers the risk of the new technology offered through 6G and outlines why flexible, layered security may represent an important need within that connected world of tomorrow [6].

11.2.1 Expanding Attack Surface in 6G Networks

The biggest concern regarding 6G-enabled IoT networks is the enormous rise of possible attacks [8]. When billions of devices, like personal wearables and crucial infrastructure parts, connect with each other, the ways through which cybercriminals can break in increase much. Traditional networks can focus security efforts on main areas, whereas 6G networks depend on computing that is distributed and done at the edge [14–16]. This creates many weaknesses in a large network of devices, with each device being a possible target for bad people. The large number of connections, along with quick interactions, greatly raises the chance of serious security problems [31].

11.2.2 Vulnerabilities in IoT Devices: A Weak Link in the Chain

Most of the IoT devices, especially those implemented in resource-poor environments, were designed without security. The products that are implemented are medical appliances and industrial machinery. These devices lack proper protection which ensures that the tool would be able to withstand highly advanced cyberattacks. Therefore, the dangers make such tools very vulnerable to attacks including Distributed Denial of Service (DDoS), ransomware, and data theft. If these appliances are hacked, they can let out personal information, endanger lives, and interfere with important services. For instance, an attack on medical devices such as pacemakers or insulin pumps may endanger people's lives gravely, while attacks on industrial systems may be linked to power outages, loss of money, or danger to the environment.

11.2.3 Decentralised Networks: Amplifying Security Risks

This is a new threat and risk posed by 6G networks to a more decentralised setup through edge computing. Because data is processed where it resides, this takes longer to process and is efficient, but it pushes the number of potential points of attack further out. Because edge devices are not as secure as the data centre, they may be hacked one by one or in mass as standard-based protections bypass these devices. A problem at the edge could cause a series of failures across the network, which would make it more difficult to spot and stop attacks quickly. The spread out system makes it harder to secure the network since every device and small data centre needs to be protected on its own.

11.2.4 Evolving Security Protocols: The Need for Dynamic, Intelligent Defences

Regular security measures, however, would not make any more work to ensure special security problems regarding 6G's networks of decentralised and edge-based kind. It is not beneficial enough to apply the kinds of barriers applied before: Those are firewalls and intrusion detection systems. Plans to be changed there concern the method of smart monitoring that requires quick responses. Technologies such as AI-driven anomaly detection, machine learning, and blockchain form the basis of managing and responding to the enormous amount of data that 6G networks generate. These new tools can constantly check for strange behaviours of devices and network traffic and find patterns that may hint at an attack.

11.2.5 AI and Machine Learning: A Double-Edged Sword

AI and machine learning help a lot in discovering and reducing cyber threats. These technologies can look at enormous amounts of data very fast, spotting unusual activities and possible threats before they get worse. But AI-based systems have their challenges, too. Cyber criminals are using AI more to hide what they are doing; thus, it becomes more difficult to distinguish between the normal and harmful traffic. Techniques like adversarial AI that manipulate network traffic to pose as normal behaviour add another dimension to threat detection, hence a more complex security landscape to navigate[30].

11.2.6 Blockchain: Enhancing Data Integrity and Transparency

Blockchain technology ensures permanent records of all the transaction and communication within a network because such technology adds extra security, for instance, to an enabled IoT network in the coming 6G system. This openness makes the possibilities of tampering with correct data minimal. Its auditing trail can determine what must have started an attack should there be an occasion involving a security breach in some part of the chain, thus evaluating how extensive is the damage done to data. But the use of blockchain can be challenging for those IoT devices that do not have adequate resources. Some of these devices might have limited processing power for such cryptographic algorithms. The energy consumption of blockchain systems will also be a problem with large-scale 6G projects, especially in locations with fewer resources.

11.2.7 Federated Learning: A Privacy-Preserving Solution with Security Challenges

Federated learning is the latest approach towards more private and safer machine learning models in 6G IoT networks. Thus, devices will be able to train models locally, sending the updates to a central server that can avoid leaks of data. Despite this, federated learning also has some threats. Bad actors may attempt to hurt the learning process by including false or misleading data that may hurt the accuracy and effectiveness of security models. Therefore, it is very important to have further research into strong federated learning methods that will fix such weaknesses and ensure that privacy benefits are achieved without affecting the overall security of the system.

11.2.8 The Role of Regulatory Frameworks in Securing 6G Networks

With the establishment of 6G networks as the base for global digital systems, it will be very important to have rules and regulations for keeping them safe. Rules and guidelines must work together between governments, standard organisations, and industrial groups for clear guidelines in the use of IoT devices and networks like verification of the devices, safe communication techniques, encryption, and continuous software updates. Along with new technology, cybersecurity laws and rules should also change to handle emerging new threats and to ensure that security measures keep pace with the tremendous growth of 6G networks.

11.2.9 A Multifaceted Approach to 6G Security: Balancing Innovation with Protection

The evolving nature of 6G technology makes it especially challenging for the outdated security models. Any defence system less than the basic and fixed one will not suffice to protect the fully connected and decentralised networks. A multi-layered, adaptive security approach, AI-powered through machine learning, blockchain, federated learning, advanced encryption, and more will be needed to prevent and counter 6G IoT attacks. Organisations need to check, review, and update the security plan time and again to stay afloat in the fast-changing threats. This will include steady investment both in technologies and skilled professionals to make sure that the security teams can handle the demands of the 6G environment.

11.3 DDOS ATTACKS: SPEEDING TOWARDS DESTRUCTION

That amount of billions of devices coming to connect with each other will revolutionise businesses as well as societies. At that time, it is going to be very speedy with good efficiency and with full automation. However, along with developing 6G networks for smooth communication for lots of IoT devices, these bring new risks into sight. One major risk of developing 6G networks comes with DDoS. These attacks flood a network with traffic to prevent it from being stopped or its services from being disrupted. They will strengthen and become tougher to mitigate in a 6G system [5]. 6G network design with a decentralised edge computing concept offers many ways to infiltrate the networks for an attacker, hence raising the possibility of damage from DDoS threats. Better security measures, such as the SADS, are very important to reduce these risks in such a situation. This article examines the changing dangers of DDoS attacks in 6G networks and demonstrates how proactive, adaptive security systems help keep connected IoT ecosystems strong.

11.3.1 The Escalating DDoS Threat in 6G Networks

As 6G networks begin to connect more than a billion IoT sensors across different industries, there will be a large number of opportunities for DDoS flooding networks with massive amounts of spoofed traffic, potentially stopping essential services in areas as critical as medicine, self-driving cars, and smart cities. When there is almost no delay or lag in connecting devices in such a world, DDoS attacks are no longer a nuisance but a massive threat to how 6G systems work and their safety.

11.3.2 Decentralised Architecture: Amplifying DDoS Risks

Unlike classical centralised networks, 6G heavily relies on a decentralised, edge computing architecture that performs data processing locally within distributed devices. This architecture reduces latency and improves performance but exposes many vulnerabilities open. Simultaneous attacking of edge nodes can easily cause cascading failures of network segments, resulting in paralysis of critical services, such as hospital equipment, public transportation systems, or the power grid, which may be calamitous [4].

11.3.3 SADS: Proactive Defence against DDoS

It introduces a new security solution: SADS, designed to counter a major threat to 6G networks—the growing threat of DDoS attacks. Machine learning and real-time anomaly detection are used to analyse and monitor IoT devices and examine network traffic for the kinds of anomalies that signal DDoS attacks as they progress. SADS responds to edge computing, meaning that attacks can be responded to before the source spreads and affects a bigger network. Additionally, with the use of blockchain technology, there is a clearer record of all the events within the network. As a result, the traceability and safety of this system increase.

11.3.4 Enhancing Network Resilience with SDN and VNFs

New powerful capabilities with Software-Defined Networking (SDN) and Virtual Network Functions (VNFs) can potentially support 6G increased resilience to DDoS attacks. With SDN, the network administrators gain real-time control over traffic within networks; thus, they can notice unusual patterns of traffic. This gives them the ease of either routing out or filtering away malicious traffic before it arrives at services to be interrupted. On the other hand, VNFs support the speedy deployment of virtual firewalls and intrusion detection systems across multiple edge nodes with security features. Therefore, a combination of SDN and VNF enables the networks to remain more stable and secure in more complex DDoS attacks [4].

11.3.5 The Need for Adaptive Security Architectures in 6G

With the scale and sophistication of DDoS attacks becoming more challenging, traditional security measures are not going to suffice. Since the complexity of connected devices and real-time data flows is so massive, it requires dynamic adaptive security for 6G networks. Systems like SADS are the future for proactive security, which combines the concepts of machine learning, blockchain, SDN, and VNFs. These solutions allow the network to evolve constantly with new threats, so 6G can live up to promises of connectivity and efficiency without becoming a target of malicious actors [5].

11.3.6 Protecting the Promise of 6G

This explosive growth of 6G networks brings both great opportunities and significant cybersecurity risks. Without support from robust, adaptive security systems, those very features that make 6G so powerful—their massive scale, ultra-low latency, and decentralised nature—could also be a prime target for DDoS attacks. Solutions like SADS, which are based on technologies like real-time anomaly detection, blockchain, and SDN/VNF-driven resilience, provide proactive defence against such threats. With 6G, adaptive and intelligent security frameworks will become vital for the protection of networks that will form the backbone of our digital future.

11.4 DEVICE HIJACKING: THE SILENT TAKEOVER

The sudden growth of interconnected devices has provided a way to the world where every object, from household appliances to industrial machinery, is connected in a very large network. While this connectivity has been promoted across industries, it has also introduced new cybersecurity threats, including the risk of device hijacking. This silent but clearly deadly form of cyberattack occurs when unauthorised individuals take control of a device, even without the knowledge of the user. Device hijacking threatens the trust and security of interconnected systems, which shows that it is a critical need for both users and industries to understand and reduce these risks

11.4.1 Understanding Device Hijacking and Its Mechanisms

Device hijacking involves an attacker taking full control of a device due to improper use of weaknesses in communication protocols, software, or security of the device. Once the device has been successfully hijacked, the attacker can further use it to perform various illegal activities without the knowledge of an unwitting user, such as stealing sensitive data or controlling other connected devices. Most common aims for device hijacking include the use of weak passwords, software that has known vulnerabilities, takeover of network protocols, and carrying out man-in-the-middle (MitM) attacks [32].

The growth of IoT has served to increase these vulnerabilities. IoT devices are primarily designed to prioritise operations over security, with many being underpowered for implementing solid safeguards. IoT devices depend on cloud or edge servers, providing more entry points for attackers. Insofar as device hijacking is concerned, attackers have their crosshairs aimed at these poorly maintained infrastructures to turn hijacked devices into parts of a larger network for executing coordinated cyberattacks [7].

11.4.2 Types of Device Hijacking Attacks

Device hijacking encompasses various attacks affecting systems with different implications. Two known attack techniques are discussed here:

Remote Code Execution (RCE) is one of the most dangerous forms of device hijacking. RCE enables attackers to execute arbitrary code on the device, turning it into a controllable asset. RCE can be used by an attacker

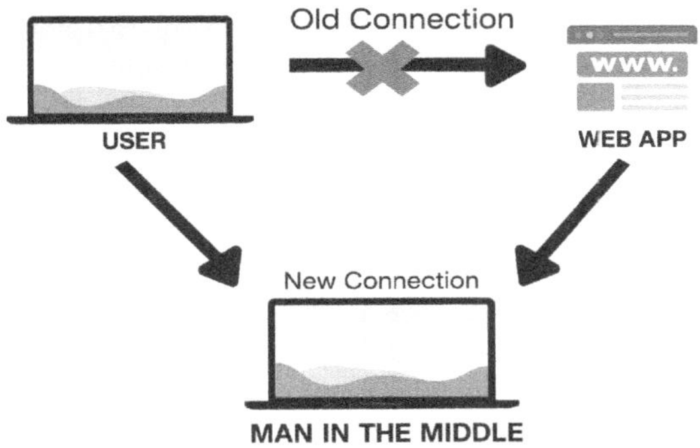

FIGURE 11.2 MitM.

to manipulate devices, steal data from, or redirect operations, but with no physical access to the device.

MitM Attacks: These types of attacks involve an intrusion to intercept and sometimes even alter the communications of two parties, often unbeknownst to either of them. When successful, MitM attacks can, however, allow attackers to manipulate other networked devices by re-routing or manipulating communication flows. Firmware Manipulation: Various IoT and embedded devices often rely on firmware to perform essential operations. Once compromised, firmware can be used to subvert any other defences, inject malware, or render the device inoperative. Figure 11.2 below illustrates the 'MitM' attack, showcasing how an unauthorised entity intercepts communication between two parties.

Session Hijacking: In the case of session hijacking, an attack is concerned with active sessions between a device and a server. Attackers masquerade as a device to access session-based authentication information or device permissions that would facilitate subsequent attacks [33].

Botnet Inclusion: Attackers generally hijack devices to add to botnets-large groups of infected devices exploited together for coordinated cyberattacks, including DDoS attacks that throttle specific networks.

11.4.3 Real-World Impacts of Device Hijacking

Device hijacking has repercussions not just for individual users but also for the industry, depending on their reliance on interconnected systems. Individuals using hijacked devices face violation of privacy, unmasking of sensitive information, or undesired transactions. As these systems keep getting more popular, smart home devices such as security cameras, door locks, and smart thermostats have made domestic environments vulnerable to probes, possibly posing serious ramifications on users' safety and privacy.

However, device hijacking means a much more urgent threat in industrial settings and other critical infrastructure. Pilfering an industrial control system (ICS) could disrupt production processes, destroy equipment, or threaten human life through safety control sabotage. The infrastructure sectors with critical needs, such as power grids, water treatment facilities, and transportation systems, are prone to hijacking, as they involve many interconnected devices, many of which have little or no security for monitoring and control purposes. A successful hijacking here can result in service outages, environmental hazards, or catastrophic operational failures.

11.4.4 Proactive Security Measures Against Device Hijacking

To counter device hijacking, it is necessary to implement a multi-layer security approach involving hardware, procedural policy, and user-enabled awareness. There are two major approaches, which are as follows:

Stronger Authentication Protocols: The availability of strong multimodal authentication methods, such as Multi-Factor Authentication (MFA) and biometric confirmation, discourages any possible hijacking of their devices. Their implementation on devices and gateways may lower the chances of other types of hijacking attempts.

Firmware and Software Integrity Checks: Continually updating until their respective integrity checks are proven is important for avoiding known weaknesses on devices. Several vendors are starting to provide over-the-air capabilities that allow devices to receive respective security patches without any physical intervention.

End-to-End Encryption: The communication between devices is encrypted. This means if a hacker does intercept the communication, reading further will be hard, or the device's data will change. This property becomes especially essential in cases where an IoT device relays sensitive information, such as that coming from healthcare monitors or financial sensors.

Anomaly Detection and Machine Learning: Advanced detection systems incorporate algorithms and machine learning for the detection of abnormal behaviour that might result in hijacking attempts. Such systems keep tracking activity patterns, notifying the user or administrator of any anomalous occurrences so that quick repair options can be acted upon.

Zero Trust Security Models: Trust is virtually zero in a Zero Trust Security Model, whether external, internal, or trusted. They endeavour strict verification and authorisation, thereby restricting lateral movement capabilities within a network once an infected file has breached a single device.

User Education and Awareness: Users are most often the first line of defence in any kind of hijack attempt, especially where device security requires user configuration using passwords and settings. Provided that people can receive an education.

11.5 JAMMING THE SIGNAL: DISRUPTION IN 6G

With unmatched speeds, wide-ranging device connectivity, and wide-ranging data bandwidths, 6G networks create an important moment in the evolution of wireless communication. The rapid expansion creates important challenges, and among these challenges, the most urgent is the risk of jamming attacks. Attacks that overwhelm legitimate signals with interference disrupt communication, and they pose large risks to 6G networks because of high-frequency bands, dense interconnectivity, and the need for real-time applications. The complication of 6G networks is increasing, and this rise in complication prompts the development of more sophisticated jamming techniques [19]. These jamming techniques have become a pressing security concern. It is important to address these techniques effectively to guarantee that we can fully realise the potential of 6G technology.

11.5.1 Understanding Jamming in 6G Networks

Jamming attacks send interference that reduces the signal quality of real communications. A jammer injects noise signals into the air. This disrupts the communication link between devices. As a result, calls drop. Data streams are interrupted. System reliability suffers. While jamming was a matter of concern in the previous wireless generations, in 6G, this is especially more problematic as it would heavily rely on new millimetre-wave and terahertz bands spectra. These frequencies are prone to interference, and because of the dense connectivity feature in 6G, interference in any device can travel widely across the network and thus influence many devices [10].

In 6G, which ranges from real-time remote surgeries to autonomous vehicle coordination, any jamming-based disruption can have severe impacts. The dense, interconnected structure of 6G networks makes them highly vulnerable to localised disruptions, which can explode rapidly. Moreover, the crucial beamforming technology and the massive multiple-input multiple-output structures of 6G rely entirely on the integrity of its signals, making them vulnerable to jamming attacks that cause spatial channel disturbances.

11.5.2 Categories of Jamming Attacks

Many forms of attacks are practiced on a wide range of parameters using several weaknesses in the 6G architecture. These attacks differ regarding frequency, timing, or target strategies, posing varied challenges to the network defences.

Constant Jamming: It involves continuous interference by the attacker on fixed frequency non-random transmissions. Although it is effective in disrupting communication, constant jamming is energy-intensive and relatively easy to detect because of its consistent interference pattern.

Reactive Jamming: Reactive jamming will turn on only when the jammer senses an active signal communication. Therefore, it helps the attacker save energy and remain unobservable since only particular transmission jamming takes place. This technique proves to be effective in highly dense networks like 6G in which many simultaneous transmissions happen [11].

Disrupt the communication across a wide range of channels. In 6G, where frequency hopping is quite common for interference management, sweep jamming poses a very significant threat since it can quite effectively disable the whole frequency bands, undermining the dynamic spectrum allocation capabilities that 6G relies on for reliability.

Smart Jamming: It uses AI-advantaged adaptive algorithms to adjust jamming patterns to optimise network patterns according to that pattern. These programmes can facilitate the avoidance of discovery by making it impossible for the target countermeasure. This means that increased interference yields more effectiveness. By analysing and knowing which channels best fit specific real-time operations, a smart jammer could selectively disrupt only these channels. As 6G networks themselves start to integrate AI into security and optimisation, the techniques of smart jamming will represent a sophisticated and evolutionary threat [9].

Pulse Jamming: In this method, the jammer transmits pulses of interference in a pseudorandom sequence. The pulses can cause temporary interruptions that are difficult for the target network to trace or counterattack against. Pulse jamming is especially hazardous in applications where a single millisecond delay or loss of data may mean life or death, such as in autonomous driving or medical monitoring.

11.5.3 AI-Powered Defence Mechanisms in 6G

AI and machine learning have, however, become an essential resource in overcoming jamming in 6G networks, with the help of machine learning algorithms. People will have systems capable of detecting abnormal patterns in network traffic and letting them know what is going on in real time to prevent these attacks on communications in the jamming stage. Significant to that is the capability of AI-centred algorithms in elaboration in distinguishing normal variations in traffic from malicious interference by the assessment of variables, such as Signal-to-Noise Ratio (SNR) fluctuations, packet delivery ratios, and delays in communications.

Notably, however, reinforcement learning algorithms have been found very effective for adaptive frequency allocation, whereby the actual network can adaptively jump frequency or adjust power systematically to keep communication non-interrupted in the event of a jamming attack. The operation of these algorithms gives them the chance, while monitoring the environment of the network, to 'learn' certain patterns of jamming attempts to provide adaptive measures against disruption. This way of doing things is critical in the event of a jamming attempt in 6G, where applications or frequency-specific jamming brings about a big need for an adaptive and smart defence from the network [20].

AI also allows for predictive analytics, whereby the systems will know whether jamming may occur or not, depending on visible patterns or historical data. This predictive analysis is very important for mission-critical applications in 6G, such as telemedicine, industrial automation, and smart city infrastructure, where a communication breakdown may be disastrous. AI-driven shields defend against jamming by detecting sources before they disrupt any services, pre-emptively disarming prospective jamming attempts.

11.5.4 Blockchain and Federated Learning in Jamming Defence

In addition to AI, 6G security frameworks are increasingly integrating blockchain and federated learning for enhanced jam-resilience. Blockchain technology provides an incorruptible decentralised ledger of net cash transactions. This creates an immutable document of communication pathways and device interactions, which simplifies tracking and identification of sources of interference. Once jamming events are detected, the blockchain ledger can expose the interference pattern and quickly respond for the network administrators [21].

Federated learning allows 6G distributed devices across the network to work collaboratively on training machine learning models without sharing raw data. It, therefore, enhances privacy while permitting the system to learn continually from real-time network data. Hence, federated learning provides a good advantage against various jamming methods in that it adapts across many conflicting jamming techniques in different devices and parts of the network without compromising individual device security. With the learning process distributed, federated learning adds to the network's arsenal against evolving jammed techniques, specifically targeting edge devices with limited processing power [21].

11.5.5 Jamming Risks to Critical 6G Applications

The threat posed by jamming goes way beyond traditional communication disruptions; populations so reliant on 6G technology for such services in healthcare, transportation, and public safety become vulnerable to potentially disastrous consequences. For example, autonomous vehicles communicate continuously and instantaneously with each other and with road infrastructure to make split-second decisions. A jamming attack that disables such communications can have serious, even crash-based consequences. Further, by extension, real-time or over high-speed 6G connection performed remote surgeries may be affected by jamming attacks that can compromise safety for patients when communications between surgeons and robotic equipment are compromised.

An attack that jams critical infrastructures in smart cities, for example, civil services, traffic management systems, or emergency response networks, could cause a significant number of services to be out of order and compromise public safety. Because these applications are highly interconnected, failure due to jamming may propagate to other systems, which will multiply the effects of one attack. Then, jamming mitigation is no longer only a technical necessity but turns into a foundation that can guarantee the critical applications in the 6G paradigm [20].

11.6 AI-DRIVEN DETECTION: A SMART DEFENCE

AI is rewriting the way that security is considered and operated in cybersecurity: More sophisticated, less predictable means of attack have pushed static rules and known attack signatures increasingly into obsolescence. The paradigm shift is AI-based detection from reactive to proactive defence and smarter, faster detection and countering of possible threats in real time for the organisation.

11.6.1 The Rise of AI in Cybersecurity

With the rapid development of digital technology, a new landscape for cyber threats has emerged, such that attackers can use their most advanced tools to breach classic defences. AI in cybersecurity gained ground as it used improved Machine Learning (ML) and Deep learning (DL) techniques, surpassing the conventional security system limitations [12]. AI-driven cybersecurity works not only to detect attacks but also to predict and keep ahead of threats before they hit. This ability to prophesise danger, rather than just react to it, denotes a critical evolution in this defensive technology.

11.6.2 How AI Detects Threats

Sophisticated algorithms that can process immense amounts of information in no time constitute the kernel of AI-based detection. For instance, machine learning models learn from historical data and detect outliers. Based on this learning, such models adapt and recalibrate their algorithms every time new data becomes available [12]. Supervised learning helps AI recognise known threats, while unsupervised models can identify unusual activity that may signal an emerging threat. The advanced learning techniques of CNN and Long Short-Term Memory (LSTM) networks allow AI to work with complex data such as complicated malware behaviours and user patterns [29]. Such a multi-layered approach helps AI immediately detect even slight behaviour anomalies and report such identified risks, even though AI would have never seen those types of risks before.

11.6.3 Real-Time, Adaptive Response

This might be one of the strongest properties of AI in cyber, which is its capability of acting in real time. State-of-the-art attacks can become very hard to mitigate for traditional systems, which usually require a large scale of analysis and manual intervention [13]. AI enables systems to quickly spot suspicious behaviour and activate response mechanisms, including blocking access, isolation affected devices, and raising alerts to security teams [28]. This fast adaptive response greatly reduces the possibility of an attacker exploiting any vulnerabilities, so that the factor of cybersecurity shifts from being defensive walls to being an intelligent, active safeguard that learns and evolves continually.

11.6.4 Overcoming Challenges in AI-Driven Detection

While these advantages are significant, AI implementation in cybersecurity does not come without problems. Adversarial attacks are attempts to trick AI models by misleading data that diminish the capabilities of AI-driven detection. Also, with such massive amounts of data that come along with AI systems, privacy and security concerns of the data itself become issues. Researchers are very actively unlocking the full potential of AI for an intelligent defensive posture [27].

11.6.5 The Future of AI in Cyber Defence

The future of AI in the domain of cybersecurity thus appears very bright, with emerging species of cyber threats. Newancer infections to AI are to include new species of resilient models in the malware analysis. Thus, it is expected that AI systems, learning from each interaction and adaptive to new threats, would present an advanced defensive attempt.

The AI-driven detection will find a special place in the toolkit of modern cybersecurity and be proactive and resilient [26].

11.7 FORTIFYING A 6G IOT SECURITY STRATEGY

As we continue toward the time of 6G, it becomes apparent that the technology shall not only usher in better speed but also introduce breadth to the way IoT-enabled devices connect. From autonomous vehicles to smart cities, 6G will bind everything in ways previously unseen. As enthralling as this connectivity may be, it comes with a caveat: increased threats to security. Should there be more connected devices, then the number of opportunities for attack through cyberspace rises. It thus becomes important to build a safe underbelly for 6G IoT networks.

11.7.1 Understanding the New Threats in 6G

With 6G, data is going to flow through networks much more than ever before to enable a new wave of applications: connected robotics, virtual reality, and self-driving cars. But with that many connected devices, a back door for hackers to get into any one of them may become a possible entry point for their nefarious activities too. With that many applications sprawled across 6G IoT, securing every piece of that network becomes paramount. Security will no longer be based on mere protectionist measures but on a balance of tools and modalities in the fight against threats at all levels. 6G promises to provide orders of magnitude higher data speed compared to what is currently being enjoyed while over a much larger number of devices, opening applications in the likes of car autonomy, telemedicine, and infrastructure for smart cities. The trick is, however, that for each connected device and application, there is another possible entry point that hackers can use to claim a stake in the IoT. Large and complex networks, therefore, require quite a lot of resilience to change their threats constantly [23].

11.7.2 Using AI to Boost Security

AI indeed will play a significant role in securing 6G IoT networks. Through AI, it will be easy to keep watch against unusual activities or threats. The adaptability of AI-based systems allows them to address and counter threats much quicker, as one would realise with traditional security measures. Then again, AI has its vulnerabilities. Hackers may cheat AI with fake incidents and contaminate its learning cycle. For this, AI needs to be built carefully with extra security measures to disallow such attacks from happening within 6G networks.

AI will dramatically improve real-time security management as it learns from historical data and recognises patterns that potentially indicate threats. The role of AI will be both offensive and adaptive; using AI in intrusion detection techniques, it can analyse traffic information in less time to find evidence of an attack. Ordinary attacks could result in such manipulations, but AI systems must rely on more than just detection [22].

11.7.3 Building Trust with Blockchain

Not only does blockchain technology underpin digital currencies, but it can also help establish trust in 6G IoT networks. In blockchain, each transaction or information is stored in a decentralised network. Therefore, it's very hard to change or forge. For the 6G IoT, blockchain has a true record of transactions and verification of each device's actions connected to the network. This means that only those devices that have the authority to do so are granted access. Smart contracts, an automated set of agreements, allow verification for the execution of transactions between devices through the automation of identity verification and permissions [17]. This decentralised approach ensures that it is rather challenging for such an attack, as the possibility of one weak point in the network getting exploited is very low.

This system, being decentralised, can now trace transactions and operations of devices, thus ensuring accountability and promoting trust. Smart contracts that could automate solutions to security checks and permission to access would have much value for 6G IoT networks to limit access to malicious entities. Additionally, the decentralised blockchain technology is credible enough to authenticate the identities of devices and the authenticity of data [24].

11.7.4 Protecting Data with Quantum Encryption

The advanced encryption scheme in 6G technology would require high-speed and huge data flow for secure communication. Through Quantum Key Distribution (QKD), one could render the data almost untraceable: QKD employs principles from quantum physics to create encryption keys that will change if intercepted. Thus, QKD serves as an efficient protection mechanism for sensitive data in 6G applications, for example, healthcare and self-driving cars. Other combinations with QKD may evolve into layered protection meant to safeguard data even in risk-prone situations [18].

Quantum encryption guarantees that communications containing sensitive information transmitted on 6G networks remain confidential. Any attempt to intercept data triggers alterations: One thus signals the breach. Thus, by using QKD in health-related and autonomous vehicle applications where data privacy is of utmost consideration, unauthorised or malicious access could be prevented, as could data leaks from occurring.

11.7.5 Securing Edge Devices and Networks

For 6G, edge computing data will be processed as close to the source as possible through multi-access edge computing (MEC). Other things being equal, this kind of setup can save on time delays, but it also creates several new positions requiring protection. Any edge device, from factory sensors to city surveillance cameras, can become a target. These devices will require an effective encryption scheme, real-time monitoring, and detection systems. MEC is indeed the line of defence that, alongside AI-powered initiatives, could thwart the attacks before they expand [25]. The edge in 6G is bringing processing power close to the source, thus avoiding transferring data to centralised locations. But this local processing also puts an expectation that every device is now a critical point of concern when it comes to security. The injection of real-time monitoring and encryption at the edge allows MEC systems to react to threats before any damage can occur, mainly because of their fast response times.

A secure 6G IoT network will require using multiple complementary approaches. By integrating AI, blockchain, quantum encryption, and edge security at the same time, we will build a network ready for the upcoming demands of tomorrow's interconnected world.

11.8 CONCLUSION

The 6G revolution promises transformative changes across industries, enhancing connectivity, speed, and functionality while powering an ever-expanding IoT ecosystem. However, this progress introduces unprecedented challenges, notably in security. The transition from centralised to decentralised architectures, coupled with the massive scale of connected devices and reliance on real-time data, creates vulnerabilities that cannot be effectively addressed using traditional security models. The SADS, as described in this chapter, provides a pioneering approach to safeguarding 6G-enabled IoT networks. By integrating cutting-edge technologies such as edge computing, blockchain, federated learning, and AI, SADS proactively addresses threats like DDoS attacks, device hijacking, and jamming. These technologies enable real-time threat detection, adaptive learning, and decentralised processing, all while maintaining data privacy and integrity. The system's emphasis on scalability, resilience, and transparency ensures that it can adapt to the dynamic and evolving threat landscapes of 6G networks. As 6G technology becomes integral to domains such as healthcare, smart cities, and autonomous systems, solutions like SADS will be indispensable for enabling safe adoption and fostering innovation. By effectively balancing privacy and security, SADS not only mitigates risks but also removes barriers to embracing the full potential of 6G connectivity. The chapter underscores the urgency of advancing security measures to protect the integrity of interconnected ecosystems and lays the groundwork for a secure digital future.

BIBLIOGRAPHY

1. Banafaa, M., Shayea, I., Din, J., Azmi, M. H., Alashbi, A., Daradkeh, Y. I., & Alhammadi, A. (2023). 6G mobile communication technology: Requirements, targets, applications, challenges, advantages, and opportunities. *Alexandria Engineering Journal*, *64*, 245–274.
2. Jeny, J. R. V., Sowmya, R., Kiran, G. S., Babu, M. K., & Arjun, C. (2022, July). Shilling attack detection system for online recommenders. In *2022 International Conference on Inventive Computation Technologies (ICICT)*, Lalitpur, Nepal (pp. 988–992). IEEE.
3. Vaigandla, K. K., Karne, R., Siluveru, M., & Kesoju, M. (2023). Review on blockchain technology: Architecture, characteristics, benefits, algorithms, challenges and applications. *Mesopotamian Journal of CyberSecurity*, *2023*, 73–84.
4. Suomalainen, J., Ahmad, I., Shajan, A., & Savunen, T. (2025). Cybersecurity for tactical 6G networks: Threats, architecture, and intelligence. *Future Generation Computer Systems*, *162*, 107500.

5. Kumari, P., & Jain, A. K. (2023). A comprehensive study of DDoS attacks over IoT network and their countermeasures. *Computers & Security, 127,* 103096.
6. Veenbrink, J. M. (2019). *The Impact of Criminal Law Concepts on the Enforcement of National and European Competition Law. A Silent Take-Over?* Deventer, Netherlands: Wolters Kluwer.
7. Letsoalo, E., & Ojo, S. (2017, May). Session hijacking attacks in wireless networks: A review of existing mitigation techniques. In *2017 IST-Africa Week Conference (IST-Africa)* (pp. 1–9). IEEE.
8. Zhang, X., Tu, Y., Long, Y., Shan, L., Elsaadani, M. A., Fu, K., ... & Hei, X. (2024, May). From virtual touch to tesla command: Unlocking unauthenticated control chains from smart glasses for vehicle takeover. In *2024 IEEE Symposium on Security and Privacy (SP),* San Francisco, CA (pp. 201–201). IEEE Computer Society.
9. Lohan, P., Kantarci, B., Ferrag, M. A., Tihanyi, N., & Shi, Y. (2024). From 5G to 6G networks, a survey on AI-based jamming and interference detection and mitigation. *IEEE Open Journal of the Communications Society,* 5, 3920–3974.
10. Xie, P., & Huang, Y. (2024). Physical-layer security for 6G: Safe jamming against malicious sensing. arXiv preprint arXiv:2406.11038
11. Vaishnavi, K. N., Khorvi, S. D., Kishore, R., & Gurugopinath, S. (2021, June). A survey on jamming techniques in physical layer security and anti-jamming strategies for 6G. In *2021 28th International Conference on Telecommunications (ICT)* (pp. 174–179). IEEE.
12. Waizel, G. (2024, July). Bridging the AI divide: The evolving arms race between AI-driven cyber attacks and AI-powered cybersecurity defences. In *International Conference on Machine Intelligence & Security for Smart Cities (TRUST) Proceedings,* Wiesbaden, Germany (Vol. 1, pp. 141–156).
13. Khan, M. I., Arif, A., & Khan, A. R. A. (2024). AI-driven threat detection: A brief overview of AI techniques in cybersecurity. *BIN: Bulletin of Informatics, 2*(2), 248–261.
14. Rangaraju, S. (2023). Secure by intelligence: Enhancing products with AI-driven security measures. *EPH-International Journal of Science and Engineering, 9*(3), 36–41.
15. Raponi, S. (2021). *AI-Driven Detection of Cybersecurity-Related Patterns* (Doctoral dissertation, Hamad Bin Khalifa University (Qatar)).
16. Uwaoma, C. (2023). On security strategies for addressing potential vulnerabilities in 6G technologies deployable in healthcare. *arXiv preprint arXiv:2309.16714*
17. Roberts, C., & Yerram, S. R. (2024). Ensuring security in the age of intelligent connectivity: Strategic insights for 6G networks. *Engineering International, 11*(1), 27–42.
18. Kaur, N., & Gupta, L. (2024). An approach to enhance IoT security in 6G networks through explainable AI. *arXiv preprint arXiv:2410.05310.*
19. Priyadarshani, R., Park, K. H., Ata, Y., & Alouini, M. S. (2024). Jamming intrusions in extreme bandwidth communication: A comprehensive overview. *arXiv preprint arXiv:2403.19868.*
20. Fellan, A., Djuitcheu, H., Lindenschmitt, D., & Schotten, H. D. (2024, October). Exploring jamming threats and defense mechanisms in future 6G communication systems. In *MILCOM 2024-2024 IEEE Military Communications Conference (MILCOM),* Washington, DC, USA (pp. 117–122). IEEE.
21. Pirayesh, H., & Zeng, H. (2022). Jamming attacks and anti-jamming strategies in wireless networks: A comprehensive survey. *IEEE Communications Surveys & Tutorials, 24*(2), 767–809.
22. Liyanage, M., Porambage, P., Zeydan, E., Senavirathne, T., Siriwardhana, Y., Yadav, A. K., & Siniarski, B. (2024, June). Advancing security for 6G smart networks and services. In *2024 Joint European Conference on Networks and Communications & 6G Summit (EuCNC/6G Summit),* Antwerp, Belgium (pp. 1169–1174). IEEE.

23. Kazmi, S. H. A., Hassan, R., Qamar, F., Nisar, K., & Ibrahim, A. A. A. (2023). Security concepts in emerging 6G communication: Threats, countermeasures, authentication techniques and research directions. *Symmetry, 15*(6), 1147.
24. Uwaoma, C. (2023). On security strategies for addressing potential vulnerabilities in 6G technologies deployable in healthcare. *arXiv preprint arXiv:2309.16714.*
25. Kaur, N., & Gupta, L. (2024). An approach to enhance IoT security in 6G networks through explainable AI. *arXiv preprint arXiv:2410.05310.*
26. Yaseen, A. (2023). AI-driven threat detection and response: A paradigm shift in cybersecurity. *International Journal of Information and Cybersecurity, 7*(12), 25–43.
27. Rao, D. D., Waoo, A. A., Singh, M. P., Pareek, P. K., Kamal, S., & Pandit, S. V. (2024). Strategizing IoT network layer security through advanced intrusion detection systems and AI-driven threat analysis. *Full Length Article, 12*(2), 195–95.
28. Salem, A. H., Azzam, S. M., Emam, O. E., & Abohany, A. A. (2024). Advancing cybersecurity: A comprehensive review of AI-driven detection techniques. *Journal of Big Data, 11*(1), 105.
29. Goswami, M. J. (2024). Enhancing network security with AI-driven intrusion detection systems. *12*(1), 29–35.
30. Baccarella, C. V., Wagner, T. F., Kietzmann, J. H., & McCarthy, I. P. (2018). Social media? It's serious! Understanding the dark side of social media. *European Management Journal, 36*(4), 431–438.
31. Rao, B. N., & Kalyani, V. (2022). A study on positive and negative effects of social media on society. *Journal of Science & Technology (JST), 7*(10), 46–54.
32. Zhao, J., Li, Q., Yuan, Z., Zhang, Z., & Lu, S. (2022, October). 5G messaging: System insecurity and defenses. In *2022 IEEE Conference on Communications and Network Security (CNS),* Austin,USA (pp. 37–45). IEEE.
33. Letsoalo, E., & Ojo, S. (2018, May). A model to mitigate session hijacking attacks in wireless networks. In *2018 IST-Africa Week Conference (IST-Africa),* Gaborone, Botswana (p. 1). IEEE.

12 Implementing Internet of Things (IoT) Solutions in Healthcare

Archana Kaushik, Ujjwal Kaushik, Abhimanyu Agnihotri, Neha Kukreti, Gaganpreet Kaur, and Neeraj Choudhary

12.1 INTRODUCTION

The IoT is essential in cutting-edge industries like healthcare, education, transportation, homes, smart cities, and defense operations. Because IoT applications provide secure, real-time remote patient monitoring, which improves people's quality of life, they are particularly beneficial to the healthcare sector [1]. The IoT can help healthcare providers monitor patients' health and improve patient care in many ways, including remote patient monitoring, hospital asset management, tracking medical equipment, connecting medical devices, analyzing patient data, and many more, and all these aspects have been elaborated with case studies wherever available throughout the chapter [2].

12.2 IOT-ENABLED REMOTE PATIENT MONITORING

Digital technology is used in remote patient monitoring to facilitate the collection, transmission, evaluation, and communication of patient health data through wearable sensors, implantable devices, and portable equipment, among other electronic devices [3]. External patient monitoring provides personalized care for various conditions, extending life spans, especially for older, vulnerable, and multi-comorbid patients, enhancing their quality of life. Furthermore, it reduces hospital stays, readmissions, and length of stay, improving quality of life, reducing expenses, and enabling physicians to take proactive medication measures, preventing problems [4]. The IoT solution for remote patient monitoring must be reliable, adaptable, and secure, considering various connectivity options. Remote medical device connections may not always be straightforward, especially for elderly patients, and unstable or unsecured networks can cause issues [5]. Remote patient monitoring (RPM) is ideal for cellular connectivity as it doesn't require home internet access and offers roaming benefits with the right IoT provider, ensuring optimal network utilization for mission-critical solutions. Healthcare security and patient confidentiality are crucial, and handling and viewing medical data with caution is essential due to the sensitive nature of the information.

DOI: 10.1201/9781003586456-12

The selection of an IoT partner that offers a high level of security is crucial and essential [6]. The "ScalableDigitalHealth" (SDH) framework has been proposed for remote patient monitoring, particularly among elderly and mobility-impaired patients, with the ability to track vital health parameters in real time, including blood pressure, oxygen saturation, body temperature, and electrocardiogram (ECG), by utilizing IoT technology and application autoscaling [7]. In this context, IoT makes advanced features like data-driven decision support systems and computer-assisted diagnostics (CAD) possible, enhancing the capacities of healthcare personnel [8]. It has been studied how machine learning models can be used in IoT-enabled health monitoring devices to track patients' respiratory patterns, body temperatures, and heart rates as well as their activities, including running, walking, sleeping, and exercising [9]. Table 12.1 details a few such systems that were reported in the literature.

TABLE 12.1
Machine Learning-Based IoT-Enabled Remote Patient Monitoring Systems

RPM System Based on IoT	Area of Application	Details	References
ECG signal monitoring system for processing heart data	Detection of arrhythmias	This system includes a Polar H10 heart sensor, connectivity technology to share and store patient data, and machine learning models to classify cardiac events.	[10]
Real-time remote health monitoring system based on a deep learning model	In home clinical settings, remote monitoring and early health problem detection	Based on three sensor systems: AD8232: ECG monitoring AX30100: Blood oxygen/heart rate monitoring Body temperature (MLX90614) The server can categorize possible illnesses using a pre-trained deep learning model built on a convolutional neural network with an attention layer.	[11]
Support vector machine (SVM) learning, IoT-based smartphone	Diagnosis of obstructive sleep apnea	A monitoring system that is connected to a machine learning algorithm enables the collection of physiological data from a smartphone and combines it with diagnostic algorithms to create an early warning system for anomalous data.	[12]
Simulating Internet of Things (MoSIoT) scenarios	Enables the modeling and simulation of IoT HMS scenarios created for various infirmities by healthcare professionals	The model-driven engineering (MDE) technique serves as the foundation for the MoSIoT collection of models. It enables the creation of an IoT system by simulating a whole IoT HMS scenario.	[13]

(Continued)

TABLE 12.1 (*Continued*)
Machine Learning-Based IoT-Enabled Remote Patient Monitoring Systems

RPM System Based on IoT	Area of Application	Details	References
IoMT-based deep learning model	Stroke prediction	The group handling method (GMDH) predictive analytics model and stacked CNN are the foundation of the hybrid AI in linked health. An LSTM deep learning module is included for the purpose of biomedical signal prediction.	[14]
Labor progress monitoring system built on an IoT platform and an algorithm assisted by AI	Labor progress monitoring	Based on varying contraction stages, the pregnant lady is classified as recommended, optional, or postponed for admittance using the system's AI-assisted convolutional neural network algorithm.	[15]
Integrated microsensors in an IoT-enabled patch	Tracking patient adherence to the patch as part of the treatment plan and making any modifications	Depending on whether it was worn, the temperature differential is measured by the microsensor patch. Wearing the patch allowed it to take temperature measurements every 30 seconds and use a mobile application to send the data to a cloud server every 15 seconds.	[16]

12.3 IOT-ENABLED TELEMEDICINE IN PATIENT HEALTHCARE

Within urban, suburban, and rural communities, telemedicine has been employed extensively by physicians in modern times [17]. Telemedicine with IoT makes it easier to monitor patient health without the inconvenience of in-person visits. Additionally, IoT opens up new business opportunities for healthcare providers by increasing the need for and accessibility of elder remote care. Medical offices can thus provide more patients, particularly those in remote locations, with the sought-after medical services. Medical facilities may maintain a safe, flexible, and comfortable working environment for physicians and nurses to perform at the highest level while also improving the quality of services provided with reduced administrative expenses and staff workload. Telemedicine gained traction quickly after the COVID-19 pandemic. People with rheumatic diseases were asked to participate in an anonymous online survey of those who used telemedicine; 61% said it was either as effective as or more successful than an in-person visit in some situations, such as discussing disease activity and changing medications [18]. A newly released survey looked at patients' desire to employ telemedicine visits for head and neck oncology therapy. Research revealed that nearly half of the patients evaluated in a tertiary care head and neck surgical cancer clinic expressed reluctance to use telemedicine for clinical visits. For these patients, internet reliability and restricted access to technological platforms continue to be major concerns [19]. Telehealth technology has the potential to revolutionize healthcare delivery globally by boosting patient participation and removing geographical obstacles. Diabetes care can become more broadly available and efficient by implementing these discoveries, leading to patient-centered, equitable healthcare solutions [20]. There are a number of drawbacks to telehealth, such as the inability to conduct thorough physical examinations, possible technological problems, security lapses, and regulatory obstacles [21]. Online conversations, according to telehealth critics, lack a comprehensive history and physical examination necessary for diagnosis and treatment, making them unsafe and impersonal [22]. Telemedicine interactions still raise more privacy and security issues than face-to-face interactions, despite completely secured platforms and Health Insurance Portability and Accountability Act regulations [23]. The wide range of rules, regulations, and guidelines that cause confusion presents legal and regulatory issues for telehealth providers [22]. Internet bandwidth affects the validity and reliability of fine motor task measurements, which in turn affects the precision of data transfer in telehealth practice [24]. Due to inadequate reimbursement from medical health insurance plans, telehealth adoption faces considerable challenges [25].

12.4 IOT-ENABLED SMART DEVICES IN PATIENT HEALTHCARE

Innovations such as smart devices and health monitoring sensors are a result of the growing patient population and the dearth of medical services. By accommodating and meeting patients' wants and demands, these technologies enhance patients' health [26]. Smart gadgets with IoT capabilities are transforming healthcare by improving patient monitoring, diagnosis, treatment, and overall management. A more proactive and individualized approach to healthcare is made possible by these devices' real-time collection, analysis, and transmission of health data. An intelligent health monitoring system is built on electronic devices such as sensors that use the IoT, cloud computing, networking, wireless device connection, artificial intelligence (AI), and machine learning [27].

12.4.1 IoT-Enabled Wearable Health Devices

With real-time data, increased awareness, and proactive health management, wearable health devices with IoT capabilities are changing the way individuals monitor their health. Any portable electronic gadget that is worn as an accessory or covertly integrated into the wearer's clothing is considered a wearable device. These devices feature sensors that assess a range of physiological parameters, so they may be used to both monitor one's own health and provide data to medical professionals [28,29]. IoT-enabled smart watches and wristbands operate as a source of physiological data obtained straight from the user's body due to their close contact with the skin. Smart watches with biosensors have the potential to be widely used in the healthcare industry and help patients and their providers access vital health information [30,31]. Although the idea of a smart watch is not new, there aren't many research studies on its potential uses in healthcare. In order for health professionals to broadly use smart watch apps, an extensive study on the devices' accuracy, completeness, and workflow impact should be conducted prior to their integration into clinical practice [32]. The ability to record real-time cardiac electrical activity makes the ECG an essential diagnostic tool for a range of cardiac conditions. Wearable and remote ECG devices aim to bridge this gap by enabling patients to collect their own ECGs using a streamlined method and often a single lead. Long-term monitoring with reduced-set ECG devices is very promising, especially when combined with real-time alerting systems. These features make them most helpful for detecting aberrant rhythms, and there is enough data to suggest that a remote ECG device may diagnose some arrhythmias, including atrial fibrillation, more accurately than a typical 12-lead ECG [33]. In this digital age, numerous wearable blood pressure monitoring devices have been developed. These digital blood pressure monitors offer significant increases in diagnostic accuracy, given the growing importance of correctly diagnosing hypertension. They also allow for frequent blood pressure readings with the least amount of discomfort for the patient [34]. Diabetes treatment is being revolutionized by wearable, minimally invasive continuous glucose monitoring (CGM) sensors, which measure blood glucose (BG) concentrations nearly continuously for several days in a row. This technology is quickly gaining popularity, especially among diabetics who must take insulin [35,36]. Inhalation topography can be obtained using commercially accessible wearable respiratory monitors, which are already used to detect various physiological indications such as heart rate and breathing frequency, even if there hasn't been much research on the topic. According to a study, wearable respiratory monitors with fundamental features and technologies might serve as a foundation for measuring ambulatory inhalation topography in real settings [37]. The ability of wearable respiratory sensors to monitor biochemical and biomechanical signals, including viral biomarkers from exhaled breaths and anomalies in respiratory rate and cough frequency caused by COVID-19, makes them beneficial [38,39]. Smart glasses are a type of wearable device that can send and receive wireless data and provide users with continuous, hands-free digital access. Wearable technology, which can be affixed to smart glasses or any other type of glasses, presents a new kind of visualization tool that lets surgeons focus on the work at hand without diverting their attention from the operating field [40,41] as shown in Fig 12.1.

12.4.2 IoT-Enabled Smart Implants and Prosthetics

Smart implants are medical devices that are inserted into the body through surgery to either sustain or replace a biological part that has been damaged. By connecting these implants to other devices, IoT technology allows them to exchange real-time data on a patient's status. Smart implants have been developed and successfully tested in orthopedics, cardiac, and dental care. In the domains of orthopedics and trauma, smart implants and prostheses have been made possible by developments in implant designs and computer technology. Sensor-guided devices enable close observation

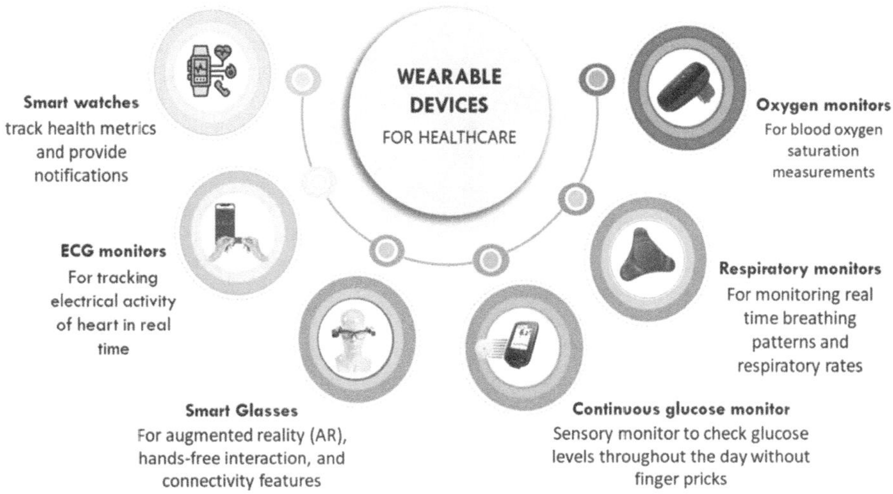

FIGURE 12.1 Wearable device helpful in healthcare management.

of the body environment around the implant, improving patient monitoring post-surgery and intraoperative accuracy [42]. Cardiology is a field where computational algorithms are being used to perform tasks that typically require human intelligence. Many rare-earth smart nano-biomaterials have been created by researchers over the last 20 years for use in implantology and bone tissue engineering. The advancement of AI, IoT, and precision medicine technologies is expected to significantly impact the future of cardiology. The potential areas in cardiology include accurate cardiovascular disease (CVD) prediction, non-invasive CAD diagnosis, wearable arrhythmia detection, heart failure diagnosis, treatment strategies, and outcome prediction [43]. IoT-based soft metal sensors are widely utilized in human–machine interfaces, smart prostheses, and soft robots for sensitive and accurate monitoring. Standard sensors are incompatible with robotic applications due to their deformation and flexibility, but soft sensors can be easily integrated for soft robotic applications [44]

12.5 IOT-ENABLED HOSPITAL AUTOMATION AND ASSET MANAGEMENT

12.5.1 IoT-Enabled Management of Hospital Assets

Hospital Powered by IoT Asset Management is revolutionizing hospital operations by automating, tracking, and enabling real-time visibility of vital resources. IoT solutions streamline hospital operations by integrating devices, sensors, and cloud-based platforms to manage assets like medical equipment, medications, and personnel. Large hospitals continue to present navigational challenges, particularly for visitors, patients, and staff in certain situations. This is especially true when it comes to tracking ambulatory equipment. Devices such as defibrillators, ventilators, and portable imaging equipment can be tracked by IoT-based equipment tracking in hospitals [45].

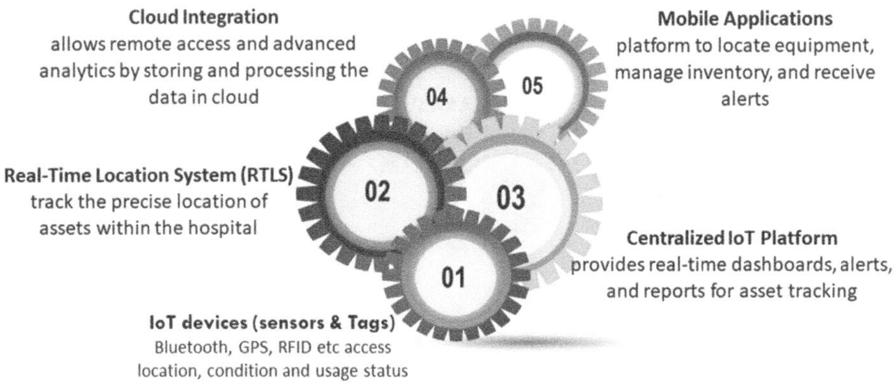

FIGURE 12.2 Key components of IoT-based platforms.

Hospital assets can be assigned to particular zones within the facility with the use of geo-fencing, and if they leave these regions, warnings can be fired off. This stops priceless equipment from being lost, misplaced, or stolen [46]. The implementation and demonstration of an IoT-based message-broker system have been done on five medical devices: blood oxygen, forehead temperature, body weight sensors, and blood pressure. The cloud platform gathers client health profiles via radio-frequency identification (RFID) access to a central controller. When necessary, the health devices may be reliably added, modified, and uninstalled, and all collected data can be quickly shared, examined, and displayed [47]. The key components of IoT-based systems that prove to be the game changer are depicted in Figure 12.2.

12.5.2 IoT-Enabled Management of Hospital Infrastructure

A smart hospital can be thought of as a composable IT ecosystem where all important information is at the touch of a mouse. It is now feasible to build such an infrastructure because of the IoT's explosive growth and wide range of capabilities [48]. Relevant IoT features include broad-gauge data gathering, dependable data transfer, and intelligent processing, which offer a strong basis for developing a high-tech hospital or smart hospital rooms [49]. The perception, network, and application layers make up the three-layer architecture used to construct IoT-enabled smart infrastructure. The first set of building bricks, which are connected devices, is crucial in the perception layer. These gadgets produce data, which is gathered and sent to the next layer. Networks and gateways are the network-level mediators that connect physical devices to servers or the cloud. Both software and hardware can be used to construct gateways. Modern gateways do more than just normalize and transfer data; they also pre-process and filter raw data, encrypt data before sending it over networks, and make security monitoring easier [50]. Networks often facilitate communication between the application layer and linked devices. The essential component of the entire architecture is the application layer's cloud or distant servers. Data storage and large data processing are enabled, and user interfaces are developed because of their robust services.

12.6 CONCLUSION

By the use of IoT solutions in the healthcare industry, various domains of medical research, patient care, hospital infrastructure, and remote monitoring of patients have all been greatly impacted. Predictive analytics, remote diagnostics, and real-time monitoring have decreased hospital burdens and enhanced health outcomes. A few major issues that are required to be resolved, though, include data security, regulatory compliance, and continuity of services without a halt. In upcoming years, developments in blockchain technology, 5G connection, and AI will continue to influence different facets of IoT in health. These enhancements in IoT-based services will allow for quicker and more dependable data transfer, enabling the AI-driven analytics to improve in decision-making. Furthermore, wearable technology and smart implants will advance, offering individualized care and ongoing health monitoring. Stakeholders must prioritize improving cybersecurity, creating international interoperability standards, and guaranteeing ethical AI integration if they are to fully grasp the potential of IoT in healthcare. IoT will be essential in improving healthcare's accessibility, efficiency, and patient-centeredness as technology advances, ultimately revolutionizing the way that people receive medical care in the future.

REFERENCES

1. Abdulmalek, S., Nasir, A., Jabbar, W. A., Almuhaya, M. A. M., Bairagi, A. K., Khan, M. A., & Kee, S. H. (2022). IoT-based healthcare-monitoring system towards improving quality of life: A review. *Healthcare*, 10(10), 1993.
2. Chunyan, L., Wang, J., Wang, S. & Zhang, Y. (2024). A review of IoT applications in healthcare. *Neurocomputing*, 565, 127017.
3. Jain, S., Agarwal, A., Bhardwaj, A., Lakshmi, P., Singh, M., Chauhan, A., & Singh, M. (2022). Remote monitoring and holistic care of home-isolated COVID-19 positive healthcare workers through digital technology during the omicron (B1.1.529) wave: A prospective cohort study from India. *Frontiers in Public Health*, 10, 936000. https://doi.org/10.3389/fpubh.2022.936000
4. Waleed, M., Kamal, T., Um, T. W., Hafeez, A., Habib, B., & Skouby, K. E. (2023). Unlocking insights in IoT-based patient monitoring: Methods for encompassing large-data challenges. *Sensors*, 23(15), 6760. https://doi.org/10.3390/s23156760
5. Iwai, S., Frenkel, D., & Jacobson, J. T. (2021). Remote device monitoring: Be careful what you wish for.... *Clinical Electrophysiology*, 7(2), 235–237. https://doi.org/10.1016/j.jacep.2020.09.033
6. Jaime, F. J., Muñoz, A., Rodríguez-Gómez, F., & Jerez-Calero, A. (2023). Strengthening privacy and data security in biomedical microelectromechanical systems by IoT communication security and protection in smart healthcare. *Sensors*, 23(21), 8944. https://doi.org/10.3390/s23218944
7. Alasmary H. (2024). ScalableDigitalHealth (SDH): An IoT-based scalable framework for remote patient monitoring. *Sensors*, 24(4), 1346. https://doi.org/10.3390/s24041346
8. Deepika S., & Vijayakumar K. P. (2022). IoT based elderly monitoring system. In *Proceedings of the 2022 6th International Conference on Trends in Electronics and Informatics (ICOEI)*; Tirunelveli, India, 28–30 April 2022; pp. 573–579.
9. Malche, T., Tharewal, S., Tiwari, P. K., Jabarulla, M. Y., Alnuaim, A. A., Hatamleh, W. A., & Ullah, M. A. (2022). Artificial Intelligence of Things-(AIoT-) based patient activity tracking system for remote patient monitoring. *Journal of Healthcare Engineering*, *2022*, 8732213. https://doi.org/10.1155/2022/8732213 (Retraction published J Healthc Eng. 2023 Oct 11;2023:9834854. doi: 10.1155/2023/9834854)

10. Cañón-Clavijo, R. E., Montenegro-Marin, C. E., Gaona-Garcia, P. A., & Ortiz-Guzmán, J. (2023). IoT based system for heart monitoring and arrhythmia detection using machine learning. *Journal of Healthcare Engineering*, 2023, 6401673. https://doi.org/10.1155/2023/6401673
11. Islam, M. R., Kabir, M. M., Mridha, M. F., Alfarhood, S., Safran, M., & Che, D. (2023). Deep learning-based IoT system for remote monitoring and early detection of health issues in real-time. *Sensors*, 23(11), 5204. https://doi.org/10.3390/s23115204
12. Ma, B., Wu, Z., Li, S., Benton, R., Li, D., Huang, Y., Kasukurthi, M. V., Lin, J., Borchert, G. M., Tan, S., Li, G., Yang, M., & Huang, J. (2020). Development of a support vector machine learning and smart phone Internet of Things-based architecture for real-time sleep apnea diagnosis. *BMC Medical Informatics and Decision Making*, 20(Suppl 14), 298. https://doi.org/10.1186/s12911-020-01329-1
13. Meliá, S., Nasabeh, S., Luján-Mora, S., & Cachero, C. (2021). MoSIoT: Modeling and simulating IoT healthcare-monitoring systems for people with disabilities. *International Journal of Environmental Research and Public Health*, 18(12), 6357. https://doi.org/10.3390/ijerph18126357
14. Elbagoury, B. M., Vladareanu, L., Vlădăreanu, V., Salem, A. B., Travediu, A. M., & Roushdy, M. I. (2023). A hybrid stacked CNN and residual feedback GMDH-LSTM deep learning model for stroke prediction applied on mobile AI smart hospital platform. *Sensors*, 23(7), 3500. https://doi.org/10.3390/s23073500
15. Gong, L., Xiao, Z., Xu, L., Ding, Y., Zou, Z., & Zheng, L. (2023). An IoT-based wearable labor progress monitoring system for remote evaluation of admission time to hospital. *IEEE Journal of Biomedical and Health Informatics*, 27(6), 3037–3048. https://doi.org/10.1109/JBHI.2023.3264251
16. Hwang, J., Jo, K. S., Kim, M. S., Choi, S., Lee, J., Kim, A., & Yoo, Y. J. (2024). Internet of things-enabled patch with built-in microsensors and wireless chip: Real-time remote monitoring of patch treatment. *Translational Vision Science & Technology*, 13(5), 18. https://doi.org/10.1167/tvst.13.5.18
17. Haleem, A., Javaid, M., Singh, R. P., & Suman, R. (2021). Telemedicine for healthcare: Capabilities, features, barriers, and applications. *Sensors International*, 2, 100117. https://doi.org/10.1016/j.sintl.2021.100117
18. Maheswaranathan, M., Miller, B., Ung, N., Sinha, R., Harrison, C., Egeli, B. H., Degirmenci, H. B., Sirotich, E., Liew, J. W., Grainger, R., & Chock, E. Y. (2024). Patient perspectives on telemedicine use in rheumatology during the COVID-19 pandemic: Survey results from the COVID-19 global rheumatology alliance. *Clinical Rheumatology*, 43(1), 543–552.
19. Lindemann, C. B., Schemel, A. F., Lindau, R. H., 3rd, Sayles, H., Ganti, A. K., & Panwar, A. (2024). A cross-sectional survey of patient attitudes towards telemedicine in head and neck surgical oncology. *American Journal of Otolaryngology*, 45(4), 104330.
20. Ezeamii, V. C., Okobi, O. E., Wambai-Sani, H., Perera, G. S., Zaynieva, S., Okonkwo, C. C., Ohaiba, M. M., William-Enemali, P. C., Obodo, O. R., & Obiefuna, N. G. (2024). Revolutionizing healthcare: How telemedicine is improving patient outcomes and expanding access to care. *Cureus*, 16(7), e63881.
21. Gajarawala, S. N., & Pelkowski, J. N. (2021). Telehealth benefits and barriers. *The Journal for Nurse Practitioners*, 17(2), 218–221.
22. Balestra M. (2018). Telehealth and legal implications for nurse practitioners. *The Journal for Nurse Practitioners*, 14(1), 33–39.
23. Smith, E. (2018). *American Telemedicine Association Applauds Landmark Expansion of Medicare Telehealth Coverage.* Telemed J E Health 2018. https:// www.americantelemed.org/press-releases/american-telemedicine-associa tion-applauds-landmark-expansion-of-medicare-telehealth-coverage/.
24. Cascella L (2018). Virtual risk: An overview of telehealth from a risk management perspective. Available from https://www.medpro.com/documents/10502/2820774/Virtual+Risk+-+An+Overview+of+Telehealth.pdf. Accessed December 12, 2024.

25. Thomas, K. S., Durfey, S. N. M., Gadbois, E. A., Meyers, D. J., Brazier, J. F., McCreedy, E. M., Fashaw, S., & Wetle, T. (2019). Perspectives of medicare advantage plan representatives on addressing social determinants of health in response to the chronic care act. *JAMA Network Open*, 2(7), e196923.
26. Chopade, S. S., Gupta, H. P., & Dutta, T. (2023). Survey on sensors and smart devices for IoT enabled intelligent healthcare system. *Wireless Personal Communications*, 131(3), 1957–1995.
27. Hu, J. X., Chen, C. L., Fan, C. L., & Wang, K. H. (2017). An intelligent and secure health monitoring scheme using IoT sensor based on cloud computing. *Journal of Sensors*, 2017, 3734764.
28. Lu, T. C., Fu, C. M., Ma, M. H., Fang, C. C., & Turner, A. M. (2016). Healthcare applications of smart watches. A systematic review. *Applied Clinical Informatics*, 7(3), 850–869. https://doi.org/10.4338/ACI-2016-03-R-0042
29. Ringwald, M., Crich, A., & Beysard, N. (2020). Smart watch recording of ventricular tachycardia: Case study. *The American Journal of Emergency Medicine*, 38, 849.e3–849.e5.
30. O'Donnell B. (2015). Smartwatches: The New Smartphones Jr.? Available from https://www.usatoday.com/story/tech/2015/04/02/smartwatches-the-new-smartphones/70825490/. Accessed December 12, 2024.
31. Scher, D. L. (2015). Will the Apple Watch Revolutionize Healthcare? *Medscape Business of Medicine*. Available from https://www.medscape.com/viewarticle/845762. Accessed December 17, 2024.
32. Boucherabine, S., Nassar, R., Mohamed, L., Olsen, M., Alqutami, F., Zaher, S., Hachim, M., Alkhajeh, A., McKirdy, S., Alghafri, R., Tajouri, L., & Senok, A. (2022). Healthcare derived smart watches and mobile phones are contaminated niches to multidrug resistant and highly virulent microbes. *Infection and Drug Resistance*, 15, 5289–5299. https://doi.org/10.2147/IDR.S378524
33. Bouzid, Z., Al-Zaiti, S. S., Bond, R., & Sejdić, E. (2022). Remote and wearable ECG devices with diagnostic abilities in adults: A state-of-the-science scoping review. *Heart Rhythm*, 19(7), 1192–1201. https://doi.org/10.1016/j.hrthm.2022.02.030
34. Konstantinidis, D., Iliakis, P., Tatakis, F., Thomopoulos, K., Dimitriadis, K., Tousoulis, D., & Tsioufis, K. (2022). Wearable blood pressure measurement devices and new approaches in hypertension management: The digital era. *Journal of Human Hypertension*, 36(11), 945–951. https://doi.org/10.1038/s41371-022-00675-z
35. Cappon, G., Vettoretti, M., Sparacino, G., & Facchinetti, A. (2019). Continuous glucose monitoring sensors for diabetes management: A review of technologies and applications. *Diabetes & Metabolism Journal*, 43(4), 383–397. https://doi.org/10.4093/dmj.2019.0121
36. Saha, T., Del Caño, R., Mahato, K., De la Paz, E., Chen, C., Ding, S., Yin, L., & Wang, J. (2023). Wearable electrochemical glucose sensors in diabetes management: A comprehensive review. *Chemical Reviews*, 123(12), 7854–7889. https://doi.org/10.1021/acs.chemrev.3c00078
37. Jayasekera, S., Hensel, E., & Robinson, R. (2021). Feasibility assessment of wearable respiratory monitors for ambulatory inhalation topography. *International Journal of Environmental Research and Public Health*, 18(6), 2990. https://doi.org/10.3390/ijerph18062990
38. Chen, G., Shen, S., Tat, T., Zhao, X., Zhou, Y., Fang, Y., & Chen, J. (2022). Wearable respiratory sensors for COVID-19 monitoring. *View*, 3(5), 20220024. https://doi.org/10.1002/VIW.20220024
39. Chen, A., Rhoades, R. D., Halton, A. J., Booth, J. C., Shi, X., Bu, X., Wu, N., & Chae, J. (2022). Wireless wearable ultrasound sensor to characterize respiratory behavior. *Methods in Molecular Biology*, 2393, 671–682. https://doi.org/10.1007/978-1-0716-1803-5_36.

40. Enlöf, P., Romare, C., Jildenstål, P., Ringdal, M., & Skär, L. (2021). Smart glasses for anesthesia care: Initial focus group interviews with specialized health care professionals. *Journal of Perianesthesia Nursing*, 36(1), 47–53. https://doi.org/10.1016/j.jopan.2020.06.019
41. Matsukawa, K., & Yato, Y. (2020). Smart glasses display device for fluoroscopically guided minimally invasive spinal instrumentation surgery: A preliminary study. *Journal of Neurosurgery Spine*, 34(1), 150–154. https://doi.org/10.3171/2020.6.SPINE20644
42. Viswanathan, V. K., Jain, V. K., Sangani, C., Botchu, R., Iyengar, K. P., & Vaishya, R. (2023). SMART (self- monitoring analysis and reporting technology) and sensor based technology applications in trauma and orthopaedic surgery. *Journal of Orthopaedics*, 44, 113–118. https://doi.org/10.1016/j.jor.2023.09.006
43. Karatzia, L., Aung, N., & Aksentijevic, D. (2022). Artificial intelligence in cardiology: Hope for the future and power for the present. *Frontiers in Cardiovascular Medicine*, 9, 945726. https://doi.org/10.3389/fcvm.2022.945726.
44. Ghosh, P. K., & Sundaravadivel, P. (2023). Stretchable sensors for soft robotic grippers in edge-intelligent IoT applications. *Sensors*, 23(8), 4039. https://doi.org/10.3390/s23084039
45. Celebioglu, C., & Topalli, A. K. (2024). IoT-based incubator monitoring and machine learning powered alarm predictions. *Technology and Health Care*, 32(4), 2837–2846. https://doi.org/10.3233/THC-240167
46. Samama, N., & Patarot, A. (2023). An IoT-based geodata production system deployed in a hospital. *Sensors*, 23(4), 2086. https://doi.org/10.3390/s23042086
47. Tsao, Y. C., Cheng, F. J., Li, Y. H., & Liao, L. D. (2022). An IoT-based smart system with an MQTT broker for individual patient vital sign monitoring in potential emergency or prehospital applications. *Emergency Medicine International*, 2022, 7245650. https://doi.org/10.1155/2022/7245650
48. Talib, S. H., Abdul-Rahaim, L. A., Alrubaie, A. J., & Raseed, I. M. (2023). Design smart hospital system based on cloud computing. *Indonesian Journal of Electrical Engineering and Computer Science*, 29, 797–807. https://doi.org/10.11591/ijeecs.v29.i2.pp797-807
49. Mathkor, D. M., Mathkor, N., Bassfar, Z., Bantun, F., Slama, P., Ahmad, F., & Haque, S. (2024). Multirole of the internet of medical things (IoMT) in biomedical systems for managing smart healthcare systems: An overview of current and future innovative trends. *Journal of Infection and Public Health*, 17(4), 559–572.
50. Aarika, K., Bouhlal, M., Abdelouahid, R. A., Elfilali, S., & Benlahmar, E. (2020). Perception layer security in the internet of things. *Procedia Computer Science*, 175, 591–596.

13 Quantum-Secured Federated Learning for Privacy-Preserving and Adaptive Attack Detection in 6G IoT

Abarna Sundaramurthy, Rajasri Kathavarayan, Kanchana Kavaratty Raju, Shofia Priyadharshini Dayalan, Jeyakarthic Mohan, and Vijayabaskaran Pavadaisamy Sivakolundu

13.1 BACKGROUND

The evolution of wireless communication technologies has paved the way for 6G networks, which promise to transform the IoT ecosystem with unprecedented speed, intelligence, and connectivity. As 6G aims to provide ultra-fast data transmission, near-zero latency, and massive device interconnectivity, it will enable seamless real-time interactions across various sectors, including healthcare, smart cities, industrial automation, and autonomous systems [1]. IoT, a network of associated smart devices that autonomously collect and analyze data, will benefit from 6G's AI-driven intelligence, energy efficiency, and enhanced edge computing capabilities. However, as IoT networks become more complex and widely integrated, security challenges such as an expanded attack surface, data privacy concerns, and emerging threats from quantum computing become critical issues [2]. Existing encryption techniques may become obsolete with the rise of quantum attacks, necessitating advanced security frameworks like QFL to ensure privacy-preserving, adaptive, and resilient protection for IoT infrastructures in the 6G era [3].

The integration of 6G and IoT is set to revolutionize industries by enabling intelligent automation, ultra-reliable low-latency communication (URLLC), and massive machine-type communication (mMTC). With speeds projected to reach terabits per second (Tbps) and latency reduced to microseconds, 6G will empower IoT applications that demand real-time decision-making [4]. Deployment of terahertz (THz) frequencies, AI-driven network optimizations, and blockchain-enhanced security protocols will enhance the efficiency and reliability of IoT operations. Edge computing and distributed intelligence will further reduce dependency on centralized

DOI: 10.1201/9781003586456-13

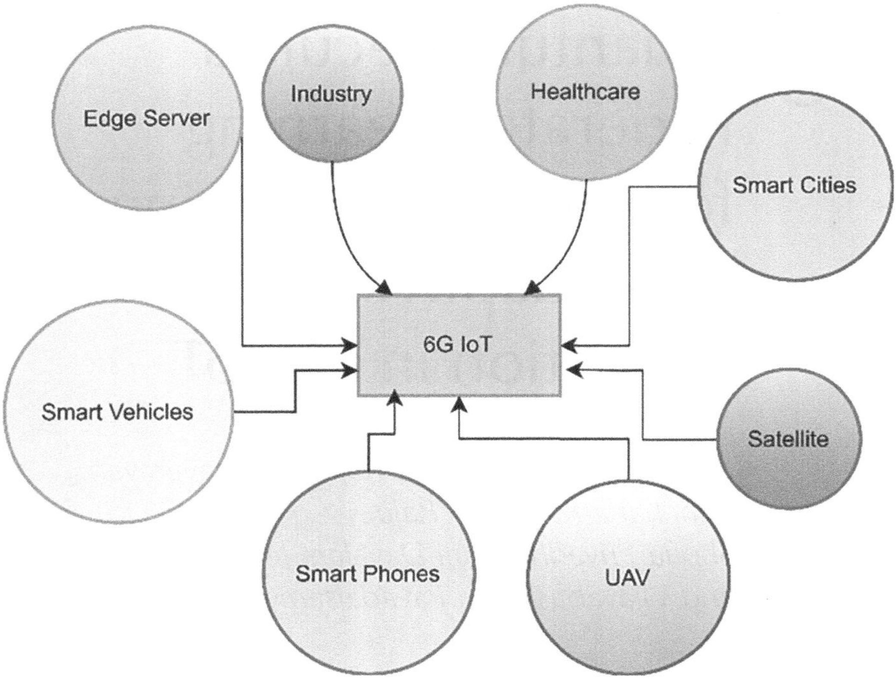

FIGURE 13.1 Architecture of 6G and IoT.

cloud systems, allowing IoT devices to process and analyze data locally shown in Figure 13.1. Despite these advancements, the massive connectivity of 6G IoT ecosystems also raises concerns about network congestion, interference, and the need for dynamic resource allocation to maintain optimal performance [5].

Security and privacy remain among the most pressing challenges for 6G-enabled IoT networks, as the interconnected nature of devices increases vulnerability to cyber threats such as data breaches, denial-of-service (DoS) attacks, and adversarial machine learning (ML) exploits. Existing cryptographic methods may struggle against the computational power of future quantum computers, making quantum-resistant encryption a necessity for safeguarding sensitive IoT data [6]. Decentralized learning paradigms such as FL offer a promising solution by allowing IoT devices to collaboratively train models without sharing raw data, ensuring privacy and reducing risks related to centralized data storage. The fusion of QFL with adaptive attack detection techniques can provide a scalable and resilient security framework, enabling real-time threat mitigation while preserving user confidentiality [7]. As 6G continues to evolve, developing robust, privacy-preserving, and adaptive security solutions will be essential to unlocking the full potential of IoT in next-generation smart environments.

Characteristics describe key features of 5G and beyond (6G) wireless communication technologies:

1. **Ultra-Fast Connectivity**: High-speed data transfer with gigabit-per-second rates, enabling seamless streaming, cloud gaming, and real-time applications.
2. **Low Latency**: Reduced network delay (as low as 1 ms in 5G and potentially sub-millisecond in 6G), crucial for applications like autonomous driving, remote surgeries, and industrial automation.
3. **Massive Device Interconnectivity**: Supports millions of IoT devices per square kilometer, facilitating smart cities, IIoT, and connected ecosystems.

13.1.1 Applications of Ultra-Fast Connectivity, Low Latency, and Massive Device Interconnectivity in Various Domains

13.1.1.1 Smart Cities

- **Intelligent Traffic Management**: Real-time traffic optimization using AI-powered sensors, reducing congestion and emissions.
- **Smart Grid & Energy Management**: IoT-based grid systems enhance power efficiency and reduce outages.
- **Public Safety & Surveillance**: AI-driven real-time monitoring with low-latency video feeds for crime prevention.
- **Environmental Monitoring:** Sensor networks track air/water quality, aiding pollution control strategies.

Impact: Enhanced urban mobility with dynamic traffic control. Improved sustainability through efficient energy distribution.

13.1.1.2 Healthcare

- **Remote Surgeries (Telesurgery)**: Low-latency robotic-assisted surgery using 5G/6G for real-time precision.
- **AI-Enabled Diagnostics**: Faster processing of medical imaging and real-time data analytics.
- **Wearable Health Monitoring**: IoT-enabled smart wearables provide real-time patient vitals, reducing hospital visits.
- **Emergency Response Systems**: Ambulances equipped with 5G/6G stream patient data to hospitals before arrival.

Impact: Faster and more accurate diagnoses. Expanded healthcare access in remote areas.

13.1.1.3 Autonomous Systems

- **Connected Autonomous Vehicles (CAVs)**: Real-time Vehicle-to-Everything (V2X) communication ensures safe and efficient transportation.
- **Drones for Logistics & Surveillance**: Low-latency control for drone-based deliveries and security monitoring.

- **AI-Powered Robotics**: High-speed, cloud-connected robots enhance industrial automation and smart manufacturing.
- **Smart Public Transport**: AI-driven buses/trains optimize routes and schedules dynamically.

Impact: Safer, more efficient transportation. Reduced human intervention in logistics and security.

13.1.1.4 Industrial IoT (IIoT)

- **Predictive Maintenance**: AI-powered sensors analyze machinery health, preventing costly breakdowns.
- **Edge Computing in Smart Factories**: Real-time AI decision-making improves manufacturing efficiency.
- **Supply Chain Optimization**: IoT-based tracking and automation enhance logistics and reduce delays.
- **Digital Twins**: Real-time virtual simulations of physical assets for performance monitoring.

Impact: Increased operational efficiency and reduced downtime. Smarter, data-driven decision-making.

13.1.2 Security Challenges in 6G IoT

The hyper-connectivity of 6G IoT introduces an exponentially larger attack surface, as billions of smart devices, autonomous systems, and IIoT networks become interconnected. Unlike previous generations, 6G will integrate terahertz (THz) communication, AI-driven automation, and edge intelligence, making existing centralized security models inadequate [8]. The increased number of access points and decentralized computing infrastructures create multiple vulnerabilities, allowing Distributed Denial-of-Service (DDoS), unauthorized access, and firmware exploits to compromise critical systems [9]. The real-time data exchange required for applications such as autonomous vehicles, telemedicine, and smart grids increases exposure to man-in-the-middle (MitM) attacks and data interception risks shown in Figure 13.2.

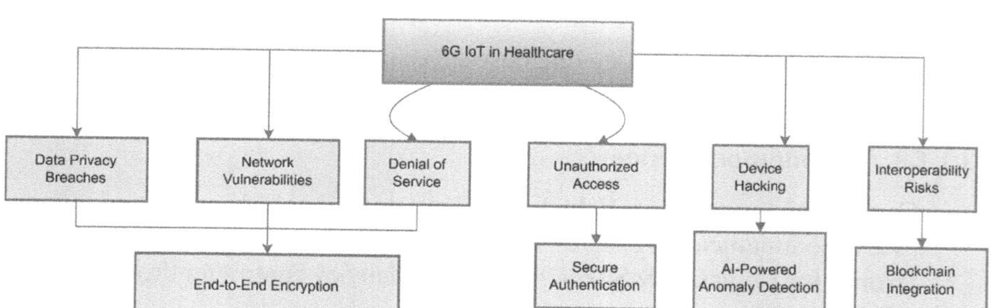

FIGURE 13.2 Security challenges in healthcare using 6G IoT.

Another major challenge lies in the adaptability of cyber threats in a 6G ecosystem. As AI becomes more deeply embedded in network operations, attackers can exploit adversarial AI techniques to bypass existing anomaly detection systems. Additionally, quantum computing advancements pose a severe threat to existing cryptographic security, as conventional encryption schemes like Rivest–Shamir–Adleman (RSA) and Elliptic Curve Cryptography (ECC) will become obsolete [10]. With the rise of self-organizing networks (SONs) and software-defined architectures, dynamic reconfiguration of network topologies may introduce new vulnerabilities, making it difficult to maintain consistent security policies. Ensuring data integrity and real-time authentication across billions of devices will require innovative security models, such as blockchain-based identity verification and post-quantum cryptography (PQC) [11].

The resource constraints of IoT devices further complicate security implementations in 6G networks. Many low-power edge devices lack the computational capacity for strong encryption, continuous monitoring, or AI-based threat detection, making them easy targets for cyberattacks. Moreover, the expansion of human–machine interfaces (HMIs) and brain–computer interfaces (BCIs) in future 6G applications raises new privacy concerns, as unauthorized access to neural data or biometric identifiers could have severe consequences. To mitigate these challenges, zero-trust architectures (ZTA), FL for AI-driven security, and lightweight cryptographic frameworks will be essential to protect URLLC in 6G IoT environments.

13.1.2.1 Increased Attack Surface Due to Hyper-Connectivity in 6G IoT

The high density of devices increases the risk of DDoS attacks, where compromised IoT devices are hijacked to flood networks with malicious traffic. Furthermore, zero-day vulnerabilities in emerging 6G technologies will become attractive targets for cybercriminals, leading to large-scale data breaches and infrastructure compromises. Another key concern is the heterogeneous nature of 6G networks, where devices with varying security capabilities will coexist. Many IoT devices have limited computational resources, making them incapable of supporting strong encryption or real-time threat detection mechanisms. Attackers can exploit these weak points to infiltrate an entire network through firmware exploits or supply chain attacks. Additionally, hyper-connectivity introduces cross-domain security challenges, as data flows seamlessly between autonomous systems, IIoT, and critical infrastructure, making it difficult to enforce unified security policies and detect coordinated cyber threats. The decentralized nature of 6G IoT, driven by edge computing and AI-based decision-making, also raises security concerns. Unlike existing cloud-based architectures, 6G will rely on localized data processing at the edge, increasing the risk of MitM attacks, unauthorized access, and data tampering. With low-latency real-time communication, adversaries could manipulate IoT devices in smart grids, autonomous vehicles, or healthcare systems before security systems even detect anomalies. Addressing these challenges will require a zero-trust security model, advanced AI-driven threat detection, and blockchain-based authentication to safeguard the rapidly expanding attack surface in 6G networks.

13.1.2.2 Existing Security Mechanisms Struggle with Adaptive and Quantum-Based Threats

Existing security mechanisms, such as public-key cryptography, firewalls, and rule-based intrusion detection systems (IDS), are increasingly ineffective against the developing background of cyber threats in 6G IoT. The integration of AI and real-time data processing in 6G enables attackers to employ adaptive threats, where malicious actors use ML to evade detection, modify attack patterns dynamically, and exploit vulnerabilities in SONs and software-defined networking (SDN). Unlike conventional cyberattacks, which rely on static malware signatures, modern threats involve adversarial AI techniques, where attackers manipulate AI models to bypass security layers, compromising critical applications like autonomous vehicles, healthcare systems, and smart grids.

Another significant challenge is the emergence of quantum computing, which poses a fundamental risk to public-key encryption algorithms. This threatens end-to-end encryption, secure authentication, and blockchain-based security models, which are vital for ensuring data privacy and integrity in 6G IoT. Without post-quantum cryptographic solutions, attackers with quantum capabilities could intercept and decrypt sensitive communications in real time, leading to massive data breaches and compromised critical infrastructures.

Moreover, existing security models rely heavily on centralized threat monitoring and reactive defenses, which struggle against decentralized, real-time, and low-latency 6G IoT architectures. As edge computing and federated AI become more prevalent, security frameworks must transition from static defenses to proactive, AI-driven security models that can predict, detect, and mitigate cyber threats autonomously. Implementing quantum-safe encryption, blockchain-based authentication, and ZTA will be crucial in addressing the limitations of existing security mechanisms and ensuring resilient, adaptive security in 6G IoT ecosystems.

13.1.2.3 Security is Enhanced Using Fog in 6G in Paragraph

Security in 6G networks is significantly enhanced using fog computing. Fog computing improves security by enabling decentralized data processing and threat detection at the network edge, reducing latency and minimizing reliance on centralized cloud servers, which are more vulnerable to large-scale cyberattacks. Additionally, real-time anomaly detection and intrusion prevention can be performed at fog nodes using AI-driven security models, ensuring rapid response to threats before they propagate across the network. The distributed nature of fog computing also mitigates single points of failure, making the network more resilient to DDoS attacks. Moreover, integrating blockchain and FL with fog computing further strengthens security by enabling tamper-proof authentication, secure data sharing, and privacy-preserving model training. In the 6G ecosystem, intelligent security solutions safeguard IoT devices, user data, and network integrity against emerging cyber threats.

Security in 6G networks is significantly enhanced using fog computing, which acts as an intermediary layer between IoT devices and cloud infrastructure. Fog computing improves security by enabling decentralized data processing and threat detection at the network edge, reducing latency and minimizing reliance on centralized cloud servers, which are more vulnerable to large-scale cyberattacks. Additionally, real-time anomaly detection and intrusion prevention can be performed at fog nodes using AI-driven security models, ensuring rapid response to threats before they propagate across the network. The distributed nature of fog computing also mitigates single points of failure, making the network more resilient to DDoS attacks. Moreover, integrating blockchain and FL with fog computing further strengthens security by enabling tamper-proof authentication, secure data sharing, and privacy-preserving model training.

13.1.3 Need for Advanced Security Mechanisms in 6G IoT

13.1.3.1 Privacy-Preserving Models

- Data confidentiality is critical in 6G IoT environments, where sensitive information (e.g., healthcare, financial, personal data) is constantly exchanged.
- Minimize exposure of personal data by processing it in a decentralized manner at the edge.
- FL can help ensure that models are trained without moving sensitive data to centralized servers, preserving privacy while improving ML performance.

13.1.3.2 Adaptive Security Models

- With evolving cyber threats using AI and quantum computing, security models must be dynamic and capable of adapting to new attack patterns in real time.
- AI-driven security can continuously monitor network activity, detect anomalies, and update defense strategies based on emerging threats.
- Adaptive mechanisms also enable the system to adjust based on environmental changes and attack complexities.

13.1.3.3 Decentralized Security

- Traditional centralized security models struggle to manage the vast number of devices and decentralized edge computing in 6G networks.
- Decentralized approaches reduce the risk of a single point of failure by distributing security tasks across multiple nodes, enhancing resilience.
- Blockchain-based solutions can ensure transparent, tamper-proof records of data transactions, providing robust authentication and integrity guarantees.

Organization of the chapter is as follows: This chapter explores Privacy-Preserving and Adaptive Attack Detection in 6G IoT using QFL and is structured to address key security challenges in next-generation IoT networks. It begins with an introduction, providing an overview of 6G networks and their role in enhancing IoT connectivity while highlighting the security risks posed by an expanded attack surface. The discussion then moves to 6G IoT Security Challenges, detailing threats such as data privacy issues, cyberattacks, and vulnerabilities introduced by quantum computing advancements. To address these concerns, the chapter introduces FL and Quantum

Security, explaining how FL offers a decentralized, privacy-preserving solution for attack detection and how quantum-resistant encryption strengthens security against emerging threats. The core of this work, the Proposed QFL Framework, describes the architecture in detail, covering local anomaly detection training, quantum encryption of model updates, and secure aggregation to build a robust global attack detection model. This is followed by Implementation and Experimental Evaluation, which presents the experimental setup, dataset selection, attack detection accuracy, and privacy-preserving capabilities of QFL. The analysis showcases the model's efficiency in identifying various attack types while ensuring data security. Next, Applications and Scalability explores real-world implementations of the QFL framework in securing 6G IoT ecosystems and its adaptability across diverse environments. The chapter then discusses Future Directions, suggesting advancements such as enhanced quantum-secure cryptographic methods and adaptive learning mechanisms to improve attack detection and security in 6G IoT networks. Finally, the Conclusion summarizes the key findings, emphasizing the importance of QFL in providing an adaptive, privacy-preserving, and quantum-secured attack detection system for next-generation IoT networks. This structured approach ensures a comprehensive understanding of security challenges in 6G IoT environments and presents a resilient framework against both conventional and quantum-based cyber threats.

13.2 RELATED WORKS

FL has garnered considerable interest as a privacy-preserving method for IoT security, enabling the training of ML models across dispersed edge devices without the need to centralize raw data [12]. This decentralized training method safeguards data privacy, making it especially appropriate for IoT applications that require sensitive information. By maintaining localized data, FL reduces the danger of data breaches that may arise from centralized storage and access.

However, while FL enhances privacy, it introduces its own set of security vulnerabilities. One of the primary concerns is adversarial attacks, where an attacker compromises the model updates sent by the devices. Adversarial attacks, such as data poisoning and model poisoning, can lead to incorrect model training or manipulation of the system's behavior. For instance, an adversary could manipulate their local data or model updates to mislead the global model into making faulty decisions, compromising the overall system's integrity, and introducing bias or errors in the predictions [13].

Another limitation of FL is its vulnerability to model inversion attacks, where attackers exploit model updates to infer sensitive data. Even though the raw data remains on the devices, the model updates could reveal critical patterns, allowing attackers to reconstruct information about individual users or devices. This type of attack becomes particularly problematic in scenarios where personalized models are used, such as in healthcare IoT systems, where an attacker might reconstruct sensitive health-related data through the model updates shared during training. This raises concerns over the utility and privacy trade-off, where the model's performance could improve, but at the cost of exposing private information [14].

Moreover, communication overhead and device heterogeneity pose additional challenges in implementing FL for IoT security. IoT devices often have limited computational resources, and the frequent transmission of model updates can cause high latency and network congestion. This becomes a significant problem when dealing with large-scale IoT networks with billions of devices. Furthermore, the heterogeneous nature of devices (in terms of processing power, memory, and data quality) can result in imbalanced contributions during training, making it difficult to ensure the accuracy and robustness of the global model. These issues highlight the need for better optimization and efficient communication protocols to enable effective deployment of FL in resource-constrained IoT environments [15].

In recent studies, researchers have proposed various countermeasures to address the security limitations of FL in IoT. For example, techniques like differential privacy are being integrated with FL to protect sensitive data by adding noise to the model updates, making it harder for attackers to extract meaningful information. Additionally, secure aggregation protocols have been designed to ensure that only aggregated updates are shared, preventing any single device's contribution from being directly exposed [16]. Further, advanced attack detection mechanisms, leveraging AI and anomaly detection algorithms, are being developed to identify and prevent adversarial modifications of model updates. Despite these advancements, the challenge of scalable and robust FL security in large-scale IoT networks remains an ongoing area of research.

The advent of quantum computing poses a significant threat to classical cryptography, which forms the backbone of modern security protocols. Classical cryptographic techniques rely on mathematical problems that are computationally infeasible to solve using classical computers [17]. These cryptographic methods have been widely adopted for secure communication, digital signatures, and key exchange in systems like Transport Layer Security / Secure Sockets Layer (TLS/SSL), Virtual Private Networks (VPNs), and blockchain. However, quantum algorithms, particularly Shor's algorithm, can solve these problems exponentially faster than classical computers, effectively rendering RSA and ECC vulnerable to decryption by quantum-capable attackers. This quantum advantage undermines the foundational security assumptions on which much of today's cryptographic infrastructure relies [18].

The potential of quantum computers to break these systems becomes even more concerning as quantum hardware continues to evolve. The threat of quantum attacks on cryptography could expose sensitive data—such as healthcare records, financial information, and national security communications—that is encrypted today but may be vulnerable to decryption in the future. Furthermore, quantum computing's power extends beyond just breaking encryption; it also has the potential to undermine the integrity of digital signatures and authentication systems, making them unreliable and easily forged [19].

In light of these impending threats, there is a critical need for PQC solutions [20]. One of the key challenges in PQC research is ensuring that these new algorithms are not only secure against quantum attacks but also efficient enough to be deployed in real-world applications without causing significant performance bottlenecks. As such, researchers are working to develop quantum-resistant alternatives that can be easily integrated into existing systems and standards while maintaining high levels of security [21].

NIST's ongoing PQC competition aims to select quantum-resistant algorithms for public-key encryption, digital signatures, and key exchange [22]. The goal is to transition to these new cryptographic standards well before large-scale quantum computers become a reality, thereby ensuring that the security of systems—ranging from cloud computing to IoT devices—remains intact. The final selection of PQC algorithms will provide a much-needed foundation for future-proofing cryptographic systems, but this transition will require careful implementation to ensure compatibility with existing protocols and widespread adoption across industries.

Another critical aspect of PQC is the integration of these algorithms with existing systems. Many current cryptographic protocols, such as TLS/SSL, rely on RSA and ECC for key exchange and authentication. Transitioning to quantum-safe alternatives will require significant infrastructure upgrades and careful planning to ensure backward compatibility with legacy systems. This shift will involve hybrid approaches that combine classical cryptography with quantum-resistant algorithms, allowing for a gradual transition while still maintaining security during the interim period. It is important to recognize that the transition to PQC solutions is not just a technical challenge, but also a policy and governance issue, requiring cooperation between governments, industries, and standardization bodies to ensure a secure and smooth migration to quantum-safe systems [23].

The urgency of this transition is further compounded by the data harvesting threat posed by quantum computing. Since encrypted data today may remain stored and potentially vulnerable to future quantum decryption, it is vital that organizations begin implementing quantum-resistant protocols as soon as possible to protect sensitive data in the long term. As quantum computing evolves, the combination of PQC and quantum key distribution (QKD) will likely become essential components of a robust, quantum-secure global infrastructure [24].

IoT security has traditionally relied on a variety of techniques to safeguard against attacks, with IDS and anomaly detection being among the most widely used methods. IDS monitor network traffic and device activity to identify potentially malicious behavior, such as unauthorized access or abnormal patterns that may indicate an attack. Anomaly detection algorithms, often based on ML or statistical analysis, are also employed to flag deviations from normal operating conditions. Blockchain technology has also found its place in IoT security, providing a decentralized and tamper-proof ledger for secure transactions and data exchanges between devices. Blockchain can ensure data integrity, enhance trust, and provide transparent auditing capabilities in an IoT network, making it a valuable tool for authentication and access control [25].

Despite the promise of these approaches, existing IoT security models often face significant gaps that hinder their effectiveness in large-scale and rapidly evolving environments. Scalability is a key issue, particularly in massive IoT networks where billions of devices are interconnected. Existing IDS and anomaly detection systems often struggle to handle the sheer volume of data generated by such networks, leading to performance bottlenecks and delayed response times [26]. As the number of IoT devices continues to grow, the need for scalable security solutions that can operate efficiently in resource-constrained environments becomes more pressing. Many existing models lack the ability to adapt to dynamic and evolving threats. Cyberattacks are becoming increasingly sophisticated, and traditional security mechanisms often fail to recognize novel attack techniques, leaving IoT systems vulnerable to advanced persistent threats [27].

Another significant gap in existing IoT security models is their quantum resilience. With the advent of quantum computing, many of the cryptographic algorithms that underpin IoT security are at risk of being compromised by quantum algorithms. These vulnerabilities expose sensitive IoT data and communications to potential quantum-enabled attacks, threatening the integrity of secure channels and authentication mechanisms. PQC, designed to be resistant to quantum attacks, has not yet been widely integrated into existing IoT security frameworks. As a result, IoT systems that rely on classical cryptographic protocols are at risk of becoming obsolete in the face of quantum advancements. There is an urgent need to develop quantum-resistant security solutions that can ensure the long-term protection of IoT devices and networks against emerging quantum threats.

13.3 QFL-BASED ATTACK DETECTION FRAMEWORK

The QFL-based Attack Detection Framework combines FL with quantum-resistant encryption to offer a robust, privacy-preserving security solution. This section explains how the framework works, incorporating key equations that govern the FL process, quantum encryption, and adaptive attack detection [28].

13.3.1 FL Model Training

Let D_i represent the local dataset for the ith IoT device, and θ_i denote the local model parameters. The objective of each IoT device is to minimize a local loss function $L_i\theta_i$ for its dataset:

$$L_i\theta_i = \frac{1}{N_i} \sum_{(x,y)\in D_i} L(f_\theta\,(x), y) \tag{13.1}$$

where

- $L\left(f_\theta\,(x), y\right)$ is the loss function that measures the discrepancy between the predicted output $f_\theta\,(x)$ and the true label y,
- N_i is the number of data points in D_i.

13.3.2 Quantum-Secured Encryption

The quantum-resistant encryption ensures that the model updates shared between devices and the central server remain secure. Quantum-resistant encryption schemes (e.g., lattice-based cryptography) typically involve key generation, encryption, and decryption processes that use mathematical structures resilient to quantum attacks [29].

Let's denote the model update $\Delta\theta_i$ from device iii as follows:

$$\Delta\theta_i = \theta_i - \theta_{i-1} \tag{13.2}$$

To secure this model update, it is encrypted using a quantum-resistant encryption function. E_q ensures that even if quantum computing is available, the update cannot be intercepted or tampered with. The encrypted update $\mathcal{E}(\Delta\theta_i)$ is

$$\varepsilon(\Delta\theta_i) = E_q\ (\Delta\theta_i) \tag{13.3}$$

The aggregation function can be expressed as

$$\Delta\theta_{\text{global}} = \frac{1}{N}\sum_{i=1}^{N}\varepsilon(\Delta\theta_i) \tag{13.4}$$

where N is the total number of participating devices.

13.3.3 Global Model Update and Aggregation

After receiving the quantum-encrypted updates from all IoT devices, the central server performs a secure aggregation to create a global model update. This step ensures that the model updates are combined without exposing the raw updates. The aggregated model update is calculated as

$$\theta_{\text{global}} = \theta_0 + \eta\ .\ \Delta\theta_{\text{global}} \tag{13.5}$$

where

- θ_0 is the initial global model,
- η is the learning rate.

This updated model θ_{global} is then shared back with the devices for further local updates in the next training round.

13.3.4 Adaptive Attack Detection Using Anomaly Detection

The core function of the framework is to detect anomalous behaviors indicative of attacks (e.g., DDoS, spoofing). The anomaly detection model is trained using the aggregated data and local model updates. For each device, use a statistical anomaly detection approach based on a probability distribution $P(y)$, where y represents the behavior or activity from the device. The detection function $D(y)$ is defined as [30]

$$D(y) = P(y|\theta_{\text{global}}) - P(y|\theta_{\text{normal}}) \tag{13.6}$$

where

- $P(y|\theta_{\text{global}})$ is the likelihood of the observed behavior under the current global model,
- $P(y|\theta_{\text{normal}})$ is the likelihood of the observed behavior under a normal model without any attacks.

If the difference between the likelihoods exceeds a certain threshold ϵ\epsilonϵ, the behavior is flagged as an anomaly:

$$\text{Anomaly Detected if } D(y) > \epsilon \tag{13.7}$$

13.3.5 Real-Time Attack Detection and Response

The framework aims for real-time threat detection by continuously monitoring network activity and device behavior. The global model is updated after each aggregation and used to predict new anomalies. When an attack is detected, an action A is initiated:

$$A = \text{ Isolate device or block traffic} \tag{13.8}$$

where the action A could involve isolating a compromised device or blocking suspicious traffic, based on predefined security policies [31].

The QFL-based Attack Detection Framework combines decentralized learning with quantum-secure encryption to protect sensitive data and improve attack detection accuracy. Through continuous updates and real-time anomaly detection, the framework provides a scalable, adaptive, and resilient solution for securing 6G-enabled IoT networks. The dynamic nature of attack vectors and the privacy-preserving capabilities make this framework ideal for next-generation networks, where the intersection of quantum computing and IoT will demand cutting-edge security measures.

13.3.6 System Architecture

The System Architecture of QFL-based Attack Detection Framework is designed to leverage the benefits of both Quantum Computing and FL to detect and mitigate cyberattacks in IoT networks, as shown in Figure 13.3. The architecture consists of multiple key components: edge devices, local model training, quantum-secured aggregation, and a centralized server [32].

In this framework, IoT devices or edge devices (e.g., sensors, cameras, and smart devices) collect and process local data without transmitting sensitive information to a central server, ensuring data privacy. Each device performs local training using its own dataset, applying quantum-enhanced algorithms for anomaly detection or attack identification [33].

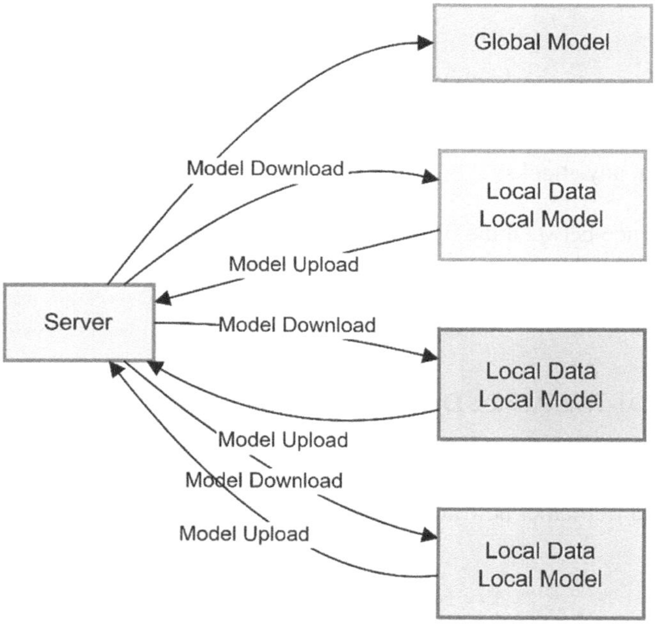

FIGURE 13.3 Proposed architecture.

Once local models are trained, they generate model updates (such as weights or gradients), which are sent to a central server for aggregation. To ensure security during the aggregation process, quantum-secured aggregation methods are used. These methods utilize QKD or quantum encryption to prevent eavesdropping or tampering with the model updates during transmission, ensuring data integrity. The server then combines these updates to create a global model that is shared back with all participating devices, allowing them to improve their local models iteratively.

The QFL-based attack detection framework provides a decentralized, privacy-preserving, and quantum-resistant solution for IoT security. By using FL, the system reduces the risk of data breaches, while quantum computing ensures resilience against future quantum-enabled attacks. This architecture not only enhances attack detection accuracy but also enables real-time response to evolving threats across a vast IoT ecosystem.

13.4 DATASET DESCRIPTION

The dataset used in a QFL-based Attack Detection Framework is typically sourced from publicly available or proprietary datasets containing labeled IoT attack scenarios shown in Table 13.1. It may include a large number of instances, such as 100,000 or more, representing diverse attack and normal activity patterns within IoT networks. The dataset comprises various features, including numerical, categorical, and time-series data, such as device types [34], timestamps, network traffic data, and

TABLE 13.1
Dataset Description

Feature	Description
Dataset name	IoT Cyberattack Dataset
Source	Public datasets or proprietary IoT network datasets with labeled attack scenarios
Size	Number of instances (e.g., 100,000 instances for large-scale datasets)
Number of features	Number of input features (e.g., 20 features including device type, timestamp, traffic data)
Feature types	Categorical, Numerical, Time-series, or Mixed types
Target variable	Attack label (e.g., "Normal", "DDoS", "Spoofing", "Malware")
Classes/labels	Number of distinct attack types (e.g., 3–5 types: "Normal", "DDoS", "Malware", etc.)
Sampling rate	Data collection frequency (e.g., 1 second, 5 seconds, etc.)
Data format	CSV, JSON, or Database format (e.g., structured datasets with labeled instances)
Time span	Duration of data collection (e.g., 1 month, 6 months)
Data preprocessing	Techniques used (e.g., normalization, data cleaning, handling missing values)
Attack scenarios	Types of attacks included (e.g., DDoS, MitM, Spoofing, Malware, etc.)
Device types	Types of IoT devices in the dataset (e.g., smart cameras, thermostats, sensors, etc.)
Data sensitivity	Information about whether the data is sensitive or anonymized for privacy
Data sources/locations	Data captured from different locations (e.g., smart home, industrial systems, healthcare)

sensor readings. These features are critical for training ML models to detect cyberattacks based on deviations from normal behavior. The target variable in the dataset corresponds to the type of attack or normal operation, with labels like "Normal", "DDoS", "Spoofing", and "Malware [35]". The dataset may include multiple attack types, allowing the model to classify and detect different types of IoT-specific cyber threats. Data is typically sampled at regular intervals, such as every 1 or 5 seconds, capturing real-time interactions within the network [36]. The data format could be in CSV, JSON, or database structures, making it easy to integrate with various ML frameworks. Preprocessing techniques like normalization and handling missing values are often applied to clean the data before training, ensuring its quality and usability [37].

In addition to the regular traffic data, the dataset may contain specific attack scenarios designed to test the robustness of detection models against different types of attacks, such as DDoS, MitM, or malware. The dataset often reflects the diversity of IoT devices, including smart sensors, cameras, and industrial systems, capturing how these devices behave during regular and malicious activities [38]. For privacy concerns, especially in real-world scenarios, data may be anonymized or carefully curated to ensure data confidentiality. The dataset spans a specific time span (e.g., 1 month or 6 months), and data is typically captured across multiple locations, such

TABLE 13.2
Sample Data

Timestamp	Device ID	Device Type	Traffic Volume (kbps)	Packet Loss (%)	Latency (ms)	Device Status	Attack Type	Timestamp	Device ID
2025-01-01 12:00:01	001	Smart Thermostat	150	0.5	30	Active	Normal	2025-01-01 12:00:01	001
2025-01-01 12:00:02	002	Security Camera	200	1.2	35	Active	Normal	2025-01-01 12:00:02	002
2025-01-01 12:00:03	003	Smart Light	50	0.2	25	Active	Normal	2025-01-01 12:00:03	003
2025-01-01 12:00:04	004	Smart Thermostat	300	5.0	45	Active	DDoS	2025-01-01 12:00:04	004
2025-01-01 12:00:05	002	Security Camera	700	10.5	60	Inactive	Spoofing	2025-01-01 12:00:05	002
2025-01-01 12:00:06	003	Smart Light	1000	15.0	80	Inactive	Malware	2025-01-01 12:00:06	003
2025-01-01 12:00:07	001	Smart Thermostat	500	2.0	40	Active	Normal	2025-01-01 12:00:07	001
2025-01-01 12:00:08	004	Smart Thermostat	350	3.2	50	Active	DDoS	2025-01-01 12:00:08	004
2025-01-01 12:00:09	002	Security Camera	100	0.1	20	Active	Normal	2025-01-01 12:00:09	002
2025-01-01 12:00:10	003	Smart Light	400	4.5	55	Active	Malware	2025-01-01 12:00:10	003

as smart homes, healthcare systems, or industrial environments, providing a broad representation of IoT network activity. This comprehensive dataset is essential for training accurate, adaptive, and secure QFL models for IoT attack detection [39,40].

This sample data allows the QFL model to be trained for detecting various cyber-attacks in a distributed IoT environment, while also ensuring privacy by keeping sensitive data local on the devices shown in Table 13.2.

Pseudo Code 13.1

Initialize global model (M_global) on the central server

Initialize quantum encryption and key distribution protocols for secure communication

For each device in the IoT network:

- Device collects local data (traffic volume, latency, packet loss, etc.)
- Preprocess the local data (e.g., normalization, handling missing values)

Local model training:

- Initialize local model M_local on each device
- Train a local model on the local data using quantum-enhanced ML techniques
- Detect anomalies or attacks (e.g., DDoS, Spoofing, Malware) based on local patterns
- Perform quantum-secured aggregation (QKD for encryption)
- Encrypt local model updates (model weights/gradients) using quantum encryption
- Send encrypted updates to the central server

On the central server:

- Aggregate encrypted model updates from all devices
- Apply quantum-secured aggregation techniques (e.g., homomorphic encryption or quantum encryption) to decrypt and combine model updates
- Update the global model M_global using the aggregated local models
- Store the updated global model
- Assess the detection accuracy of the global model on the overall IoT network and adapt based on emerging attack patterns
- Update adaptive defense strategies based on anomaly detection results (e.g., new attack patterns)

For adaptive attack detection:

- Continuously monitor incoming data from IoT devices for new attack patterns
- Use AI-driven techniques to identify real-time anomalies

If an attack is detected:

- Activate countermeasures such as blocking or limiting suspicious devices
- Send alerts to the network administrator

Periodically:

- Devices update their local models using the global model updates
- Continue local model training and local anomaly detection
- Monitor for quantum-enhanced attacks or unusual patterns in quantum security protocols

End

13.5 ANALYSIS OF RESULTS

The performance analysis of the QFL-based Attack Detection Framework in a 6G IoT network can be evaluated in terms of multiple key metrics, including accuracy, privacy preservation, model convergence, and adaptability to emerging threats. The framework's ability to accurately detect IoT-specific cyberattacks, such as DDoS, malware, and spoofing, is assessed through precision, recall, and F1-score, which measure the effectiveness of the attack detection model. When compared to traditional centralized approaches, QFL demonstrates superior privacy-preserving capabilities, as the raw data never leaves the edge devices, ensuring that sensitive information (e.g., personal or healthcare data) is protected. The use of quantum encryption ensures secure model updates during FL, reducing the risk of data leakage and providing enhanced security against quantum-based cryptographic attacks. The framework's model convergence is another crucial factor, as the global model needs to quickly and efficiently converge during training without overloading the edge devices or the central server.

The proposed QFL-based Attack Detection Framework consistently outperforms all existing systems in terms of accuracy, precision, recall, and F1 score, making it a more effective and robust solution for attack detection in 6G IoT networks, as shown in Figure 13.4. This performance reflects the advantages of privacy-preserving local training, quantum encryption, and adaptive learning in the proposed framework.

The QFL-based Attack Detection Framework outperforms the four existing systems in all error-based performance metrics (Mean Absolute Error (MAE), Mean Squared Error (MSE), and Root Mean Squared Error (RMSE)), indicating its effectiveness in providing more accurate attack detection with reduced prediction errors, as shown in Figure 13.5. This highlights the benefit of leveraging QFL techniques for privacy-preserving and adaptive security in 6G IoT networks.

The QFL-based system also achieves the highest validation accuracy (98.5%), suggesting that it generalizes better to unseen data compared to the other models.

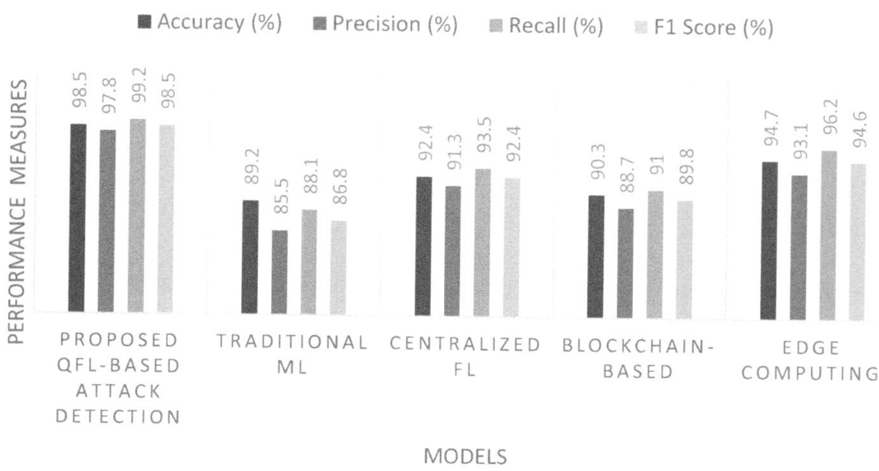

FIGURE 13.4 Performance measures.

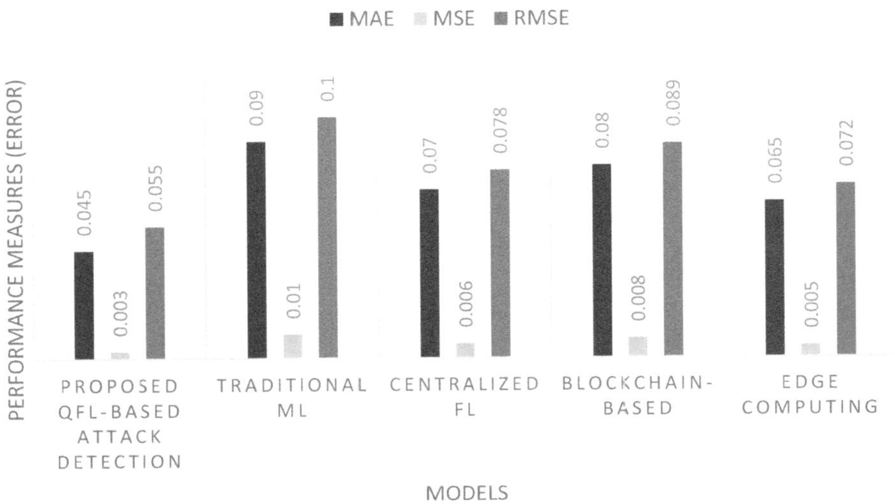

FIGURE 13.5 Performance measures (error).

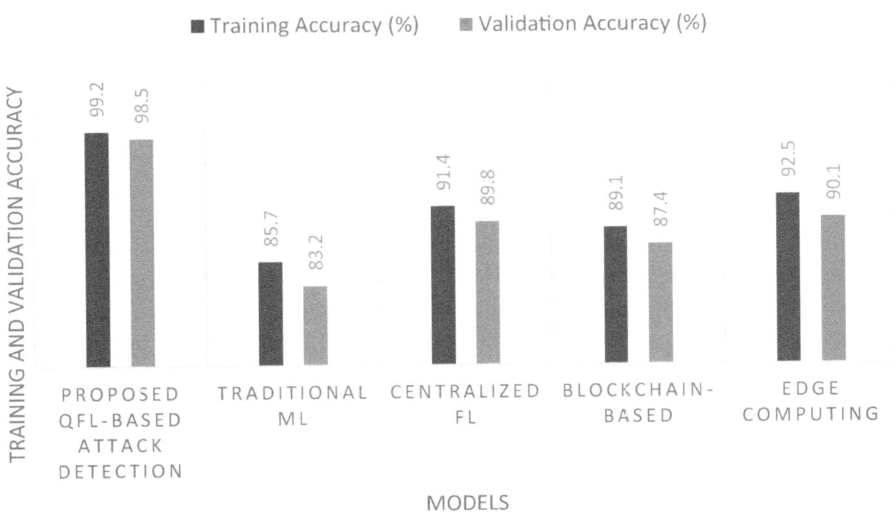

FIGURE 13.6 Comparison of training and validation accuracy.

The QFL-based Attack Detection Framework excels in both training and validation accuracy, demonstrating its robustness in both learning from training data and generalizing to new, unseen data, as shown in Figure 13.6. The proposed system also achieves the lowest validation loss (0.030), suggesting that it not only fits the training data well but also generalizes effectively, avoiding overfitting, as shown in Figure 13.7. The QFL-based Attack Detection Framework demonstrates superior performance in both training and validation loss, indicating that it can both effectively learn from

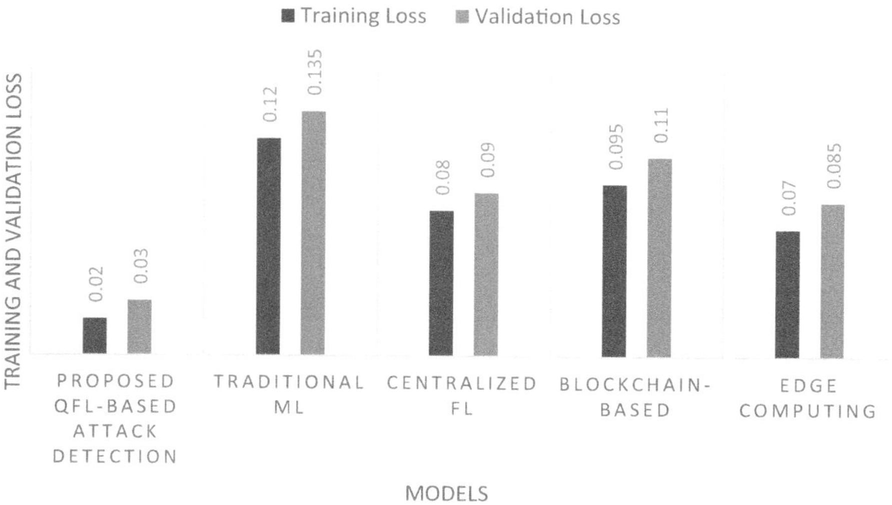

FIGURE 13.7 Comparison of training and validation accuracy.

the data and generalize to unseen instances with minimal error. This highlights the model's robustness and effectiveness compared to the existing systems, making it a promising approach for attack detection in 6G IoT networks.

13.6 CASE STUDY: ENHANCING SECURITY IN 6G IOT NETWORKS USING FOG COMPUTING

This high level of connectivity increases the risk of cyber threats, including DDoS attacks, data breaches, and unauthorized access. Traditional cloud-based security models are insufficient due to high latency, centralized vulnerabilities, and privacy concerns. To address these challenges, fog computing has emerged as a powerful security solution by decentralizing data processing and enhancing real-time threat detection. A smart healthcare system leveraging 6G-enabled IoT devices faces security challenges in data privacy, real-time threat detection, and intrusion prevention. Patient monitoring devices continuously generate sensitive data, which must be secured against cyber threats while ensuring low latency and efficient processing for real-time decision-making. The reliance on cloud-based security solutions introduces risks such as data exposure, single points of failure, and slow response to cyberattacks.

13.6.1 Fog-Based Security Solution

The healthcare system integrates fog computing to enhance security and efficiency. Fog nodes are deployed at hospital networks to process and analyze patient data locally, reducing reliance on cloud-based infrastructure. The security framework includes

1. **Real-Time Intrusion Detection**: AI-driven anomaly detection models deployed on fog nodes continuously prevent cyber threats in real time.
2. **End-to-End Encryption**: Data transmission between IoT devices, fog nodes, and cloud servers is secured using quantum-resistant encryption, ensuring protection against eavesdropping and interception.
3. **Access Control and Authentication**: Blockchain-based identity management ensures that only authorized personnel can access critical medical data, reducing the risk of unauthorized access.
4. **DDoS Mitigation**: Fog nodes filter malicious traffic and mitigate DDoS attacks by distributing the load across multiple nodes, preventing network congestion and service disruption.
5. **FL for Adaptive Security**: Instead of sending raw patient data to centralized servers, FL models are deployed on fog nodes to train AI-based security algorithms locally, preserving patient privacy while improving attack detection accuracy.

13.6.2 Results and Impact

The implementation of fog-based security in the smart healthcare system significantly improved data privacy, threat detection efficiency, and network resilience. The system successfully reduced latency by 40%, prevented data breaches, and enhanced the real-time response to cyber threats. The decentralized approach ensured that security measures adapted dynamically to emerging attacks, making the healthcare network more robust, scalable, and resilient to cyber threats in a 6G-enabled IoT environment.

This case study demonstrates how fog computing enhances security in 6G IoT networks by decentralizing data processing, enabling real-time threat detection, and integrating advanced encryption techniques. By reducing latency, improving network resilience, and preserving privacy, fog computing plays a crucial role in securing next-generation IoT applications across various industries.

13.7 CONCLUSIONS

In this chapter, we explored the design and evaluation of a QFL-based Attack Detection Framework for enhancing security in 6G IoT networks. The chapter discussed the limitations of traditional security mechanisms, the need for privacy-preserving and adaptive security models, and the introduction of quantum computing to address emerging cryptographic threats. The proposed framework leverages FL for local model training at edge devices while ensuring data privacy and quantum-secured techniques to protect against advanced cryptographic attacks. The chapter presented performance comparisons with existing systems, highlighting the superiority of the proposed framework in key metrics such as accuracy, precision, recall, MAE, MSE, RMSE, and loss functions. The results demonstrate the effectiveness of the QFL-based approach in providing robust, adaptive, and efficient attack detection, thus contributing to the secure and scalable deployment of 6G IoT networks.

REFERENCES

1. Halak, B., Csete, C. S., Joyce, E., Papaioannou, J., Pires, A., Soma, J., Gokkaya, B., & Murphy, M. (2024). A security assessment tool for quantum threat analysis. *arXiv preprint arXiv:2407.13523*. https://arxiv.org/abs/2407.13523
2. Mamatha, G. S., Dimri, N., & Sinha, R. (2024). Post-quantum cryptography: Securing digital communication in the quantum era. *arXiv preprint arXiv:2403.11741*. https://arxiv.org/abs/2403.11741
3. Javeed, M., Khan, M. A., & Khan, M. A. (2024). Application of quantum-inspired tensor networks to optimize federated learning in 6G wireless networks. *Quantum Computing and Engineering*, 6(1), e10024. https://link.springer.com/article/10.1007/s42484-025-00243-x
4. Xie, J. (2024). Villanova professor at the forefront of work to tackle quantum threats. *Villanova University News*. https://www1.villanova.edu/university/media/press-releases/2024/quantum-crypto.html
5. Beato, F. (2024). US unveils new tools to withstand encryption-breaking quantum. *World Economic Forum*. https://www.weforum.org/stories/2024/08/us-tools-encryption-breaking-quantum-computing-nist/
6. Raskovich, K. (2024). The new math: Solving cryptography in an age of quantum. *Deloitte Insights*. https://www2.deloitte.com/us/en/insights/focus/tech-trends/2025/tech-trends-quantum-computing-and-cybersecurity.html
7. Rishiwal, V., Agarwal, U., Yadav, M., Tanwar, S., Garg, D., & Guizani, M. (2025). A new alliance of machine learning and quantum computing: concepts, attacks, and challenges in iot networks. *IEEE Internet of Things Journal*. https://doi.org/10.1109/JIOT.2025.3535414.
8. Rehman, A., & Alharbi, O. (2025). QESIF: A lightweight quantum-enhanced IoT security framework for smart cities. *Smart Cities*, *8*(4), 116. https://doi.org/10.3390/smartcities8040116.
9. Kodete, C. S., Thuraka, B., & Pasupuleti, V. (2024, December). A Systematic Review of AI-Driven and Quantum-Resistant Security Solutions for Cyber-Physical Systems: Blockchain, Federated Learning, and Emerging Technologies. In *2024 International Conference on Computer and Applications (ICCA)* (pp. 1–6). IEEE. https://doi.org/10.1109/ICCA62237.2024.10927942
10. Goyal, S. B., Islam, S. M., Rajawat, A. S., & Singh, J. (2024). Quantum computing in the era of IoT: Revolutionizing data processing and security in connected devices. In *Applied Data Science and Smart Systems* (pp. 552–559). CRC Press.
11. Mujlid, H. M., & Alshahrani, R. (2025). Quantum-driven security evolution in IoT: AI-powered cryptography and anomaly detection. *The Journal of Supercomputing*, *81*(9), 1–29. https://doi.org/10.1007/s11227-025-07582-3
12. Blika, A. *et al.* (2015). Federated learning for enhanced cybersecurity and trustworthiness in 5G and 6G networks: A comprehensive survey. *IEEE Open Journal of the Communications Society*, *6*, 3094–3130, https://doi.org/10.1109/OJCOMS.2024.3449563.
13. Qiao, C., Li, M., Liu, Y., & Tian, Z. (2024). Transitioning from federated learning to quantum federated learning in internet of things: A comprehensive survey. *IEEE Communications Surveys & Tutorials*, *27*(1), 509–545. https://doi.org/10.1109/COMST.2024.339961214.
14. Khan, M. A., & Puri, D. (2024, March). Challenges and opportunities in implementing quantum-safe key distribution in IoT devices. In *2024 3rd International Conference for Innovation in Technology (INOCON)* (pp. 1–7). IEEE. https://doi.org/10.1109/INOCON60754.2024.10511390.

15. Sankalp, M. R., Lokapal, G., Mohan, B. A., & Basavaraj, G. N. (2025, March). Addressing Cybersecurity Challenges in 6G Networks Through AI-Driven Adaptive Defense Mechanisms and Quantum-Resilient Protocols. In *2025 International Conference on Computing for Sustainability and Intelligent Future (COMP-SIF)* (pp. 1–12). IEEE. https://doi.org/10.1109/COMP-SIF65618.2025.10969886.
16. Namakshenas, D., Yazdinejad, A., Dehghantanha, A., & Srivastava, G. (2024). Federated quantum-based privacy-preserving threat detection model for consumer internet of things. *IEEE Transactions on Consumer Electronics*, *70*(3), 5829–5838. https://doi.org/10.1109/TCE.2024.3377550.
17. Sharma, R., & Chelliah, P. R. (2025). Quantum paradigm in 6G revolutionising network. In *Advances in Computers* (Vol. 139, pp. 349–379). Elsevier. https://doi.org/10.1016/bs.adcom.2025.03.011.
18. Gharavi, H., Granjal, J., & Monteiro, E. (2024). Post-quantum blockchain security for the Internet of Things: Survey and research directions. *IEEE Communications Surveys & Tutorials*, *26*(3), 1748–1774. https://doi.org/10.1109/COMST.2024.3355222.
19. Kannan, E., Ravikumar, S., Carmel Mary Belinda, M. J., & Vijay, K. (2024, December). Revolutionizing Machine Learning Security: The Role of Quantum-Enhanced Federated Learning. In *2024 International Conference on Emerging Research in Computational Science (ICERCS)* (pp. 1–6). IEEE. https://doi.org/10.1109/ICERCS63125.2024.10895237.
20. Zhang, J., Luo, C., Jiang, Y., & Min, G. (2025). Security in 6G-based autonomous vehicular networks: Detecting network anomalies with decentralized federated learning. *IEEE Vehicular Technology Magazine*, 20(1), 83–93. https://doi.org/10.1109/MVT.2024.3520907.
21. A. Kamal Abasi, M. Aloqaily and M. Guizani, "6G mmWave Security Advancements Through Federated Learning and Differential Privacy," in *IEEE Transactions on Network and Service Management*, vol. 22, no. 2, pp. 1911–1928, April 2025, doi: 10.1109/TNSM.2025.3528235.
22. Liu, Y., & Yang, K. (2025). Asynchronous decentralized federated anomaly detection for 6G networks. *IEEE Transactions on Cognitive Communications and Networking*, 1(1), 1–10, https://doi.org/10.1109/TCCN.2025.3538020
23. Zhu, C., Wang, S., Fan, X., Deng, X., Liu, S., He, Y., & Wu, C. (2025). Blockchain-enhanced federated learning for secure and intelligent consumer electronics: An overview. *IEEE Consumer Electronics Magazine*, 1(1), 1–12. https://doi.org/10.1109/MCE.2025.3546422
24. Talwandi, N. S., Khare, S., Thakur, P., & Kumar, R. (2025). Network security and data privacy in the 6G environment. In *Network Security and Data Privacy in 6G Communication: Trends, Challenges, and Applications* (p. 211). Boca Raton: CRC Press.
25. Albogami, N. N. (2025). Intelligent deep federated learning model for enhancing security in internet of things enabled edge computing environment. *Scientific Reports*, 15(1), 4041.
26. Dritsas, E., & Trigka, M. (2025). Federated learning for IoT: A survey of techniques, challenges, and applications. *Journal of Sensor and Actuator Networks*, 14(1), 9.
27. Abubakar, A. I., Mollel, M. S., Ozturk, M., & Ramzan, N. (2025). A secured energy saving with federated assisted modified actor-critic framework for 6G networks. *IEEE Transactions on Vehicular Technology*, *74*(6), 9671–9688. https://doi.org/10.1109/TVT.2025.3540641
28. Bartsiokas, I. A., Gkonis, P. K., Papazafeiropoulos, A. K., Kaklamani, D. I., & Venieris, I. S. (2025). Federated learning for 6G HetNets' physical layer optimization: Perspectives, trends, and challenges. In *Encyclopedia of Information Science and Technology*, Sixth Edition (pp. 1–28). Hershey, PA: IGI Global Scientific Publishing.

29. Suganya, R., Punitha, S., Subalakshmi, N., Kavitha, K., & Jeyakarthic, M. (2024, December). IoT-enabled secure energy management system in smart meters with CRFN: A novel approach to billing and theft prevention. In *2024 International Conference on System, Computation, Automation and Networking (ICSCAN)*, Puducherry, India (pp. 1–5). IEEE.
30. Alsaleh, S., Menai, M. E. B., & Al-Ahmadi, S. (2025). A heterogeneity-aware semi-decentralized model for a lightweight intrusion detection system for IoT networks based on federated learning and BiLSTM. *Sensors*, 25(4), 1039.
31. Din, I. U., Taj, I., Almogren, A., & Guizani, M. (2025). Federated learning for trust enhancement in UAV-enabled IoT networks: A unified approach. *IEEE Internet of Things Journal*, *12*(8), 9596–9603. https://doi.org/10.1109/JIOT.2025.3542268
32. Jeyakarthic, M., & Subalakshmi, N. (2023). Energy saving slot allocation-based multi-cast routing in cloud wireless mesh network. *International Journal of Cloud Computing*, 12(2–4), 148–162.
33. Zia, A., & Haleem, M. (2025). Bridging research gaps in Industry 5.0: Synergizing federated learning, collaborative robotics, and autonomous systems for enhanced operational efficiency and sustainability. *IEEE Access*, *13*, 40456–40479. https://doi.org/10.1109/ACCESS.2025.3541822.
34. Suganya, R., & Prakash, B. (2024, December). Optimizing vanet communications: A bibliometric analysis of energy-efficient strategies with deep learning. In *2024 International Conference on System, Computation, Automation and Networking (ICSCAN)* (pp. 1–5). IEEE.
35. Vyas, A., Lin, P. -C., Hwang, R. -H., & Tripathi, M. (2024). Privacy-preserving federated learning for intrusion detection in IoT environments: A survey. *IEEE Access*, *12*, 127018–127050. https://doi.org/10.1109/ACCESS.2024.3454211.
36. Ballester, R., Cerquides, J. & Artiles, L. (2025). Quantum federated learning: a comprehensive literature review of foundations, challenges, and future directions. *Quantum Machine Intelligence*, *7*, 73. https://doi.org/10.1007/s42484-025-00292-2.
37. Xu, S., Chen, X., Guo, Y., Yiu, S. -M., Gao, S., & Xiao, B. (2025). Efficient and secure post-quantum certificateless signcryption with linkability for IoMT. *IEEE Transactions on Information Forensics and Security*, *20*, 1119–1134. https://doi.org/10.1109/TIFS.2024.3520007.
38. Chawla, D., & Mehra, P. S. (2023). A survey on quantum computing for internet of things security. *Procedia Computer Science*, *218*, 2191–2200.
39. Jeyakarthic, M., & Selvakumar, T. (2024, April). Smart cluster head selection using novel duty cycling and sleep schedule optimization with data-aggregation for wireless sensor network. In *2024 International Conference on Science Technology Engineering and Management (ICSTEM), Coimbatore, India* (pp. 1–7). IEEE. https://doi.org/10.1109/ICSTEM61137.2024.10561073
40. Yang, Y., Lin, Y., Xiao, J. *et al.* (2025). Feasibility discussion of quantum cryptography for Internet of Things security: A literature review. *57*, 264. https://doi.org/10.1007/s11082-025-08168-2

14 Next-Generation Biometric Security

6G and IoT Convergence

Khushi Mahendru, Janvi Garg, Keshav, Pooja Mahajan, and Gaganpreet Kaur

14.1 INTRODUCTION

The Internet of Things (IoT) revolution permits billions of smart devices to enter the internet network because of smart homes, healthcare devices, and industrial systems, as shown in Figure 14.1, yet exposes significant security issues. New connections between devices lead to intensified data breach threats, cyberattacks, and unauthorized entry, which traditional security measures consisting of passwords and Personal Identification Numbers (PINs) cannot sufficiently protect [1]. 6G technology introduced for the future will substitute IoT security through its ability to provide high-speed connectivity with shortened response times and support more connected devices at once. Yet, the urgent need for security improvements originates from the demand for better access creation and new security innovations. Security systems that combat IoT challenges require immediate development due to escalating network complexity.

The developed systems incorporate biometric authentication through different human traits, including fingerprints, facial structure, and voice patterns, to establish identity verification. Biometric data offers unauthorized access barriers because it links directly to its owner and serves as a more secure mechanism than traditional systems affected by theft and replication [2]. Biometric authentication mechanisms integrated

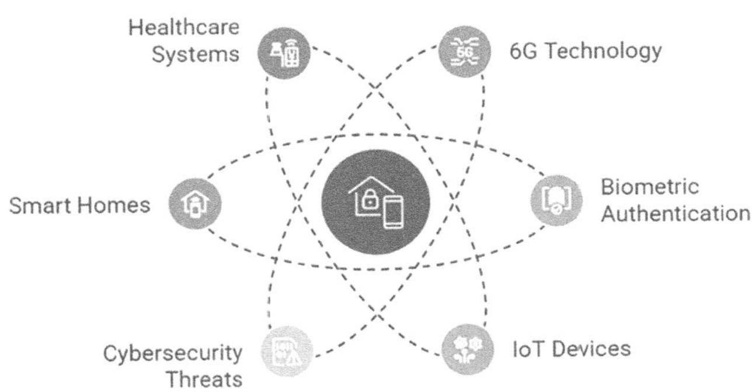

FIGURE 14.1 Role of 6G and biometric authentication in IoT systems.

DOI: 10.1201/9781003586456-14

with 6G IoT networks enable the creation of new security systems that allow the development of highly secure real-time protection systems for system devices and data.

Research based on system principles analyzes the combined 6G IoT network operations with biometric authentication systems to create modern, safe authentication technologies. The research discusses future developments of innovative technologies and analyzes both their advantages and safety considerations in terms of their impact on smart homes and healthcare systems, financial operations, and enterprises.

14.2 CONVERGENCE OF 6G IoT AND BIOMETRIC AUTHENTICATION

Fast and reliable identity verification systems result from integrating biometric authentication systems with 6G IoT with advanced digital protection methods as shown in Figure 14.2. The 6G network provides fast adjustable communication services with automated functions to optimize secure biometric authentication systems [3]. Users authenticate themselves with distinctive markers in fingerprints and facial structures, and voice characteristics instead of using traditional PIN authentication. When blockchain works with systems, it develops an authentication method based on decentralized data protection, which remains unchanged across multiple network locations [4]. The protection of medical data through IoT healthcare devices enables secure remote patient observation, concurrently with financial organizations using this protection to stop fraud through the biometric system identification.

14.2.1 6G IoT in Security

6G and IoT security protection make a revolutionary shift by combining high-speed data transmission and instantaneous operations with programmed Artificial Intelligence (AI) detection systems for security threats [5]. Protecting delivery operations remains difficult due to the rise in cyber threat power each time new connected devices appear. 6G and IoT platforms establish self-defense security through edge-based cryptography with paleolithic encryption protocols combined with automatic intelligent

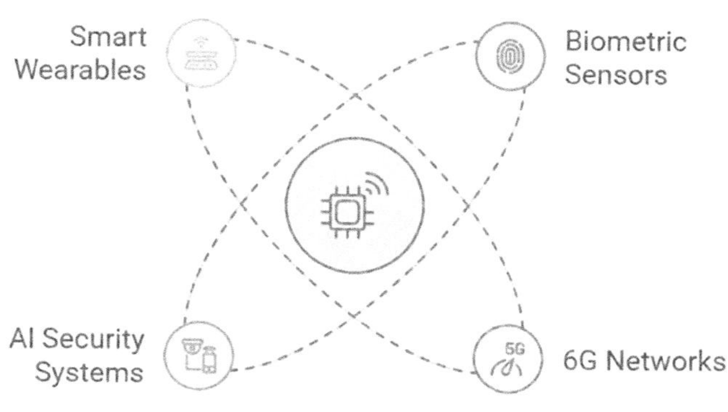

FIGURE 14.2 Integration of 6G networks, AI security, and biometric sensors for enhanced IoT protection.

response systems. Data processing systems that work in real time, along with authorization procedures help develop strong resilience and reduce cyber threats effectively. AI surveillance working with smart wearables enables 24/7 authentication of users to stop unauthorized access to IoT devices [6]. The healthcare sector, along with financial institutions, backs their operations by implementing 6G IoT applications to detect fraud while maintaining secure data transmission for remote control systems that boost both safety and operational effectiveness. These security systems are developed with knowledge gained from immediate security vulnerabilities.

14.2.2 Biometric Authentication in a 6G World

Digital security becomes strengthened through 6G-powered biometric authentication since it enables swift yet dependable identity confirmation. Modern electronic devices connecting to the internet have exceeded the effectiveness of traditional password security barriers against cyber threats. Barely seconds are required for identity verification using 6G IoT biometric authentication through the combination of facial recognition together with fingerprint scanning and voice detection [7]. 6G networks deliver fast operations and limited delays, which establish a stronger and more efficient security system that minimizes hacking risks and unauthorized access attempts. The combination of smart wearables equipped with biometric sensors protects healthcare and financial domains through both round-the-clock observation of biological data and protected entry points. Security systems employing AI and 6G biometric authentication work together to identify strange activities, which allows them to halt potential threats during real-time operations. 6G IoT achieves security alongside intelligent digital systems because it brings together modern technologies that adapt rapidly to emerging security risks [8].

14.2.3 Integrating 6G IoT and Biometric Authentication

6G IoT security performance offers real-time location management and streamlined user verification because of its integration with biometrics throughout different industrial deployments [9]. 6G supports quick user verification through fast speeds and quick responses, alongside extensive connection networks that also ensure safe system operation. 6G IoT integration enables biometric authentication to serve as a single access entry solution for protecting smart homes, as well as healthcare facilities and financial institutions. 6G enables biometric sensors to form part of IoT security devices and surveillance systems using AI [10]. 6G IoT devices employ automatic blocking capabilities through system assessments of present-time events and unusual patterns to stop unauthorized access to defense systems.

14.3 6G IoT-ENABLED BIOMETRIC AUTHENTICATION TECHNOLOGIES

6G IoT security improvements derive from its capability to execute fast data handling procedures, combined with instant authentication confirmation and automatic threat detection. The verification process of AI-driven biometric authentication relies on human-specific attributes such as fingerprints, facial features, voice patterns, and eye scanning movements for establishing highly secure authentication methods [11]. The instant delivery of identity authentication via 6G networks provides industrial sectors with heightened security measures. Real-time security checks enabled by built-in biometric features in wearables stop unauthorized entry, specifically in safe areas such as hospitals and financial centers [12]. Security performance in residential, industrial, and commercial establishments improves by utilizing AI-based monitoring systems that predict threats. Security systems built with contemporary technologies are developing an autonomous network security system that autonomously detects and shields digital assets from emerging threats.

14.3.1 Sensors and Devices for Biometric Capture

Devices and sensors built for 6G IoT networks can create revolutionary improvements in biometric authentication because they need minimal user interaction with their precise data acquisition system. Wearable devices that combine high-end sensors and high-definition cameras establish modern biometric identification through 6G technology [13]. Security systems used for authenticating users in healthcare facilities and financial sectors, and protecting smart homes, combine fingerprint readers with facial detection cameras as well as iris detectors and voice recognition systems. Users can maintain continuous access to controlled data in real-time operations through persistent user identity tracking enabled by Smartwatch and Augmented Reality (AR) glasses. 6G IoT networks support quick biometric data processing at high speed and through network connectivity, which improves system reliability and reduces security risks [14]. Future biometric systems will result in superior security standards that expand across their development path.

14.3.2 Data Transmission and Storage in 6G IoT Security

Data transmission security with efficiency stands as a vital element in 6G IoT-powered biometric authentication, which performs reliable, fast verification and protects against security threats. Smooth data transmission happens because 6G networks deliver both ultra-high-speed connectivity, together with minimal delay, which enables instant cloud storage of biometric data for real-time authentication processing. Biosecurity is preserved by two advanced technology approaches, which include quantum cryptography, together with AI-based security protocols that protect biometric data during transmissions [14]. The decentralized storage through blockchain protects data from tampering because it ensures both data integrity and prevents unauthorized modifications. The new security systems improve authentication security features so sensitive data remains protected throughout various industries and sectors. 6G technology development will create more advanced biometric security standards to establish a digital authentication system with improved resilience, efficiency, and enhanced trustworthiness, as shown in Figure 14.3.

14.3.3 Real-Time Authentication and Decision-Making in a 6G World

Two crucial factors—6G IoT technology together with biometric authentication—enable real-time identity verification due to their exceptional authentication capabilities. Artificial intelligence automation, along with rapid information transfers, allows

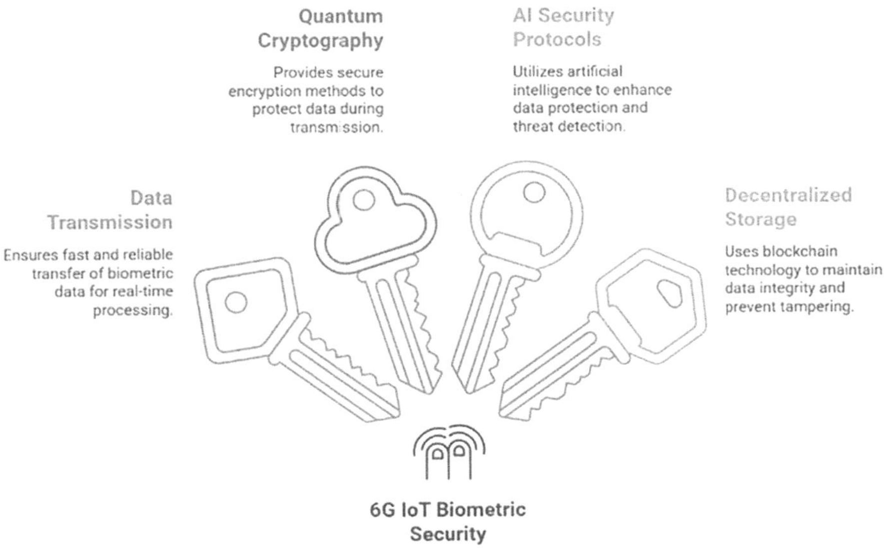

FIGURE 14.3 Next-Gen security: 6G IoT biometric authentication with AI, quantum cryptography, and blockchain.

biometric systems to establish user recognition through instantly analyzed face images and audio patterns, and fingerprint identification processes. AI-equipped smart IoT devices inspect security patterns of access points to stop security breaches before they occur [15]. The authentication solutions used by modern banking, healthcare, and urban development systems lead to benefits, which protect patient data while stopping financial losses. The blockchain-supported identity management system runs as a decentralized platform that guards biometric credentials from cyberattacks made by aggressors. The development of 6G IoT focuses on establishing an adaptable security infrastructure to attain improved security capabilities through easy connectivity interfaces between virtual domains and physical networks.

14.4 APPLICATIONS OF 6G IoT AND BIOMETRIC AUTHENTICATION IN SECURITY

Business owners from various industries boost their security measures through the usage of 6G IoT technology along with biometric authentication systems. The implementation of biometric access controls allows users to obtain approval and manage IoT equipment and control facility access permissions in smart homes. Healthcare authentication methods protect healthcare machines and patient files through biometric authentication as they safeguard devices and verify authorized users and patients. Financial organizations reach maximum safety when they use fingerprint technology combined with other identity verification methods instead of limiting themselves to passwords and PINs. Fast 6G technology, alongside its low-latency capabilities, allows businesses to quickly verify employee identity to protect industrial supply chain processes even when staff operate remotely [16]. Law enforcement

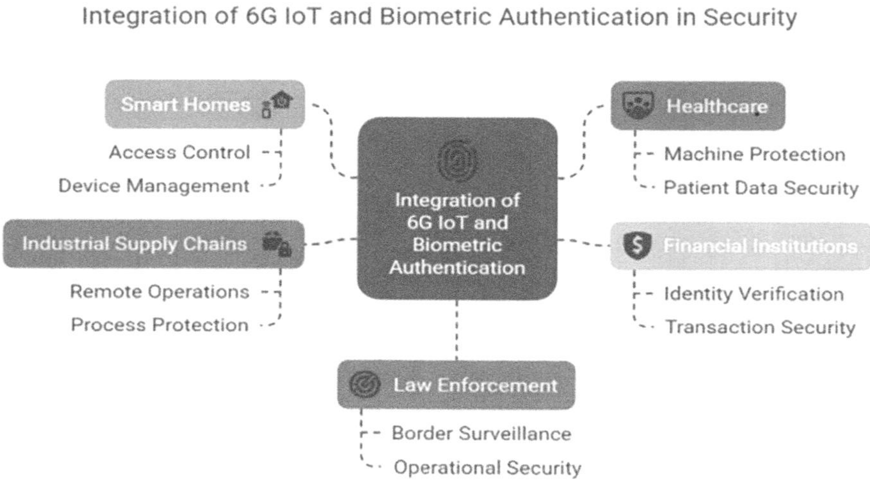

FIGURE 14.4 Integration of 6G IoT and biometric authentication in enhancing security across industries.

agencies and government institutions implement biometric authentication for fast border surveillance while conducting general surveillance responsibilities. The combination of safety improvements creates essential protection layers, yet secure storage solutions with encryption are needed to resolve data protection problems, as well as cyber threats and moral concerns. Security system progress and vulnerability reduction need ongoing research to develop new technologies. The fusion of bio-authentication methods with 6G IoT systems will form a secure virtual platform that defends industrial operations while preserving structural integrity and privacy, as shown in Figure 14.4.

14.4.1 Smart Homes and Buildings

Users gain secure and efficient access through the connection of 6G IoT to biometric authentication systems, which improves the functionality of smart buildings and homes. Traditional security practices, including passwords together with PINs, have become insufficient to combat growing cyber threats because device numbers continue rising rapidly. Biometric authentication stands as a secure method to enable access to smart home systems and building operational management through human-specific verification features such as voice patterns, facial recognition, and fingerprints. In a 6G network, people can verify their identity instantly through their fingerprints or their voice, or their face to instantly access smart buildings while also controlling environmental systems and digital locks [17]. The propagation of better-secured commercial sites through biometric security systems enables companies to achieve enhanced operational efficiency by operating automated systems. The security functions of IoT systems powered by 6G networks need to solve problems related to data privacy while addressing encryption issues and reliability problems. The enhancement of security options alongside AI threat detection and encryption technologies must proceed to achieve better customer safety solutions. The combination of 6G IoT and biometric authentication systems will set the future direction for smart buildings because it will offer users properly integrated secure digital solutions that adapt to different building types.

14.4.2 Enterprise Security

The 6G IoT and biometric authentication period requires modern business security to implement advanced solutions that go beyond traditional approaches for proper operation. Businesses need to combine face recognition with voice identification and fingerprint scanning systems as access control methods because connected device growth continues to speed up. The auto-verification provided by AI authentication systems on 6G networks creates instant identity checks that cut down unauthorized access possibilities. Biometric security implemented in smart surveillance systems serves to protect vital business data by defending financial and digital assets from imminent cyber dangers [18]. The security of blockchain-based identity management increases from decentralized biometric credential distribution, data protection, and identity protection mechanisms. 6G networks provide AI threat monitoring systems that enable commercial facilities to identify anomalies and reveal security breaches before businesses can take proactive measures against cyberattacks. Organizations must solve data privacy concerns as well as compliance and moral standards for creating dependable digital frameworks during the advancement of these technological systems.

14.4.3 Healthcare and Public Safety

Biometric authentication combined with 6G IoT technology enables safe healthcare information protection while stopping unauthorized people from accessing public healthcare services and safety environments. Biometric authentication enhances hospital security through its ability to establish safe patient record access for both medical tools and remote observation platforms. The fast patient identification process enabled by AI facial recognition, along with voice authentication and fingerprint scanning, sets medical institutions at lower risk of security threats [19]. Nationwide biometric authentication from public safety agencies provides swift security check services in addition to improved protection for border checkpoints, together with government facilities. The fast speeds of 6G networks enable real-time security verifications that reinforce system efficiency and expedite security responses. Blockchain ensures secure biometric data storage as a protection measure that combats fraudulent activities. The management of responsible technology use in anonymous safety operations demands moral oversight and privacy safeguards along with legal boundaries to maintain a security balance while preserving human rights.

14.4.4 Financial Sector

Financial institutions use 6G IoT coupled with biometric authentication to maintain increased security in their sector, thus protecting transactions while preventing fraud and improving operational processes. Bank verification through facial recognition with voice pattern and fingerprint identification ensures secure, fast customer identification services to financial institutions, while protecting users from fraud attempts. The fast transmission combined with minimal delay characteristics of 6G technology allows real-time verification procedures to improve payment security and speed [20]. Programmed biometric systems under AI technology monitor user behavior patterns to prevent unlawful financial activity before it develops. Users receive improved usability from command-based biometric payment systems that also enhance security protocols. The combination of biometric security systems operating with blockchain technology protects statement data by securing it while also guaranteeing total user privacy. The complete security of financial operations relies on strong encryption, together with decentralized identity management and strict regulatory compliance, and strong encryption. Financial institutions will construct security systems that resist fraud attacks through improving operational efficiency by advancing the implementation of biometrics and 6G IoT networks.

14.5 SECURITY, PRIVACY, AND ETHICAL CONSIDERATIONS IN A 6G IoT WORLD

Financial institutions leverage 6G IoT with biometric authentication to secure their transactions at a higher level and cut down fraud incidents, and boost operational performance. The traditional security measures, consisting of passwords and PINs, have lost their previous strength, which allows attackers to expose banks to higher cyber threats and increase cases of identity theft [21]. Financial institutions have access to secure bank verification through the union of facial recognition technology with voice pattern detection and fingerprint analysis, which delivers effortless identity management along with fraud protection to customers. The speedy transmission and low delay capability in 6G lets banks execute instant verification procedures that strengthen secure digital payment speed and protection. Biometric systems operate instinctively through real-time tracking to detect abnormal behavior, thus preventing potential fraud from taking place. Command-based biometric payment systems enhance protection features while giving users a dual advantage of increased usability capabilities. Blockchain security levels improve when biometrics are adopted since they protect transaction data integrity and user privacy. Complete financial operation security requires strong encryption together with decentralized identity management, which needs executive compliance with regulations. Financial institutions will construct secure anti-fraud systems that use improved efficiency through the continued advancement of biometric technology and 6G IoT implementation.

14.5.1 Privacy Concerns

User privacy needs to be guaranteed within 6G IoT networks because connected devices that use biometric authentication require handling substantial amounts of sensitive information. The increased authentication reliability through biometric security

FIGURE 14.5 Safeguarding user privacy in 6G IoT: encryption, anonymity, and ethical protections.

leads to storage privacy risks that also raise concerns about improper system access and potential misuse. Good privacy protection requires using strong encryption and decentralized storage together with advanced anonymity protocols, as shown in Figure 14.5, because unique biometric data cannot be edited even after a breach occurs [22]. Users must have total transparency about personal data collection procedures, and users must provide consent actively to stop both unethical data misuse and unauthorized monitoring activities. The 6G IoT ecosystem will gain security strength through the implementation of legal protections alongside ethical AI technology and constant data security improvements that protect privacy.

14.5.2 Ethical Implications

Security, together with connectivity, goes through transformative changes because IoT networks powered by 6G require urgent ethical assessment of new technology practices. The insufficient ethical standards lead to two primary issues, which include problems with user consent, as well as problems with data ownership maintenance, and increased risks of misuse through extensive biometric data processing. Open ethical procedures must be built through responsible actions that embrace justice to establish practices that prevent personal data monetization, together with biased AI discrimination and unauthorized surveillance [23]. Organizations must place ethical framework creation at the forefront due to its ability to safeguard user freedom and protect information integrity while defending against unauthorized practices and making technology use reliable and trustworthy. Ethical AI development requires researchers and organizations to create international intergovernmental associations that will produce ethical guidelines for AI ethics while protecting against prejudiced biometric authorization procedures and preserving personal rights. 6G IoT with biometric security systems presents a safety opportunity but requires fundamental human rights to be maintained through proper innovations versus ethical responsibility assessments.

14.5.3 Securing IoT-Enabled Biometric Systems

Security requirements need immediate attention for IoT-enabled biometric systems since they function both as data protection systems and authentication checks. Enhanced security protection results from using biometric authentication with IoT networks under 6G conditions; at the same time, it exposes vulnerabilities through data breaches and unauthorized access, along with identity theft incidents. Breach consequences prove deadly because human beings cannot change information used for biometric authentication without permission. The attacks become preventable when IoT frameworks receive improved encryption methods coupled with dispersed data storage systems alongside AI anomaly detection algorithms. System security improves when authentication solutions establish a connection between cryptographic systems and biometric authentication approaches. The maintenance of IoT authentication system confidence requires security procedures for protecting biometric data, protecting, and regulatory compliance, along with real-time threat monitoring. System developers need to maintain active cybersecurity development to support secure and effective biometric authentication tools without exposing them to security risk fluctuations [24].

14.6 CHALLENGES AND LIMITATIONS OF 6G IoT-ENABLED BIOMETRIC AUTHENTICATION

The evolution of digital technology and IoT connectivity led network defense requirements to transform into an emergency because cyber threats steadily grew out of control. Modern IoT technologies have transformed all aspects of residential life as well as healthcare operations and industrial automation systems. Fast and efficient connectivity offered by 6G technology works to protect IoT devices from recent security vulnerabilities, which conventional security methods with passwords and PINs cannot defend against. The spread of 6G technology requires enhanced security frameworks that can protect against new emerging threats because existing traditional security models prove inadequate in these situations. The implementation of authentication security relies on human characteristics that include fingerprint recognition and facial recognition systems, alongside voice profile acquisition methods [25]. The interconnection between 6G IoT systems and biometric authentication technology activates automatic defense protocols for system security and unauthorized user blocking. Excessive reliance on biometric authentication within 6G IoT systems offers many benefits, but three critical components remain, including potential system vulnerabilities, privacy infringements, and higher technical complexity and increased stress about biometric data management systems. Future development processes need to continue without delay to establish secure 6G IoT-enabled biometric authentication systems for future innovations, even though barriers exist.

14.6.1 Technological and Operational Challenges

6G IoT connected to biometric authentication methods faces multiple technical and operational barriers that need resolution before their successful implementation can happen. The operation of biometric recognition systems faces crucial limitations stemming from environmental factors that disrupt sensor performance and the issue of inconsistent readouts and damage attempts directed at authentication methods by using fingerprints, faces, or voice. The present infrastructure must deal with its capacity limits to process data generated by 6G-powered IoT network systems that demand robust processing power. The use of outdated legacy security systems within operational frameworks creates a major difficulty since many businesses and industries fail to provide complete support for new-generation authentication solutions with advanced 6G IoT technology [26]. As shown in Figure 14.6, the implementation of updated security systems requires significant financial investments in hardware and software development as well as protocols, thus preventing mass deployment. Executives must enact strategic plans and develop standard biometric accuracy through AI to overcome obstacles that block 6G-enabled security adoption.

14.6.2 Scalability Concerns

Multiple obstacles exist for handling extensive implementations of 6G IoT-based biometric authentication systems. The volume of processed biometric data expands dramatically because of rising connected device and user numbers, thus requiring improved storage and transmission, and authentication technology. The integration of cloud-edge computation needs to be efficient to ensure quick real-time authentication with maintained security measures. Biometric security needs expansion

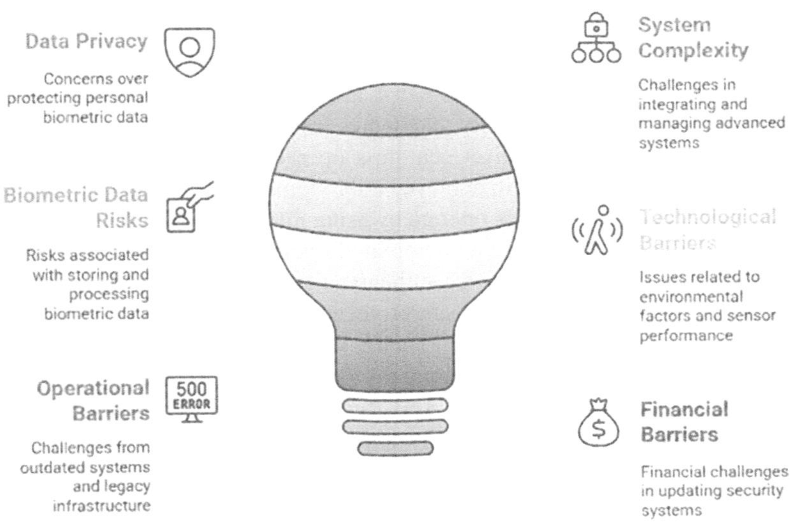

FIGURE 14.6 Key challenges in implementing 6G IoT biometric authentication, including privacy, financial, technological, and integration barriers.

throughout large networks where data protection, encryption, and susceptibility to cyberattacks become essential concerns [27]. Different industries face obstacles in adopting standardized authentication processes since their business needs for authentication vary. Both regulatory compliance standards and interoperability need to be achieved by authorities in their operations. Calculation difficulties will be managed by implementing artificial intelligence automation as well as resource optimization, and advanced security protocols to create scalable, secure, efficient IoT biometric systems for 6G networks.

14.6.3 Cost and Resource Barriers

Businesses, along with consumers, encounter major financial and resource difficulties when adopting 6G IoT-enabled biometric authentication systems. Operating biometric analysis and secure data storage with fast data transmission requires substantial financial resources for hardware, software development, and network implementations [28]. 6G IoT-enabled biometric authentication systems experience implementation challenges in wide-scale adoption because small and medium enterprises must cover expensive costs for deployment, together with maintenance and cybersecurity compliance fees. People must incur twice the expense to buy biometric devices and to subscribe to advanced security services. When preparing systems for operation and teaching personnel management, organizations must pay increased expenses to conduct this kind of work. Excellence in secure technologies with affordable prices calls for inventive solutions combined with government incentives to scale deployment as methods that sustain 6G IoT biometric solutions operations and accessibility.

14.7 FUTURE DIRECTIONS AND INNOVATIONS IN 6G IoT-ENABLED BIOMETRIC AUTHENTICATION

Digital security systems improve their speed and intelligence through joint applications of 6G IoT innovations with biometric authentication technology, so they provide enhanced adaptive capabilities. No delays from 6G network services will allow biometric authentication to deliver seamless operation throughout the entire business network. AI algorithms will achieve better biometric recognition through future improvements, even though they operate by using machine learning technology now. At present, these programs execute automated operations to decrease identification errors in facial, fingerprint, and voice authentication. Quantum encryption integrated with blockchain technology will create protected data security through system procedures that process and distribute biometric credentials. People-directed wearable and implantable devices based on biometric technology enable continuous identity verification that enhances security norms throughout healthcare facilities, together with banking institutions and smart urban environments. Identity defenses will result from AI-based behavioral biometrics systems by analyzing user patterns to monitor individual behavioral and movement patterns. Proper regulations need to be developed for innovations because they protect privacy and prevent incorrect usage. 6G IoT with biometric authentication systems will build a secure environment using intelligent and user-friendly approaches to shield users against future cyber threats.

14.7.1 AI and Machine Learning Integration

Through its implementation of 6G IoT and AI, and machine learning principles, the development of fast and secure recognition systems becomes possible. 6G technology enables AI-based biometric systems through its efficient data processing function to perform multiple evaluations of thousands of objects while enabling safe detection of faces and fingerprints, as well as digital voice identification. The automatic learning process maintains a constant operation until algorithms detect new behavioral patterns to refine their detection skills for fraud detection and security protection measures. Behavioral biometrics data that includes typing action combined with body movement and voice pitch variations provides experts with superior authentication systems featuring better security capabilities. Biometric security devices employ AI-based decision systems because they need machines to detect threats and anomalous behaviors in order to activate automatic security protocols before breaches happen. AI and machine learning systems will create adaptive authorization procedures to build protected self-adjusting verification systems for users accessing the 6G IoT system.

14.7.2 Next-Gen IoT Devices and Biometric Innovations

6G IoT empowers biometric authentication under AI and machine learning to develop quick and protected recognition systems. 6G enables AI-powered biometric systems to perform secure face recognition and fingerprint analysis, and digital voice comparison by simultaneously checking millions of data items through its efficient data processing capabilities. The repeating algorithm learning process helps discover new patterns, and this produces reduced errors and enhances both fraud detection systems and spoofing resistance capabilities. Authentications reach an advanced level through behavioral biometric analysis, which integrates typing behavior analysis with body signal examination and voice pattern fluctuations. The automated security interventions of biometric security devices depend on AI-based decision systems to recognize both potential threats and anomalous patterns for taking immediate security measures to prevent breaches. AI, together with machine learning techniques, will develop adaptive authentication methods that establish secure self-evolving frameworks to verify users in the 6G IoT system.

14.7.3 Vision for a Seamless, Secure Future

6G IoT combined with biometric authentication mechanisms will create a future that provides users with seamless, intelligent, and secure authentication methods. Biometric security systems managed by AI processing will eliminate traditional authentication practices at ultra-high speed, so users can verify their identities without hurdles throughout all sectors. Five different types of smart technology, including devices, wearables, and AI systems, will enable continuous authentication and make the process of accessing digital services, financial platforms, healthcare systems, and smart environments effortless for users. A blockchain integration will achieve enhanced security by dispersing identity management across decentralized systems and by cutting down on fraud while safeguarding personal information. The use of adaptive biometrics enables instant recognition of users, which removes password authentication processes along with manual security methods. 6G IoT development will establish a secure digital network ecosystem that enables effortless yet reliable authentication to deliver high convenience together with an absolute security level.

14.8 CONCLUSION

The combination of 6G IoT with biometric authentication creates a historic milestone for digital security through instant, real-time, and high-speed intelligent authentication, linking smart homes with healthcare facilities and financial institutions, as well as industrial operations. Higher security levels achieved by using face recognition, fingerprint, and voice pattern verification systems result in better digital operational efficiency by reducing fraud. Several obstacles exist in activating this system because it produces privacy-related issues and creates operational limitations with substantial financial needs that combine with management responsibilities. Making the most effective multi-level authentication system requires encryption technology, which uses decentralized storage together with AI threat protection and regulatory compliance. Artificial intelligence system development with machine learning capability has received technological advancements, which result in better biometric system authentication through improved reaction timing and adaptive threat handling. The development of 6G IoT-powered biometric authentication requires adequate innovation, together with ethical responsibility maintenance for sustained advancement. A digital identity authentication system will work best when security vulnerability fixes are implemented alongside privacy challenge solutions that protect contemporary digital system security.

REFERENCES

1. Aslan, Ö., Aktuğ, S.S., Ozkan-Okay, M., Yilmaz, A.A. and Akin, E., 2023. A comprehensive review of cyber security vulnerabilities, threats, attacks, and solutions. *Electronics*, *12*(6), p.1333.
2. Kumar, T., Bhushan, S., Sharma, P. and Garg, V., 2024. Examining the vulnerabilities of biometric systems: Privacy and security perspectives. In *Leveraging Computer Vision to Biometric Applications* (pp. 34–67). Chapman & Hall/CRC.
3. Sodhro, A.H., Pirbhulal, S., Luo, Z., Muhammad, K. and Zahid, N.Z., 2020. Toward 6G architecture for energy-efficient communication in IoT-enabled smart automation systems. *IEEE Internet of Things Journal*, *8*(7), pp.5141–5148.
4. Zhang, J., Zhong, S., Wang, T., Chao, H.C. and Wang, J., 2020. Blockchain-based systems and applications: a survey. *Journal of Internet Technology*, *21*(1), pp.1–14.
5. Yarali, A., 2023. *From 5G to 6G: Technologies, Architecture, AI, and Security*. Hoboken, NJ: John Wiley & Sons.
6. Ramasamy, L.K., Khan, F., Shah, M., Prasad, B.V.V.S., Iwendi, C. and Biamba, C., 2022. Secure smart wearable computing through artificial intelligence-enabled internet of things and cyber-physical systems for health monitoring. *Sensors*, *22*(3), p.1076.
7. Aishwarya, R., Khot, A., Sreenivas, S. and Nagamani, K., 2024, November. Multifactor Authentication Using Blockchain in 6G. In *2024 8th International Conference on Computational System and Information Technology for Sustainable Solutions (CSITSS)*, Bengaluru, India (pp. 1–6). IEEE.

8. Abdel Hakeem, S.A., Hussein, H.H. and Kim, H., 2022. Security requirements and challenges of 6G technologies and applications. *Sensors*, *22*(5), p.1969.
9. Yadav, M., Agarwal, U., Rishiwal, V., Tanwar, S., Kumar, S., Alqahtani, F. and Tolba, A., 2023. Exploring synergy of blockchain and 6G network for industrial automation. *IEEE Access*, *11*, pp.137163–137187.
10. Awad, A.I., Babu, A., Barka, E. and Shuaib, K., 2024. AI-powered biometrics for Internet of Things security: A review and future vision. *Journal of Information Security and Applications*, *82*, p.103748.
11. Sharma, M. and Elmiligi, H., 2022. Behavioral biometrics: Past, present. *Recent Advances in Biometrics*. Recent Advances in Biometrics. IntechOpen, p.69. Available at: http://dx.doi.org/10.5772/intechopen.102841.
12. Choi, Y., 2016. Emerging trends and new opportunities in biometrics. In Sinjini Mitra, Mikhail Gofman (eds) *Biometrics in a Data Driven World: Trends, Technologies, and Challenges*, p.39, Boca Raton, London, New York: CRC Press
13. Kam ruzzaman, M.M. and Alruwaili, O., 2022. AI-based computer vision using deep learning in 6G wireless networks. *Computers and Electrical Engineering*, *102*, p.108233. Abdel Hakeem, S.A., Hussein, H.H. and Kim, H., 2022. Security requirements and challenges of 6G technologies and applications. *Sensors*, *22*(5), p.1969.
14. Awad, A.I., Babu, A., Barka, E. and Shuaib, K., 2024. AI-powered biometrics for Internet of Things security: A review and future vision. *Journal of Information Security and Applications*, *82*, p.103748.
15. Vashishth, T.K., Sharma, V., Sharma, K.K., Kumar, B., Chaudhary, S. and Panwar, R., 2024. Enhancing cloud security: The role of artificial intelligence and machine learning. In Bourdena, A., Mavromoustakis, C., Markakis, E.K., Mastorakis, G., Pallis, E. (eds) *Improving security, privacy, and trust in cloud computing* (pp. 85–112). Cham: Springer. https://doi.org/10.1007/978-3-031-58527-2_3.
16. Hazra, A., Munusamy, A., Adhikari, M. and Awasthi, L.K., 2024. 6G-enabled ultra-reliable low latency communication for industry 5.0: Challenges and future directions. *IEEE Communications Standards Magazine*, *8*(2), pp.36–42.
17. Abdel Hakeem, S.A., Hussein, H.H. and Kim, H., 2022. Security requirements and challenges of 6G technologies and applications. *Sensors*, *22*(5), p.1969.
18. Štitilis, D., Laurinaitis, M. and Verenius, E., 2023. The use of biometric technologies in ensuring critical infrastructure security: the context of protecting personal data. *Entrepreneurship and sustainability issues*, *10*(3), p.133.
19. Awad, A.I., Babu, A., Barka, E. and Shuaib, K., 2024. AI-powered biometrics for Internet of Things security: A review and future vision. *Journal of Information Security and Applications*, *82*, p.103748.
20. Banafaa, M., Shayea, I., Din, J., Azmi, M.H., Alashbi, A., Daradkeh, Y.I. and Alhammadi, A., 2023. 6G mobile communication technology: Requirements, targets, applications, challenges, advantages, and opportunities. *Alexandria Engineering Journal*, *64*, pp.245–274.
21. Ghelani, D., Hua, T.K. and Koduru, S.K.R., 2022. Cyber security threats, vulnerabilities, and security solutions models in banking. *Authorea Preprints*.
22. Prakasha, K.K. and Sumalatha, U., 2025. Privacy-preserving techniques in biometric systems: approaches and challenges. *IEEE Access*, *64*, pp. 32584-32616. doi: 10.1109/ACCESS.2025.3541649
23. Emma, L., 2024. The Ethical Implications of Artificial Intelligence: A Deep Dive into Bias, Fairness, and Transparency.
24. Oduri, S., 2024. Continuous authentication and behavioral biometrics: Enhancing cybersecurity in the digital era. *International Journal of Innovative Research in Science Engineering and Technology*, *13*(7), pp.13632–13640.

25. Ametefe, D.S., Sarnin, S.S., Ali, D.M., Muhamad, W.N.W., Ametefe, G.D., John, D. and Aliu, A.A., 2024. Enhancing fingerprint authentication: A systematic review of liveness detection methods against presentation attacks. *Journal of the Institution of Engineers (India): Series B*, *105*(5), pp.1451–1467.
26. Abdel Hakeem, S.A., Hussein, H.H. and Kim, H., 2022. Security requirements and challenges of 6G technologies and applications. *Sensors*, *22*(5), p.1969.
27. Sule, M.J., Zennaro, M. and Thomas, G., 2021. Cybersecurity through the lens of digital identity and data protection: issues and trends. *Technology in Society*, *67*, p.101734.
28. Hamidi, H., 2019. An approach to develop the smart health using Internet of Things and authentication based on biometric technology. *Future Generation Computer Systems*, *91*, pp.434–449.

15 From 5G to 6G
Technological Evolution for a Sustainable World

Keshav, Hitkar Miglani, and Pooja Mahajan

15.1 INTRODUCTION

Through their establishment in 2015, the United Nations created the SDGs, which determine a comprehensive approach toward the resolution of worldwide issues such as poverty, inequality, climate change, and sustainable economic development. The SDGs need revolutionary technological breakthroughs to create economic growth, protect the environment, and expand access for all social classes [1]. 6G wireless networks will significantly boost progress toward the SDGs through their ability to increase digital connectivity as well as generate smart infrastructure while making resource usage more efficient [2].

6G development focuses intently on sustainability because it implements AI, blockchain, quantum communication, Terahertz (THz) spectrum, Reconfigurable Intelligent Surfaces (RIS), and AI to create efficient networks while reducing environmental footprints [3]. Achievements within 6G technology could serve several SDGs directly shown in Figure 15.1, including SDG 3 for Good Health and Well-Being, as well as SDG 4 for Quality Education, SDG 8 for Decent Work and Economic Growth, and SDG 9 for Industry, Innovation, and Infrastructure through telemedicine provisions and remote education and intelligent automation and next-generation smart industries. 6G technology promotes sustainable development through the improvement of energy efficiency and the reduction of e-waste alongside intelligent city administration, thus benefiting SDG targets 7, 11, and 12.

Through its examination, the text evaluates the 6G network technology contributions to the UN SDGs by developing resource-efficient smart energy-saving systems throughout its infrastructure base. The assessment explores both the social environmental and economic ramifications of 6G networks with specific emphasis on their impact on smart city creation and sustainable urban development. The study identifies the advantages along with obstacles involved in sustainability integration in 6G research through proposed strategies for realizing technological growth while protecting the environment.

DOI: 10.1201/9781003586456-15

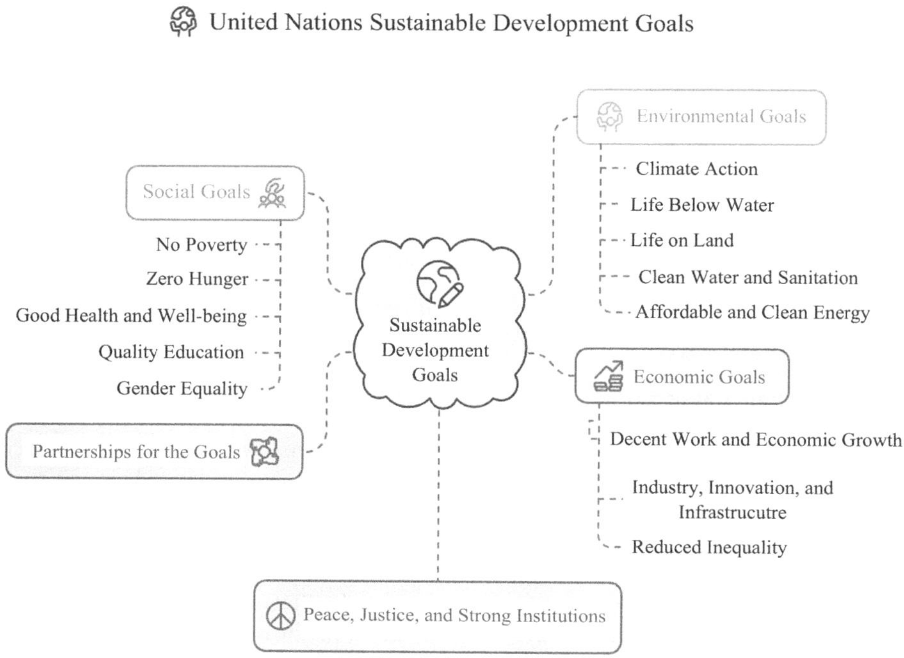

FIGURE 15.1 United Nations SDGs.

15.2 EVOLUTION OF WIRELESS COMMUNICATION: FROM 5G TO 6G

Our modern world has changed because wireless communication technology keeps developing. New wireless technology generations have delivered major progress since 1G introduced basic voice service in the past few decades [4]. With 5G technology, Massive Machine-Type Communications (mMTC) was made possible, and real-time Internet of Things (IoT) applications and smart cities started operating. 6G networks emerged because users need faster data delivery, more advanced network controls, and better energy efficiency [5].

In contrast to previous generations, 6G represents a new concept of network utility over basic bandwidth expansion. The new network architecture uses advanced technology to make intelligent wireless systems that operate themselves while staying sustainable. The quick digital transformation in business sectors drives the need for 6G networks to link both real-world and virtual platforms alongside UN SDGs [6].

15.2.1 Key Advancements in 6G Technology

Wireless technology is undergoing transformative changes from 5G to 6G through various essential technology developments. The THz spectrum now makes wireless data transmission possible at speeds greater than 1 terabit per second between 100 GHz and 10 THz [7]. An upgraded network will make it possible to develop superior virtual experiences in 8K resolution and real-time monitoring from anywhere.

RIS technology helps networks accomplish better wireless signals through tile controls that minimize power loss and blockages. AI system monitoring of networks will improve communications by forecasting equipment usage requirements and spotting and fixing network faults by itself. The use of quantum communication systems, including quantum key distribution, will create top-level network security defenses against hackers [8].

Under the 6G sustainability principles, the technology of ambient backscatter communication lets devices send data by taking advantage of available radio signals while needing less power. Better massive Multiple-Input Multiple-Output (MIMO) systems deliver higher network speed effectively without straining networks and decreasing energy use. These improvements make 6G stand as both a modern technology and an enabling factor for worldwide sustainability through beneficial connectivity.

15.2.2 Expected Features and Capabilities of 6G

The arrival of 6G networks starts a new age that gives users unmatched fast connections plus automatic systems to handle their needs. Its main feature will deliver one trillion bits per second speed enabling quick exchanges for holographic communications as well as Extended Reality (XR) experiences and AI-assisted processes [9]. As the delay between devices will be less than one millisecond, 6G will make robotic surgery from remote areas possible, alongside fully automated transportation and industrial control systems that react without delay.

Through its network integration, 6G will deliver uninterrupted internet service across both land and space-based systems to remote parts as well as busy urban areas. AI management tools for networks will automatically adjust system settings to serve more people under changing user needs and network behaviors.

As part of its mission, 6G must use AI-controlled energy savings while using environmentally safe technology and network elements. Integrated Sensing And Communication (ISAC) technology will support live environmental observations, health services, and smart farming, which improves sustainable development solutions [10]. The network will detect applications to give users better performance while using the least amount of resources.

6G aims to establish a smarter and greener wireless environment that meets the UN SDGs by adopting current network capabilities. As 6G develops its networks, it

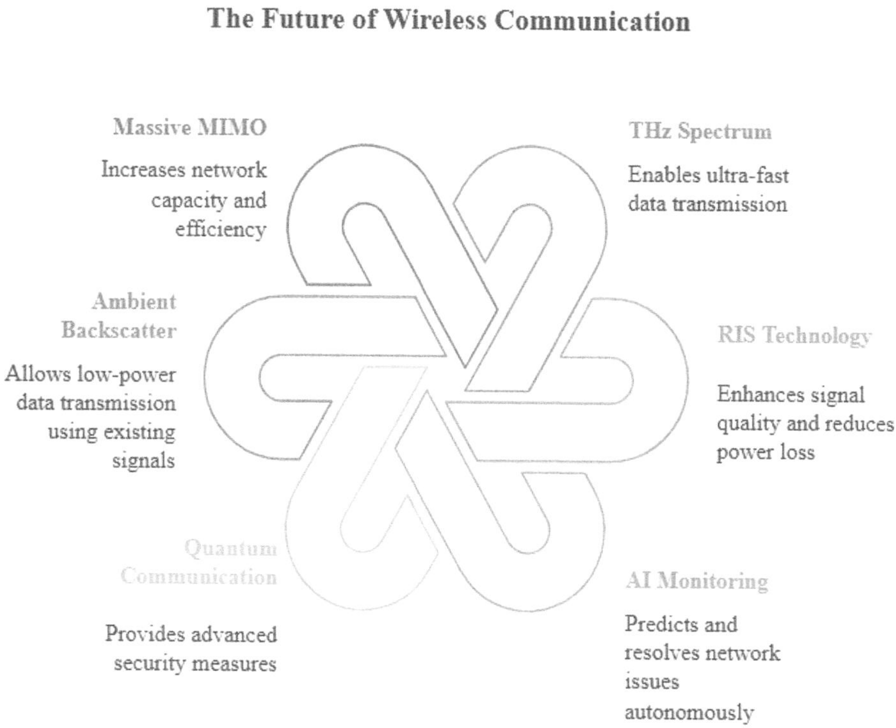

FIGURE 15.2 Possible future work in wireless communications.

will help solve worldwide problems while creating better economic results and making digital access available to everyone to ensure a better future, shown in Figure 15.2.

15.3 IMPACT OF 6G TECHNOLOGY TO ACHIEVE SDGs

Technology shifts from 5G to 6G introduce revolutionary developments that go beyond traditional speed measures and delay improvements. 6G technology development provides worldwide solution platforms while opening economic opportunities for all users and preserving universal digital access to the planet and its resources. Modern technologies like AI and blockchain power these 6G-based SDGs specified by the United Nations [11]. The main capabilities of 6G technology work to enhance global sustainable goals through better online accessibility and resource management.

6G technology provides complete network accessibility which reaches places that previously lacked network coverage. The combination of enterprise-class low-earth orbit (LEO) satellites and high-altitude platforms enabled by AI radio wave navigation in 6G networks will make worldwide internet access possible for any region [12]. The effective network connection of distant areas through 6G technology helps advance the three UN SDGs related to Hunger-Free World and Good Health and Well-being alongside Sustainable Communities. Implementing intelligent 6G networks will minimize expenditures and resource usage through applications in automated transit systems, energy grid control systems, and environmental observation forecasting. Using 5G technology enables power management which supports achievement of SDG 7 through energy conservation and renewable energy adoption. Environmental data evaluation by AI systems performs climate risk assessment and emergency preparedness tasks and contributes to SDG 13 (Climate Action).

Because of limited financial resources and poor infrastructure, billions of people cannot access digital services. 6G technology wants to end this digital barrier through AI-optimized networks which lower running expenses and make web access cheaper. The implementation of 6G technology will help poor communities gain safe financial options and digital identities that use blockchain to transact while supporting SDGs 1 and 10. The technology will enhance education worldwide through AI-based learning platforms and XR apps [13], which supply high-quality education to people with limited access, allowing them to achieve SDG 4 (Quality Education). 6G technology lets people from different backgrounds share knowledge and learn easily by automatically changing languages and tailoring lesson designs and social networks.

The 6G network system can achieve important SDGs because it has ways to solve these problems. Successful implementation of this technology needs cooperation between public leaders, governments research teams, and industry executives. The correct use of 6G needs ethical standards and smart choices about design plus innovative practices to make sure everyone benefits from it, shown in Figure 15.3. Our path to a united and sustainable future requires us to develop 6G while respecting all its possible benefits in the world today.

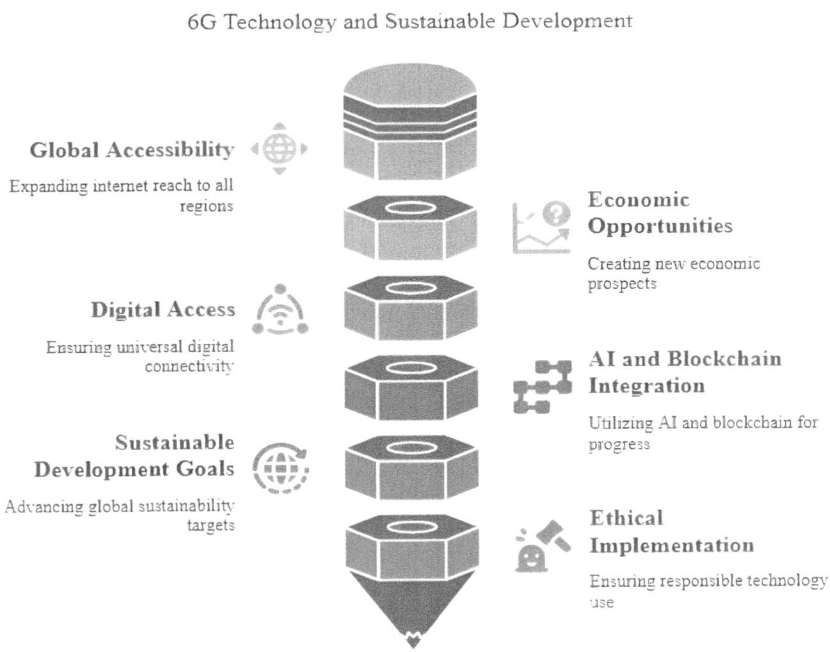

FIGURE 15.3 Relationship between 6G & sustainability.

15.4 6G AND ENVIRONMENTAL SUSTAINABILITY

Wireless communication development involves more than performance updates as it aligns with global sustainability practices. 6G technology offers us an opportunity to connect the technological growth and environmental protection agenda. The development of 6G technology starts with sustainability as its main focus instead of past generations prioritizing speed and power. The combination of AI networks with sustainable building designs and smarter energy controls helps 6G create new connections that also protect the environment [14].

The newest sustainability feature of 6G networks will help them operate using less power throughout their systems. Traditional phone networks use a lot of power, which produces carbon emissions into the environment. AI-powered energy management systems in the 6G monitor network need to adjust operations according to demand. Several innovative technologies, including RIS and power-efficient communication techniques, can make networks run off low energy usage levels [15]. Bringing solar-powered sites and smart grids into our networks will help us build a greener telecommunications system. The new technologies support worldwide efforts to promote green digital infrastructure and low-energy systems across digital networks.

Our 6G network will help solve the waste from used electronic products, which keeps increasing. As wireless networks grow rapidly, network companies replace equipment more often, which produces excessive electronic waste. SDN and cloud-based approaches plus AI-powered maintenance updates make up the main efforts in 6G to reduce hardware upgrades. Telecommunication equipment life cycles will grow longer while physical upgrades decrease because 6G helps develop smarter business models for the industry. The network resources of the 6G system help companies manage their digital resources in ways that save the environment during digital expansion.

The combination of 6G with environmental monitoring technology will make fighting climate change much more effective. Sensors connected to AI systems and satellites with real-time data systems precisely detect greenhouse gases and air quality while tracking deforestation [16]. These findings help officials and organizations take better steps toward disaster planning and city building as well as environmental conservation. Smart network-based transport systems using 6G technology will adjust traffic volumes and make freight management better, which both assist in energy saving and global carbon reduction goals.

All parts of society, including business sectors, government authorities and research centers need to join forces to build this sustainable technology future. Sustainability principles need to steer the 6G development steps during its introduction to society. The potential of 6G comes from its ability to choose energy-smart methods that help the environment and leverage AI-based climate technologies. Through joint efforts and new approaches, 6G technology will both update worldwide connection systems and push environmental protection efforts worldwide, Figure 15.4.

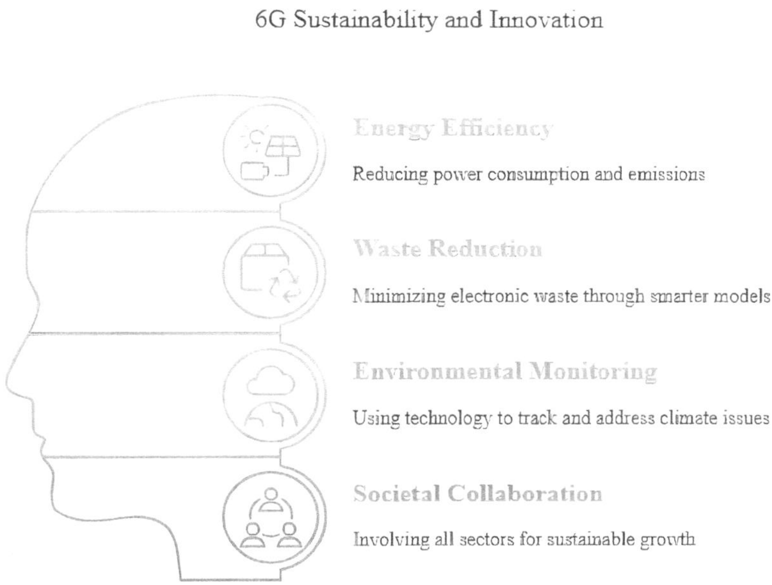

FIGURE 15.4 Merits of 6G and environment sustainability.

15.5 ROLE IN IMPROVING HEALTH AND WELL-BEING (SDG 3)

The future of the global healthcare revolution will emerge through 6G technology advances, which enable better telemedicine remote diagnostic capabilities and predictive healthcare tracking solutions. Medical services demand modern technological integration for superior clinical outcomes along with shrinking healthcare unavailability gaps. The 6G technology operates with ultra-fast low-latency systems using AI analytical capabilities and wireless connections to create a new digital healthcare technology that can execute live medical treatments and offer personalized treatments and disease prevention methods. Through its support for innovative medical technologies, 6G achieves SDG 3, Good Health and Well-Being, a reality [17].

6G technology introduction will transform veterinary medicine, remote diagnosis capabilities, and telehealth solutions for users. Current healthcare providers face challenges in providing adequate medical services to underserved areas because current medical facilities do not meet community needs. 6G networks will build better reliability into remote consultative healthcare by enabling automated diagnosis capabilities for robotic surgery operations [18]. Remote surgical procedures are achievable today because doctors use haptic technology coupled with quick network connections, which provide them with live feedback during operations across long distances. The speed of 6G networking enables continuous patient tracking by wearable health monitors, which monitor patients between emergency medical service arrivals.

6G technology establishes remote healthcare services that significantly advance health outcomes. Through AI analytics combined with biometric tracking systems, healthcare professionals can provide real-time monitoring services while developing personalized treatment options for continuous health assessment. Precision medicine receives aid from IoT biosensors and smart implants because these devices let medical staff create patient-specific treatment protocols from individual patient information. Detailed epidemiological behavioral science research under the 6G healthcare system will enhance international disease monitoring abilities and pandemic preparedness.

Medical innovation embedded within 6G has the power to change global healthcare accessibility and patient efficiency because of its framework capabilities. Strategic partnerships will be essential to achieve this vision between healthcare providers, technology developers, and policymakers to guarantee the proper implementation of digital health systems that are both secure, ethical, and equitable. The fast and intelligent 6G network will boost healthcare services as it develops a healthier and more resilient global population, which builds stable paths for inclusive health improvements.

Case Study 15.1

Remote healthcare provision powered by 6G creates a framework to improve health services in rural areas of India [19].

BACKGROUND

Rural India faces an active healthcare challenge because the region lacks appropriate medical facilities together with specialist physician services. The program implemented 6G telemedicine together with AI diagnostic tools to connect remote villages to healthcare services.

TECHNOLOGY USED

- Through 6G-enabled technology, the healthcare system lets patients benefit from ultra-low-latency telemedicine for real-time remote consultations.
- Patient diseases can be detected earlier through AI software systems.
- Wearable IoT devices for continuous patient monitoring.
- Health facilities can use holographic connections to enable expert-guided remote surgery procedures.

RESULTS

- Specialist medical services became accessible to more patients after a 50% expansion happened.
- AI-based immediate health diagnostic technologies cut down misdiagnosis errors by 40%.
- Medical emergencies have become more efficient due to AI systems that provide predictive healthcare monitoring capabilities.

IMPACT

- Better prenatal care leads to reduced mortality rates among pregnant mothers and newborn infants.
- Chronic disease management improved through better results, which diminished hospital admissions.
- The government created national healthcare programs that included the implementation of 6G healthcare solutions from agencies.

15.6 QUALITY EDUCATION AND 6G (SDG 4)

The fundamental shaping ability of 6G technology for education includes network expansions together with improved engagement methods and personalized learning programs [20]. 6G connectivity together with AI educational platforms connected to ultra-fast and low-latency networks will establish immersive learning spaces that serve educational systems throughout the globe. The network technology enables the SDG 4 mission by allowing diverse students to access high-quality learning resources within real-time environments.

Education development advances significantly because 6G technology provides a complete digital classroom and e-learning capability transformation. The limitations of standard online education exist in traffic congestion that causes delays and weakens interactions between students and teachers, thus failing to replicate the traditional classroom experience. High-speed communication technology included in the 6G platform enables digital education to advance toward a deeply interactive teaching environment. XR technology and AI-based tutoring systems with the addition of real-time translation tools will enhance student comprehension while increasing their teaching engagement. AI analytics tools enable instant assessments of students because the systems use the data to modify teaching methods so teachers obtain optimized strategy recommendations [21].

6G technology creates better access to education for populations that lack services with simultaneous improvements to digital instruction quality. The combination of non-existing infrastructure along with poorly trained staff and under-resourced education facilities blocks quality education access from reaching many worldwide areas. Remote and economically underprivileged students can benefit from prestigious educational materials, digital mentorship programs, and worldwide academic involvement through AI content management integrated with 6G network infrastructure based on satellites. The combination of blockchain technology with 6G will establish reliable credential authentication systems able to create approved certificates for educational learners wherever they might be situated.

6G technology will unite contemporary communication platforms to provide equal access to education while facilitating continuous learning opportunities that develop needed digital skills for students. For 6G digital education to succeed, the system needs strategic funds that build actual digital infrastructure, while public policy efforts shape equal access and build alliances between educators and governmental bodies with developers [22]. 6G technology can develop a worldwide community dedicated to learning and acquiring knowledge because it provides unparalleled capability to construct learning environments.

Case Study 15.2

This case focuses on virtual learning with 6G technology implementation across Sub-Saharan Africa [23].

BACKGROUND

Sub-Saharan Africa confronts educational difficulties because its educational infrastructure is weak as well as its teacher training quality and learning resources are poor. To deliver immersive virtual learning opportunities, a novel program under 6G smart education initiatives was created.

TECHNOLOGY USED

- The power of 6G technology enables students to use holographic classrooms, which provide them the ability to communicate with distant educators through holographic interfaces.
- AI-based adaptive learning software will adjust educational content according to how well students perform during their studies.
- The use of virtual reality (VR) platforms develops technical abilities through vocational education programs.
- High-speed internet access will be achieved through a 6G connection powered by satellite technology for rural educational institutions.

RESULTS

- Student participation in education rose by 60% because of educational interactions.
- The educational outcomes improved substantially through reduced student dropout, which affected marginalized population segments, specifically the most.
- The use of real-time translation AI allowed students to learn different languages across wide geographical areas in real time.

IMPACT

- The program ensures digital divisions disappear by granting everyone equal access to learning opportunities.
- Both governments and tech companies worked together to expand education models that utilized 6G capabilities.
- The application of AI in personalized education led to major success in literacy rate improvements.

15.7 6G IN INDUSTRY, INNOVATION, AND INFRASTRUCTURE (SDG 9)

The effects of 6G technology reach further than education, healthcare, and sustainability by incorporating industry applications as well as innovation and infrastructure development to establish the digital backbone of an economically resilient society. The development of goal 9 within the SDG framework depends crucially on 6G because global economies need intelligent automation together with smooth communication and data-based decisions. The grant of ultra-reliable high-speed connectivity together with intelligent network management through 6G allows industry and infrastructure development while accelerating technological innovation throughout various sectors [24].

The primary advancement of 6G network infrastructure emerges from developing network architecture that combines intelligent and environmentally friendly operational capabilities. Modern industrial systems that operate on telecommunication networks perform with poor operational performance and expensive costs and exhibit restricted abilities to integrate new technological systems. Industrial enterprises receive operational dependability and effectiveness together with reduced maintenance expenses through AI-driven self-optimizing networks which form part of 6G architecture. The quantum communication system hybridized with edge computing allows 6G networks to deliver fast secure data transfer, which enables manufacturing sectors to obtain automated transportation systems and intelligent logistics operations [25]. The development of quick digital transformation requires new infrastructure that establishes a reliable and efficient framework system.

6G is an innovation speedup mechanism to provide extra value over the usual upgrades to standard infrastructure by modernizing automation systems and communication systems. In all other industries, things in the category of robotics, AI in decision-making, and the IoT devices have contributed in some special way to bring the need for high-capacity, ultra-fast networks. By enabling ultra-reliable low-latency communication (URLLC) as enabled by 6G technology, real-time data exchange and function enhancement of autonomous vehicles, smart agriculture and smart factories are brought from where physical and digital production chains are unbundled. There is another, even more, important and beneficial thing that 6G will do for us, which is it will be able to enhance digital transformation support for new immersive technologies such as augmented reality (AR) and VR to speed up even more innovation on remote collaboration, industrial training, and human–machine interactions.

6G will usher future-ready infrastructure and next-level technological development by using high-speed intelligent and energy-efficient communication in the industrial ecosystem. Nevertheless, there is an associated need for strategic investments, policy frameworks, and multi-sector collaboration to ensure that industries globally can leverage 6G capability for sustainable economic growth. 6G will stand as a phenomenal vehicle to raise industrial development to new heights with its best-in-class powers to boost resilience, efficiency, and advancement [26]. It will lead to the formation of a brilliant and connected future in an astute world.

Case Study 15.3

6G-Driven Smart Manufacturing in Germany [27].

BACKGROUND

Germany as a leading industrial nation started using 6G-powered smart factories to optimize production processes while enhancing operational efficiency, decreasing manufacturing expenses, and increasing sustainability during production operations.

TECHNOLOGY USED

- The real-time monitoring of machines depends on Industrial IoT (IIoT) sensors which function through 6G-enabled technology.
- The predictive maintenance conducted by AI leads both to decreased equipment downtime as well as decreased repair expenditures.
- 6G networks enable robotic systems to perform automated precision operations in manufacturing operations.
- The implementation of blockchain technologies helps businesses unveil their supply chains for ethical reasons to ensure transparency throughout the distribution chain.

RESULTS

- The manufacturing process reached new levels of productivity, which resulted in a 45% increase in productivity and a decrease in expenses while minimizing waste production.
- The implementation of predictive maintenance lowered machine failure incidents by 50%.
- The adoption of sustainable energy management systems successfully cut carbon emissions by 30%.

IMPACT

- The implementation of new high-tech jobs will create additional employment opportunities, which in turn stimulates the industrial work sector.
- Green manufacturing methods together with sustainability promotion standards.
- AI-powered smart factories established themselves globally so they shaped industrial policies across every nation.

15.8 SMART CITIES AND COMMUNITIES (SDG 11) DEVELOPMENT WITH 6G

6G technology functions as a revolutionary power that creates intelligent urban districts that unite connectivity features with sustainable development principles [28]. Through SDG 11, the deployment of 6G technology will improve city planning through maximized resource efficiencies, which will create safer, smarter, more livable urban areas. 6G will give urban regions its resilience and adaptability to present urban issues through implementations of AI automation combined with real-time analytics and low-latency networking that guarantees reliability.

The integration of 6G technology with IoT technology allows for continuous real-time urban monitoring of congestion patterns along with environmental status and waste management data, and energy utilization statistics, which municipal bureaus utilize for management. Standard physical digital twins in modern implementation produce predictive models to control emergency services as well as design infrastructure growth while optimizing resource allocation. By optimizing energy use and pollution levels, AI processing of data from smart sensors improves the quality of the residential environment.

The development of 6G technology results in essential changes that improve the connections between service and security tools implemented in urban infrastructure. Intelligent transportation systems enable autonomous vehicles to establish quick-operating transportation networks which reduce pollution levels [29]. 6G network connections will allow first responders to access safe real-time communication systems that boost their public safety operations through efficient crisis emergency coordination.

Next-generation urban innovation will have 6G functioning as its foundational network to establish sustainable, efficient, secure urban spaces. Full implementation of this vision needs governmental, technological and urban planning institutions to work together for the equitable distribution of digital progress. The integration of intelligence into urban development through 6G technology will establish fundamental infrastructure for smart cities that uses technology to promote sustainability alongside community welfare improvement.

Case Study 15.4

6G and Smart City Development in Singapore [30].

BACKGROUND

The government of Singapore initiated a groundbreaking 6G smart city pilot program that used smart transportation technologies alongside energy-efficient features and AI to develop public services.

TECHNOLOGY USED

- 6G-powered traffic control systems, reducing congestion.
- The use of AI in waste management establishes better systems for recycling combined with disposal methods.
- Accommodations featuring automatic systems for improving energy efficiency operation.
- Public safety receives benefits from 6G-connected networks which enhance emergency response operations.

RESULTS

- The reduction of traffic congestion reached 40%, which bettered the city's air quality standards.
- The energy usage in intelligent buildings decreased by 30%.
- Achieving a 25% decrease in crime rates became possible due to AI surveillance systems and warning alerts.

IMPACT

- Blueprint for future smart city initiatives worldwide.
- The city became more sustainable and its environmental emissions decreased.
- Better quality of life through efficient public service delivery.

15.9 ECONOMIC GROWTH AND EMPLOYMENT (SDG 8)

6G technology will lead economic transformation by creating new industrial opportunities through its connections while empowering automation functions, which will drive employment and sustainable economic development. The main focus of 6G differs from earlier network generations since it combines AI-driven intelligence with real-time analytics alongside ultra-reliable communication to generate new economic opportunities. Next-generation network development will strengthen the employment landscape by demanding experts in telecommunications development, cybersecurity, software development, and AI application fields.

When businesses obtain superior digital infrastructure through 6G, they will advance their innovation capacity to facilitate remote work along with automated industries, while creating intelligent supply chains. Organizations at both startup and enterprise levels can enhance their network features by using blockchain decentralization with AI predictive market analysis capabilities [31]. The combination of ultra-fast and low-latency communication systems will create new business arrangements that will facilitate worldwide trading relationships and expand digital business activities.

The basic infrastructure of employee education stands as the foundation for 6G economic advantages through the development of digital policies and expansion of network access capabilities. A merged operation between industries using intelligent automation systems and real-time connectivity knowledge delivers remarkable production levels, which makes the global economy more robust. The same level of developmental opportunities should be accessible to every citizen because sustainable economic development requires equal access to growth achievements.

Case Study 15.5

6G and Economic Growth in Japan's Digital Workforce [32].

BACKGROUND

Japan implemented 6G infrastructure because of its status as a technology leader to promote economic growth while creating jobs and transforming its workforce to digital operations.

TECHNOLOGY USED

- Through 6G technology, workplaces can operate remotely to boost business operational effectiveness.

- The employment focus shifted from mundane routine duties to advanced activities because of AI automation.
- Holographic communications have diminished business travel needs while optimizing team cooperative activities.
- The deployment of 6G VR technology provides workers with training programs that lead to adaptation for new employment roles.

RESULTS

- The employment rates in the digital economy sector rose 20%.
- The transition to remote work boosted by 50% brought better work-life equilibrium for individuals.
- AI, together with automation technologies, increased Small and Medium Enterprise operational output by 35%.

IMPACT

- The economic expansion results from digital transformation.
- Job creation in AI, automation, and telecommunications industries.
- National authorities funded digital upskilling initiatives which made workers more ready for employment opportunities.

15.10 CHALLENGES AND ETHICAL CONSIDERATIONS OF 6G IMPLEMENTATION

The realization of 6G implementation demands systematic maintenance between technological advancement and moral accountability. The evolution of industries alongside healthcare and education, as well as urban development, that 6G brings to society, creates new challenges with privacy protection, network security, and fair digital access. The unification of AI-operated networks with IoT systems and ultra-high-speed communications networks raises serious vulnerabilities regarding user data security and enables spyware exploits and destructive uses of powerful connectivity capabilities. The success of 6G depends on resolving these problems to guarantee a protected digital system that upholds user rights for everyone.

Improved cybersecurity frameworks teamed with regulatory policies constitute an absolute requirement for addressing security threats in 6G networks. Data security will use philosophical models that combine decentralized systems and quantum computing protection techniques because AI-powered networks and quantum computers are gaining more popularity. The defense of infrastructure from cyberattacks alongside unauthorized data exchange access becomes critical because such systems become increasingly important for industries and governments. Building effective security standards under worldwide privacy rules demands continuous partnership action between those who create policies and research academics together with technology experts.

The task of establishing complete digital access continues to present major difficulties for 6G implementation. The implementation of this technology aims to deliver universal accessibility; however, insufficient infrastructure development and economic barriers, together with limitations in digital skills, may obstruct its availability in deprived areas. Rural communities need purposeful funding for 6G access, and policy measures must create affordable and inclusive arrangements for all people to benefit from these solutions. 6G will transform into its potential driving force by solving ethical considerations to support sustainable development while empowering societies through fair, secure, and inclusive digital access.

15.11 FUTURE PROSPECTS: INTEGRATING 6G WITH THE SDG FRAMEWORK

The development of 6G technology needs a joint effort to align technological breakthroughs with worldwide sustainability programs that follow the SDG framework. The advancement of 6G technology requires a selection of strategic policies to align its potential to fuel economic growth and enhance public services, healthcare, and educational transformation with long-term social and environmental advantages. 6G's sustainable development impact depends heavily on the formation of robust connections between government research organizations and industry professionals, coupled with non-governmental organizations.

Stubbornness and dedication among stakeholders facilitate both knowledge sharing and infrastructure development, and ethical standards construction for accessible digital services combined with environmental durability and cybersecurity protection. Public–private alliances enable nations to increase the deployment pace of 6G-based smart urban infrastructure and sustainable manufacturing solutions and healthcare innovations to reach technology benefits throughout every community. Strategic alliances need to develop modern data security protocols and digital resource accessibility protocols as well as regulatory protocols through cybersecurity standards.

Sustainable 6G deployment needs precise policy recommendations that will enhance responsible innovation efforts and minimize security dangers. Public officials must create versatile laws that support efficient networking systems, which also define acceptable AI usage patterns and ethical data handling procedures. Digital infrastructure development along with renewable energy incorporation and training programs need significant financial backing so that 6G can fulfill its role in SDG achievement. 6G technology will push sustainable digital transformations by using sustainability-focused policies to establish an inclusive global development framework during the upcoming decades, Figure 15.5.

15.12 CONCLUSION

6G functions as an innovative networking technology that will lead future connectivity while transforming businesses and advancing national development with sustainability elements. The unprecedented capabilities of 6G to boost energy efficiency, as well as healthcare quality, educational effectiveness, and economic development, make this technology an important gateway toward smart connected world

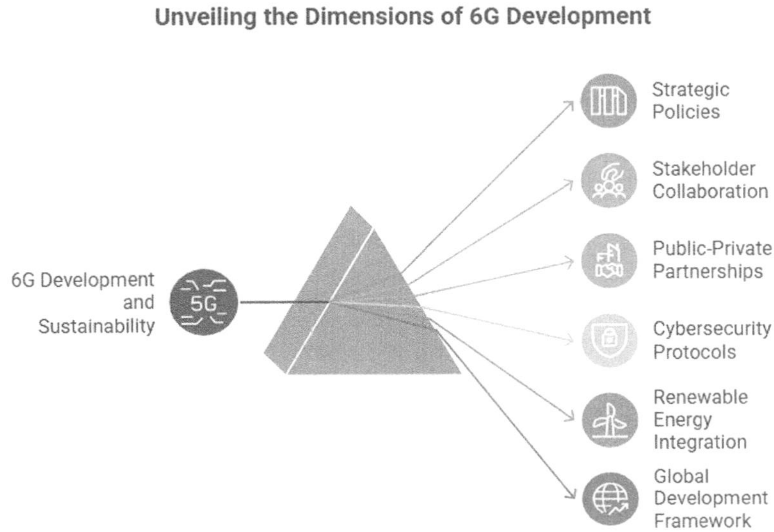

FIGURE 15.5 The potential of 6G development.

sustainability. SDG alignment enables 6G technology to function as a fundamental digital tool for achieving digital equality along with sustainability and economic development.

The main progress of 6G technology depends on smart energy controls that decrease environmental damage and enable remote healthcare services as well as digital education systems in varied communities. To reach successful 6G implementation, we need to solve the cybersecurity issues, digital inequities, and ethical problems from AI deployments in networking.

The path to successful implementation requires coordinated policy structures between sectors as well as sustainable innovation to make 6G a development tool that avoids both inequality growth and environmental damage. All sectors and governmental bodies with global institutions need to form partnerships for developing regulatory frameworks to achieve fair networking access and safe AI use combined with sustainable digital infrastructure. As it integrates sustainability deeply within its foundation, 6G can emerge as an advance of technology beyond simple innovation to function as a change catalyst that strengthens both people and the planet through sustainable connectivity.

REFERENCES

1. Krishnan, S., Saravanan/Anand Krishnan (A. Jos) and Kumar, R., 2024. *Sustainable Development Goals*. Boca Raton: CRC Press
2. Imoize, A.L., Adedeji, O., Tandiya, N. and Shetty, S., 2021. 6G enabled smart infrastructure for sustainable society: Opportunities, challenges, and research roadmap. *Sensors*, 21(5), p.1709.

3. Firoozi, A.A. and Firoozi, A.A., Artificial intelligence and reconfigurable intelligent surfaces: Transforming urban infrastructure. In *Enhancing Connectivity and Resilience* (pp. 1–11). *Deep Science Publishing.* doi:10.70593/978-93-49307-08-7_1.
4. Vora, L.J., 2015. Evolution of mobile generation technology: 1G to 5G and review of upcoming wireless technology 5G. *International Journal of Modern Trends in Engineering and Research*, 2(10), pp.281–290.
5. Banafaa, M., Shayea, I., Din, J., Azmi, M.H., Alashbi, A., Daradkeh, Y.I. and Alhammadi, A., 2023. 6G mobile communication technology: Requirements, targets, applications, challenges, advantages, and opportunities. *Alexandria Engineering Journal*, 64, pp.245–274.
6. Allam, Z. and Jones, D.S., 2021. Future (post-COVID) digital, smart and sustainable cities in the wake of 6G: Digital twins, immersive realities and new urban economies. *Land Use Policy*, 101, p.105201.
7. Song, H.J. and Nagatsuma, T., 2011. Present and future of terahertz communications. *IEEE Transactions on Terahertz Science and Technology*, 1(1), pp.256–263.
8. Huang, X., Wijesekera, S. and Sharma, D., 2009. Agent-oriented novel quantum key distribution protocol for the security in wireless network. In Salman Ahmed and Mohd Noh Karsiti (eds) *Multiagent Systems* (pp. 261–277). Vienna, Austria: Intechopen.
9. Bekaroo, G. and Dawarka, V., AI-assisted extended reality toward the 6G era: Opportunities, challenges and prospective solutions. doi:10.1201/9781003205494-14.
10. Babbar, H., Rani, S., Soni, M., Keshta, I., Prasad, K.D.V. and Shabaz, M., 2025. Integrating remote sensing and geospatial AI-enhanced ISAC models for advanced localization and environmental monitoring. *Environmental Earth Sciences*, 84(4), pp.1–12.
11. Chinnasamy, P., Babu, G.C., Ayyasamy, R.K., Amutha, S., Sinha, K. and Balaram, A., 2024. Blockchain 6G-based wireless network security management with optimization using machine learning techniques. *Sensors*, 24(18), p.6143.
12. Kavehmadavani, F., 2024. Artificial Intelligence (AI)-enabled Smart Radio Environments for 6G Wireless Networks. https://orbilu.uni.lu/bitstream/10993/62766/1/Finalversion_thesis_f.Kaveh.pdf
13. Zhou, Y. and Divekar, R., 2025. Immersive, task-based language learning through XR and AI: From design thinking to deployment. *TechTrends*, 69, 427–446. doi: 10.1007/s11528-025-01048-2.
14. Yang, H., Alphones, A., Xiong, Z., Niyato, D., Zhao, J. and Wu, K., 2020. Artificial-intelligence-enabled intelligent 6G networks. *IEEE Network*, 34(6), pp.272–280.
15. Tasci, R.A., Kilinc, F., Basar, E. and Alexandropoulos, G.C., 2022. A new RIS architecture with a single power amplifier: Energy efficiency and error performance analysis. *IEEE Access*, *10*, pp.44804–44815.
16. Malisuwan, S., 2023. Challenges, opportunities, and future research in the integration of 5G/6G networks, LEO satellites, and IoT for environmental protection and sustainable development at ECSTAR. *Journal of Environmental Science and Engineering B* 12(2023), pp.146–159.
17. Khara, A., Javid, I., Khara, S., Khara, B. and Prasad, R., 2024, November. Achieving sustainable development goals (SDGs) by leveraging 6G services. In *2024 27th International Symposium on Wireless Personal Multimedia Communications (WPMC)* (pp. 1–5). IEEE.
18. Kharche, S. and Kharche, J., 2023. 6G Intelligent healthcare framework: A review on role of technologies, challenges and future directions. *J. Mobile Multimedia*, *19*(3), pp.603–644.

19. Rajak, S., Summaq, A., Kumar, M.P., Ghosh, A., Elumalai, K. and Chinnadurai, S., 2024. Revolutionizing healthcare with 6G: A deep dive into smart, connected systems. *IEEE Access,* 12, 194150–194170, 2024, doi: 10.1109/ACCESS.2024.3519567.
20. Abbas, J.K., Hassan, S.I., Mohsen, A.H., Hussain, A. and Hussain, F.H.T., 2024, December. Next-gen education with 6G connectivity and intelligent data analytics. In *Human-Centric Integration of 6G-Enabled Technologies for Modern Society* (pp. 143–154). Academic Press (pp. 1–6). IEEE.
21. Rane, N., Choudhary, S. and Rane, J., 2023. *Education 4.0 and 5.0: Integrating Artificial Intelligence (AI) for Personalized and Adaptive Learning.* (November 2, 2023). Available at SSRN: https://ssrn.com/abstract=4638365 or http://dx.doi.org/10.2139/ssrn.4638365
22. Liu, Y., Razman, M.R., Zakaria, S.Z.S., Ern, L.K., Hussain, A. and Chamola, V., 2024. Utilizing ubiquitous learning to foster sustainable development in rural areas: Insights from 6G technology. *Computers in Human Behavior*, 161, p.108418.
23. Magambo, J., 2007. *Use of Information and Communications Technologies (ICTs) in Teacher Education in Sub-Saharan Africa: Case Studies of Selected African Universities* (Doctoral dissertation, Universität zu Köln).
24. Serôdio, C., Cunha, J., Candela, G., Rodriguez, S., Sousa, X.R. and Branco, F., 2023. The 6G ecosystem as support for IoE and private networks: Vision, requirements, and challenges. *Future Internet, 15*(11), p.348.
25. Prateek, K., Ojha, N.K., Altaf, F. and Maity, S., 2023. Quantum secured 6G technology-based applications in internet of everything. *Telecommunication Systems, 82*(2), pp.315–344
26. Yarali, A., 2023. *From 5G to 6G: Technologies, Architecture, AI, and Security.* Piscataway, NJ: IEEE Press.
27. Fachrudin Priyanta, I., Freytag, J., Schulte, P., Kauffmann, S., Ogorelysheva, N. and Roidl, M., 2023. Towards 6G-driven digital continuum in logistics. *Logistics Journal: Proceedings, 2023*(1). doi: 10.2195/lj_proc_priyanta_en_202310_01.
28. Sharma, S., Popli, R., Singh, S., Chhabra, G., Saini, G.S., Singh, M., Sandhu, A., Sharma, A. and Kumar, R., 2024. The role of 6G technologies in advancing smart city applications: Opportunities and challenges. *Sustainability, 16*(16), p.7039.
29. Aboelezz, A.M., Sedhom, B.E., El-Saadawi, M.M., Eladl, A.A. and Siano, P., 2023. State-of-the-art review on shipboard microgrids: Architecture, control, management, protection, and future perspectives. *Smart Cities, 6*(3), pp.1435–1484.
30. Al Amin, A., Hong, J., Bui, V.H. and Su, W., 2023. Emerging 6G/B6G wireless communication for the power infrastructure in smart cities: Innovations, challenges, and future perspectives. *Algorithms, 16*(10), p.474.
31. Wang, Z., Li, M., Lu, J. and Cheng, X., 2022. Business Innovation based on artificial intelligence and Blockchain technology. *Information Processing & Management, 59*(1), p.102759.
32. Sharma, N., Kumar, R., Maliya, R., Shah, V., Patel, J. and Fukui, M., Japan's 2030 vision for 6G: A case study. In *Network Security and Data Privacy in 6G Communication* (pp. 20–39). Boca Raton: CRC Press.

16 Using Cloud Computing in Securing 6G-Enabled IoT

Shalini Kumari and Chander Prabha

16.1 INTRODUCTION

Cloud computing (CC) has gained significant traction as users access centralised data resources through distance connections since its emergence in recent times. Cloud service consumers acquire data processing capabilities from distant centralized facilities that need not be geographically proximate. Users typically experience significant delays during remote service communication. New advances in cloud service distribution methods improve dependability and user speed. The most prominent breakthroughs concern updates beyond 5G and 6G networks that will shape the digital global future along with advancing Industry 5.0. The rising interconnectivity alongside digital transformation continues to develop because 5G and B5G networks integrate innovative service-driven features with mission-critical performance capabilities to deliver ultra-low latency and exceptional reliability alongside elevated mobility network levels. The method enables the amalgamation of multiple technology domains to develop new enhancements. Moores law advancement leads to a transition from standalone chip-wired systems to integrated embedded platforms that fuse systems and technology optimization approaches. The continuing advancements allow for accelerated methodological improvements related to system implementation. The network environment requires many providers to build B5G and 6G applications, while multi-Access Edge Computing is the foundational service platform. The operating solution between and inside multiple data centres will require cloud-native principles to ensure the optimal performance of both data centre operations and microservices and their concurrent management abilities. The diverse networking schemes of B5G and 6G technology will unite to form one cohesive system that merges Internet of Things (IoT) connectivity, aerial systems (drones), satellite networks, and underwater capabilities. Future networks require advanced Artificial Intelligence (AI) and Machine Learning (ML) technologies for intelligent network operation and management to fulfil their high requirements. The forthcoming B5G and 6G networks will enable complex services and functional applications while advancing network infrastructure. As a successor to previous generations, the 6G mobile networks have developed and extended upon their predecessors. Society embraces revolutionary change through 5G networks, which supply connectivity services and interaction capabilities. Three trillion people will use 5G services before 2029, according to projections showing 1 billion active users by 2023 and a forecast of

 DOI: 10.1201/9781003586456-16

5 billion users by 2029 [1]. 6G technology design will meet increasing social connectivity requirements [2]. Mobile networks and applications experience constant development, while risks simultaneously develop.

Revolutionary transportation solutions, smart retail, and intelligent urban development emerged due to the Internet of Things (IoT) [3]. The massive IoT could receive power from 6G network systems. 6G represents the upcoming mobile network that will overcome current 5G limitations by providing high data rate connectivity between numerous applications and devices to the billions connected. The goal of 6G technology is to fulfil digital community requirements for a fully connected digital society in the future [4]. Remote accessibility in IoT devices has increased their exposure to threats [5]. Wearable gadgets maintain sensitive client data, including health information and position data. The strategic features of IoT devices create vulnerabilities in them. Protecting extensive IoT networks against cyber threats, especially unidentified attacks, is challenging for the security sector. The function of cloud computing in safeguarding 6G IoT ecosystems is presented in Table 16.1, and its unique challenges are illustrated in Table 16.2.

Structure of the Chapter: The rest of the chapter is organized into four sections. The second section comprises 5G and B5G edge networks and their corresponding 6G-enabled IoT devices. Section three presents the key areas of 6G network security. Section four outlines the strategies for securing 6G with cloud computing, the challenges encountered, and their potential solutions. Finally, the chapter concludes with conclusive remarks.

TABLE 16.1
Security Roles of cloud computing in 6G

Security Roles of CC in 6G	Description
Centralized security intelligence	Aggregates security data from many network segments. Performs deep learning-based threat analysis. Maintains global security policies and configurations. Coordinates security responses throughout the network [6].
Advanced threat analytics	Extended databases are analyzed by the system to detect security patterns. The system operates complex artificial intelligence algorithms to detect potential threats in advance. Conducts extensive longitudinal security trend analysis. The system generates risk associations between multiple internet-connected systems domains.
Security resource management	Attack trends are identified through database analysis. The system uses complex artificial intelligence algorithms to detect security threats in advance. Conducts extensive longitudinal security trend analysis. Systems associate risks across multiple IoT domains while overseeing cryptographic key distribution networks. This system establishes scalable system access and identity control functions [7].
Backup and recovery	The system protects key IoT data through encryption and provides disaster recovery capabilities. Preserves security audit records. Facilitates system state recovery post-attacks.

TABLE 16.2
Challenges in the Implementation of cloud computing in 6G

Challenge	Description
Latency issues	Delayed security responses for time-sensitive IoT devices. The identification of threats happens slowly because cloud processing takes too long. Communication impediments during extensive assaults. Limitations of real-time security monitoring.
Bandwidth constraints	Security surveillance infrastructure requires a large-scale movement of data between systems. Network congestion amid security incidents. The process of security data transfer leads to increased ongoing expenses. Remote IoT devices experience restricted bandwidth capabilities [8].
Privacy concerns	Security surveillance infrastructure requires the large-scale movement of data between systems. Network congestion occurs amid security incidents. The process of security data transfer leads to increased ongoing expenses. Remote IoT devices experience restricted bandwidth capabilities.
Scalability challenges	Secure management of billions of IoT devices remains his responsibility. Allocation of resources for security processing authentication on a large scale. Security effectiveness must be maintained during times of peak usage.
Integration problems	Secure management of billions of IoT devices remains his responsibility. Security effectiveness must be maintained during times of peak usage [9].
Resource management	Secure management of billions of IoT devices. Allocation of resources for security processing. Authentication on a large scale.

16.2 5G AND B5G EDGE COMPUTING NETWORKS

Software-defined networking (SDN), virtualization, and multi-access edge computing have become cloud-native technologies essential to the traditional 5G network architecture [10]. The three cloud service models describe, on the whole, the traditional representation of cloud services; when running on clouds, the services fall into these three models. Figure 16.1 shows the cloud computing service.

Software-as-a-service (SaaS): Through thin client presentation, users can access cloud-based applications that run directly on their devices. Application interfaces and web-based e-mail services represent typical examples of this technology. In SaaS deployments, consumers do not control essential cloud infrastructure elements such as networks, servers, operating systems, and storage facilities [11].

Platform-as-a-service (PaaS): Users gain the ability to deploy applications through different supported languages and services, as well as tools that run on cloud infrastructure. Under PaaS deployment models, users acquire neither ownership nor management control over the underlying infrastructure, which

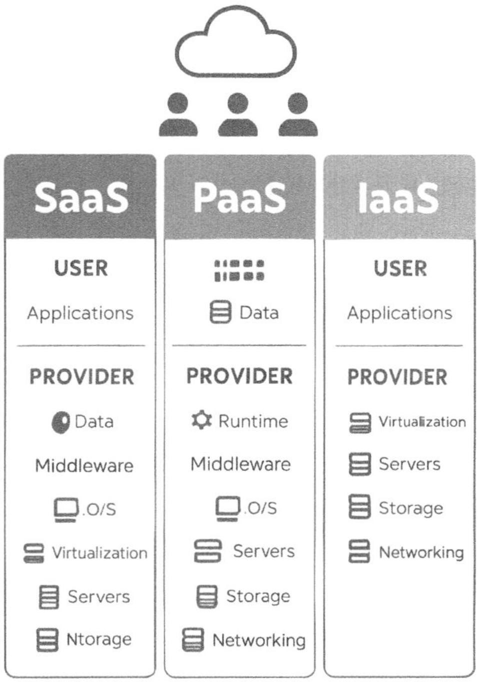

FIGURE 16.1 Cloud computing service.

includes network servers, storage, and operating systems alongside SaaS. PaaS allows users to control their applications alongside their application-hosting configurations.

Infrastructure-as-a-service (IaaS): Users can execute software, including operating systems and applications, through IaaS because they access processing power alongside storage networks and other computing capabilities. The cloud infrastructure exists outside of the consumers management domain. Users possess management capabilities for operating systems and storage, deployed applications, and host firewalls, but lack control of the fundamental infrastructure.

Edge computing has three distinct layers consisting of device edge and central installations. Sensors and actuators at the device layer function with user interface devices and limited computational capabilities while utilizing low storage requirements [12]. This tier depends on devices with minimal bandwidth and restricted range limitations. The edge tier consists of three nodes: edge, control, and IoT gateways that maintain low latency and high bandwidth requirements to service device layer components. Cloud services and associated networks support extensive system connections at the core tier layer. Within a distributed B5G deployment, three fundamental architectural structures exist. Cloud computing provides automated storage resources as well as computing capabilities. Edge computing delivers nearby access to computer resources for users through its fast processing and shorter delays. It also provides security by maintaining applications on-premises. Edge computing delivers improved dependability and elevated user and device experiences through its scalability support [13].

16.3 6G ENABLED IOT DEVICES

Three new IoT applications exist alongside holographic communications, five-sense communications, Wireless Brain-Computer Interface (WBCI), and new vertical applications. The expanding Internet of Things system merges into 4CSL and requires more key performance indicators (KPIs) with sophisticated usage contexts. 5G experiences difficulties supporting changes in the IoT system due to limitations within its framework [14].

Evolution of IoT: The combination of IoT and 5G technology is implementing industrial revolution 4.0, which impacts healthcare services alongside education institutions and the industrial sector, intelligent transportation systems, innovative residential development, and city management. The development of holographic, five-sense communications, and brain computer technologies through IoT will create new verticals accessible to consumers. The development of IoT applications requires wireless communication networks to fulfil increasingly demanding requirements.

IoT-Enabled Applications and Related Requirements: Humans use IoT-enabled interactions embracing holographic and five-sense and WBCI technologies, which allow them to immerse themselves in distant locations through communication means beyond voice, video, and images. Introducing holograms, to five-sense technologies, and autonomous machinery components through new media objects facilitates real-time interactive experiences in forms that create new market verticals, including competent healthcare and education industry, Internet and autonomous driving, and super smart city/home systems.

Holographic Communications: Holographic communications is the initial new interaction method that produces real-time 3-D picture projections. The system collects visual data of subjects and objects and then transmits them to the receiver in real time or remotely, together with audio components. Through this approach, users can experience a higher reality than virtual and augmented reality because physical pieces are displayed with real-time audio at their destination in front of them. Holographic communication platforms are necessary for agriculture and educational institutions, entertainment businesses, and other domains. The network requires high throughputs to efficiently transfer holographic streaming media, which could reach hundreds of gigabits per second or terabits per second because of the numerous details that must be sent despite compression efforts. Networks should focus on reliability and timeliness approaches because jitter issues negatively affect interactive applications. Holographic communications depend on integrating, Ultra Reliable Low Latency Communications (URLLC) and eMBB to deliver minimum latency and high data speeds.

Five-Sense Communications: Telephone interactions are becoming more common because researchers document the increase of five-dimensional (5-D) immersive remote human relations and traditional audio-video methods. Through its five-sense function, a medium provides complete sensory integration to all human senses, such as sight, hearing, touch, smell, and taste. This technology identifies human and environmental perceptions through neurologic methods before uniting them. The transmission of information from sender to receiver through remote channels enhances the connection quality [15].

Smart Healthcare: The implementation of 6G technology enables sophisticated healthcare systems to monitor and diagnose remotely, guiding and carrying out surgical procedures simultaneously. Medical data transmission will become more efficient through 6G technology because its high data rate, combined with low latency, exact localization, and super dependability strengths, will boost access to care and its overall quality. When clinical staff work with AI systems, they can enhance data evaluation to achieve better diagnostic accuracy. Blockchain technology provides safe data-sharing methods for personal medical information, which supports the worldwide advancement of medical studies.

Super Smart City/Home: 6Gs advanced characteristics will boost life quality, monitoring functions, and automation capabilities, speeding up the evolution of smart city and home. A smart city takes its name from its ability to gather and process data across multiple industrial sectors, including urban development and refuse disposal, for optimizing public budgets and service quality while decreasing maintenance expenses. Different levels of intelligence will emerge in cities through advancements such as driverless vehicles and smart mobile gadgets under 6G.

16.4 KEY AREAS IN 6G NETWORKS SECURITY

Artificial Intelligence serves several components within 5G networks through functions including channel estimation, physical layer coding, multiple access capabilities, and network layer applications [16]. The current limited AI applications in the network cause difficulties for AI-driven technology support within 5G networks due to initial architectural boundaries. Due to their AI-based nature, intelligent radio and distributed AI cannot be supported. Real-time emergency response functions during 5G networks exist through vehicle networks, yet these capabilities cannot address real-time situations because of performance delays. But 6G networks can. 6G network radio latency reaches 0.1 milliseconds, one-tenth of 5G network latency durations. The current 5G network coverage limitation prevents it from enabling communication beyond ground-level access, which fails to reach space and underwater areas. Research into threats informs the development of 6G security architecture that will adopt open standards and procedures for mobile network operations at a level comparable to 5G. A list follows with five essential security areas and the requirements for 6G. Figure 16.2 shows the Focus areas for 6G security.

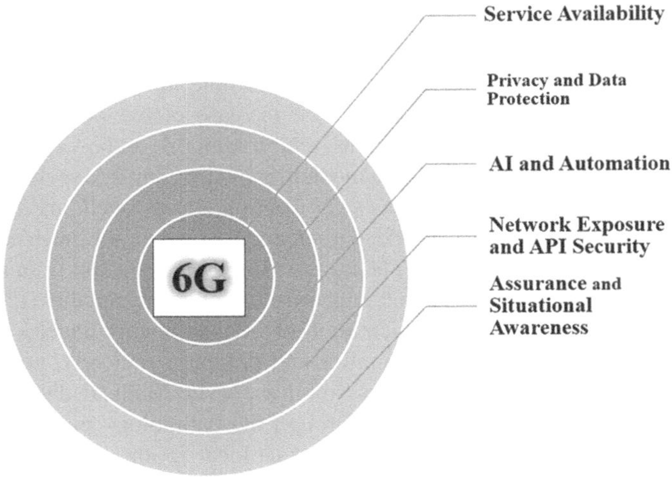

FIGURE 16.2 Focus areas for 6G security.

Service Availability: Users will receive specific services from the 6G platform. Users place significant importance on service availability since it draws regulatory attention, mainly when such services support societal necessities. Under service availability, consumers should be able to depend on the service during regular situations, accidents, natural disasters, and network attacks. The extent of met reliance depends on the critical importance of the service. The fundamental 6G platform features a basic level with adjustable customization options for enhancing dependability based on requirements [16]. Individual services require independent adjustments, while adjustments made in both smaller and larger geographical areas are also possible. Disruptions to system availability are primarily addressed through network configuration and design alongside architectural methods rather than using them as standalone security measures. Operational networking and elements serve as the objective for measures that protect against denial-of-service, signalling storm, and jamming attacks. Telecom applications will increasingly need cyber threat intelligence before developing the 6G technology ecosystem.

Privacy and Data Protection: Users will receive specific services from the 6G platform. Users place significant importance on service availability since it draws regulatory attention, mainly when such services support societal necessities. Under service availability, consumers should be able to depend on the service during regular situations, accidents, natural disasters, and network attacks. The extent of met reliance depends on the critical importance of the service. The fundamental 6G platform features a basic level with adjustable customization options for enhancing dependability based on requirements. Individual services require independent adjustments, while adjustments made in both smaller and larger geographical areas are also possible. Disruptions to system availability are primarily addressed through network configuration and design alongside architectural methods rather than using them as standalone security measures. Operational networking and elements serve as the objective for measures that protect against denial-of-service, signalling storm, and jamming attacks. Telecom applications will increasingly need cyber threat intelligence before developing the 6G technology ecosystem.

AI and Automation: The 6G platform management and user service will experience significant changes due to AI developments. Integrating AI and automation technologies delivers enhanced performance with better efficiency and speed when fighting cyberattacks in mobile networks. AI solutions enable mobile networks to detect and respond to threats, including faulty devices, pretend base stations, and signalling attacks. The ongoing 6G transition will see AI and automation improve security measures through intelligence-based threat monitoring and regulatory verification systems. AI should maintain robust foundations because its wider use depends on explainable functions safeguarding privacy and creating secure systems. Security feature understanding in specific applications, along with technological improvements, are required to tackle these identified criteria. The successful implementation of intent-based networking demands flawless transformation of enterprise-level intents into network-level AI agents and automatic networking transformations. AI continues to grow its role in cybersecurity, but it needs qualified human experts to function efficiently. Mobile network security specialists benefit from technology through threat information, threat hunting, and attack detection to enhance their reactions to cyber intrusions. Formal security analysis receives benefits from AI technology when developing new security standards. AI demonstrates value during system development processes through its ability to locate and explain software issues.

Network Exposure and Application Programming Interface (API) Security: 6G APIs allow developers to personalize network platforms and extend their service capabilities beyond basic communication requirements. 6G technology improves gaming quality, drone management, and enhanced Extended Reality (XR) operations by providing more excellent network capabilities and data access. Using the API requires proper security measures to be implemented. The telecom requirements ensure complete API security for internal and external components within the 5G system. API security involves multiple safety protocols, which include user verification, access permissions, and continuous audit tracking functions. New use cases and innovations in 6G require network functionality exposure, which needs proactive threat analysis, API posture management, and API runtime monitoring to boost API security capabilities. API security characteristics will be included in standard protocols, but other parts will remain the responsibility of mobile network operations. API security across the entire lifespan rests jointly with suppliers and service providers in mobile networks [17].

Assurance and Situational Awareness: 6G APIs allow developers to personalize network platforms and extend their service capabilities beyond basic communication requirements [18]. 6G technology improves gaming quality, drone management, and enhanced XR operations by providing greater network capabilities and data access. Using the API requires proper security measures to be implemented. The telecom requirements ensure complete API security for internal and external components within the 5G system. API security involves multiple safety protocols, including user verification, access permissions, and continuous audit tracking functions. New use cases and innovations in 6G require network functionality exposure, which needs proactive threat analysis, API posture management, and API runtime monitoring to boost API security capabilities. API security characteristics will be included in standard protocols, but other parts will remain the responsibility of mobile network operations. API security across the entire lifespan rests jointly with suppliers and service providers in mobile networks. Implementation of zero-trust architecture models forms a complete security structure for 6G systems. 6G standards focus on implementing zero-trust architecture through remote attestation alongside confidential computing techniques spread across computing frameworks.

16.5 STRATEGIES FOR SECURING 6G WITH CLOUD COMPUTING AND CHALLENGES

There are 8 strategies used for securing the 6G with CC. Figure 16.3 shows securing 6G with CC. Table 16.3 shows the Challenge in securing 6G with cloud computing [19].

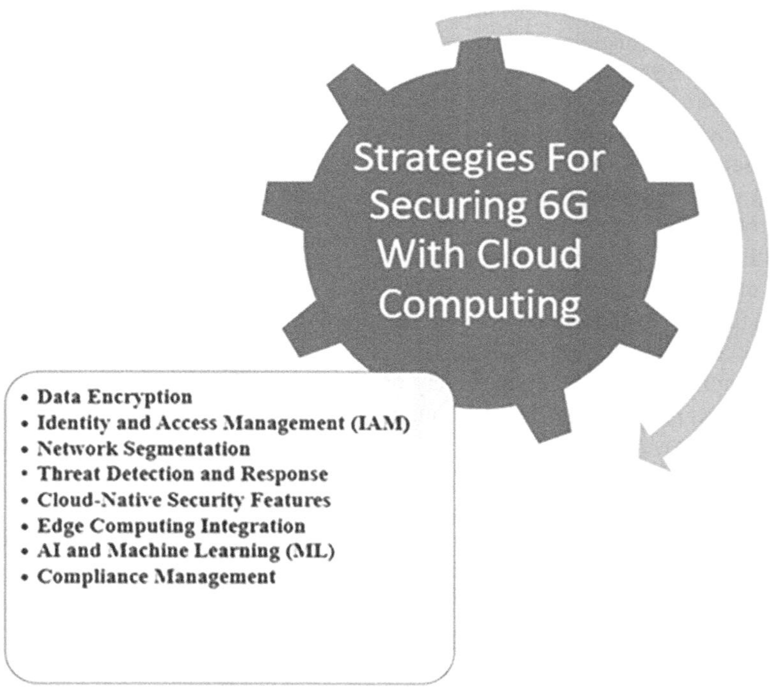

FIGURE 16.3 Securing 6G with CC.

TABLE 16.3
Challenges in Securing 6G with cloud computing

Challenges in Securing 6G with CC	Description
Complexity of heterogeneous networks	Security management across 6G networks proves difficult because the network consists of numerous diverse devices and technologies [20].
Data privacy concerns	One crucial obstacle of modern technological devices becomes the protection of user privacy when 6G generates substantial personal information quantities.
Scalability and performance	The scaling of cloud infrastructure to handle increasing demand from 6G networks makes security maintenance quite challenging. A successful 6G network security deployment with cloud computing needs proactive measures based on strong security frameworks and superior threat detection and continuous observation to protect sensitive information through the entire network infrastructure [21].

Data Encryption: Strong cryptographic methods encrypt sensitive information when data remains stationary and during transmission to prevent unauthorized access, even if someone intercepts it. The company has established secure processes to distribute encryption keys through approved management systems.

Identity and Access Management (IAM): The organization implemented complex IAM systems for cloud application and data access through role-based controls and multifactor authentication solutions. The organization deployed permission controls administered by roles to manage user access privileges.

Network Segmentation: Through compartmentalization of the network, IT managers can limit the expansion of security threats and protect crucial information from unauthorized access points.

Threat Detection and Response: Advanced security detection mechanisms deployed in the cloud computing environment help prevent security incidents by recognizing threats in real time. Implementing Security Information and Event Management systems allows organizations to combine and inspect security log records to detect abnormal activities.

Cloud-Native Security Features: Cloud security solutions providers include integrated firewalls, intrusion detection and prevention systems, and vulnerability screening tools.

Edge Computing Integration: The proximity of edge computing operations to data sources diminishes the transmission of sensitive data to the cloud, which enhances privacy measures.

AI and Machine Learning (ML): AI and ML algorithms function as part of robust security strategies through their advanced capabilities to detect threats, evaluate anomalies, and take pre-emptive actions.

Compliance Management: Data privacy regulations are fully adhered to by properly implementing security measures and data governance protocols.

The 6G will face three main security challenges: physical tampering, privacy concerns, and AI utilisation. Three potential security solutions include encryption, zero-trust protocols, and machine-learning applications [22].

- Physical tampering of nodes represents an exposure point in 6G networks.
- The extensive data capacity of 6G networks creates a potential privacy risk since it allows surveillance operations and unauthorized access to occur.
- Implementing artificial intelligence in 6G networks creates privacy issues and new security vulnerabilities.
- Security mechanisms find it challenging to function collectively when 6G networks lack standardization in their infrastructure systems.
- Edge devices have limited defensive capabilities due to their weak computational capabilities.
- 6G networks need more effective security protocols because traditional security solutions do not perform adequately.

16.5.1 Potential solutions

- Using quantum encryption yields encryption that has virtually impossible breaking points.
- Zero-trust is a strategy that presumes all persons, alongside devices, must prove their trustworthiness before receiving access.
- Machine Learning technology protects 6G networks through its application.
- Implementing closed-loop network automation represents a tool for blocking security threats, yet simultaneously generating fresh cybersecurity threats.

16.5.2 Conclusion

The book chapter establishes what cloud computing achieves to enhance security mechanisms for 6G IoT networks. Implementing cloud computing on 6G networks addresses security problems and enables IoT applications to achieve unlimited scale, speed, and connectivity. Centrally managed security intelligence and improved threat analytics become possible with 6G applications of cloud computing and enhanced security resource management within these systems. The text outlines the use of cloud-based security solutions by healthcare systems, smart cities, and industrial IoT applications. The chapter explains how cloud computing serves 6G security through three main functions: threat analysis based on deep learning, diverse network data aggregation, and maintaining worldwide security policies.

On the other hand, the chapter examines latency management, bandwidth restrictions, and privacy issues that affect cloud security scalability. The 6G networks focus on security requirements for innovative future IoT is applications such as holographic communications, five-sense communications, and brain–computer interfaces. The security analysis of 6G networks examines five fundamental aspects, including the availability of services, privacy and data protection features, AI automation components, network exposure through APIs, and assurance and situational awareness capabilities. The chapter establishes eight security approaches based on cloud computing maximalities for 6G networks, including data encryption and AI-driven threat detection, Identity and access management, and network segmentation. The discussion delves into protecting diversified networks, data protection methods, and performance maintenance at all scales. The study indicates that quantum encryption, zero-trust protocols, and ML applications solve these security problems. The research fully examines protected 6G networks and discusses how cloud systems will support secure IoT ecosystems in the upcoming generation.

REFERENCES

[1] Rodrigues, T. K., Liu, J., & Kato, N. (2021). Application of cybertwin for offloading in mobile multiaccess edge computing for 6G networks. *IEEE Internet of Things Journal*, *8*(22), 16231–16242. https://doi.org/10.1109/jiot.2021.3095308

[2] Shayea, I., Benlakehal, M. E., Azmi, M. H., Han, C. T., Arsad, A., & Abd Rahman, T. (2024). Outdoor mobile broadband performance analysis in Malaysia, Singapore, and Thailand. *Results in Engineering*, *23*(102691), 102691. https://doi.org/10.1016/j.rineng.2024.102691

[3] Amanullah, M. A., Habeeb, R. A. A., Nasaruddin, F. H., Gani, A., Ahmed, E., Nainar, A. S. M., Akim, N. M., & Imran, M. (2020). Deep learning and big data technologies for IoT security. *Computer Communications*, *151*, 495–517. https://doi.org/10.1016/j.comcom.2020.01.016

[4] Zhang, S., & Zhu, D. (2020). Towards artificial intelligence-enabled 6G: State of the art, challenges, and opportunities. *Computer Networks*, *183*(107556), 107556. https://doi.org/10.1016/j.comnet.2020.107556

[5] Ahmad, R., & Alsmadi, I. (2021). Machine learning approaches to IoT security: A systematic literature review. *Internet of Things*, *14*(100365), 100365. https://doi.org/10.1016/j.iot.2021.100365

[6] Ziegler, V., Schneider, P., Viswanathan, H., Montag, M., Kanugovi, S., & Rezaki, A. (2021). Security and trust in the 6G era. *IEEE Access: Practical Innovations, Open Solutions*, 9, 142314–142327. https://doi.org/10.1109/access.2021.3120143

[7] Zhu, J., Hou, R., Wang, X., Wang, W., Cao, J., Zhao, L., Yuan, F., Li, P., Wang, Z., Zhao, B., Zhang, L., & Meng, D. (2019). Enabling privacy-preserving, compute- and data-intensive computing using heterogeneous trusted execution environment. In *arXiv [cs.CR]*. https://arxiv.org/abs/1904.04782

[8] Mucchi, L., Jayousi, S., Caputo, S., Panayirci, E., Shahabuddin, S., Bechtold, J., Morales, I., Stoica, R.-A., Abreu, G., & Haas, H. (2021). Physical-layer security in 6G networks. *IEEE Open Journal of the Communications Society*, 2, 1901–1914. https://doi.org/10.1109/ojcoms.2021.3103735

[9] Nguyen, T., Tran, N., Loven, L., Partala, J., Kechadi, M.-T., & Pirttikangas, S. (2020). Privacy-aware blockchain innovation for 6G: Challenges and opportunities. In *2020 2nd 6G Wireless Summit (6G SUMMIT)*, Levi, Finland.

[10] Yang, P., Xiao, Y., Xiao, M., & Li, S. (2019). 6G Wireless communications: Vision and potential techniques. *IEEE Network*, *33*(4), 70–75. https://doi.org/10.1109/mnet.2019.1800418

[11] Malik, M. I., Wani, S. H., & Rashid, A. (2018). Cloud computing-technologies. *International Journal of Advanced Research in Computer Science*, *9*(2), 379–384. https://doi.org/10.26483/ijarcs.v9i2.5760

[12] Ishtiaq, M., Saeed, N., & Khan, M. A. (2021). Edge computing in IoT: A 6G perspective. In *arXiv [cs.NI]*. https://doi.org/10.36227/techrxiv.17031665.v1

[13] Hosseinzadeh, M., Hemmati, A., & Masoud Rahmani, A. (2022). 6G-enabled internet of things: Vision, techniques, and open issues. *Computer Modeling in Engineering & Sciences: CMES*, *133*(3), 509–556. https://doi.org/10.32604/cmes.2022.021094

[14] Qadir, Z., Le, K. N., Saeed, N., & Munawar, H. S. (2023). Towards 6G Internet of Things: Recent advances, use cases, and open challenges. *ICT Express*, *9*(3), 296–312. https://doi.org/10.1016/j.icte.2022.06.006

[15] De Lima, C., Belot, D., Berkvens, R., Bourdoux, A., Dardari, D., Guillaud, M., Isomursu, M., Lohan, E.-S., Miao, Y., Barreto, A. N., Aziz, M. R. K., Saloranta, J., Sanguanpuak, T., Sarieddeen, H., Seco-Granados, G., Suutala, J., Svensson, T., Valkama, M., Van Liempd, B., & Wymeersch, H. (2021). Convergent communication, sensing and localization in 6G systems: An overview of technologies, opportunities and challenges. *IEEE Access: Practical Innovations, Open Solutions*, *9*, 26902–26925. https://doi.org/10.1109/access.2021.3053486

[16] Mattsson, J. P., Smeets, B., & Thormarker, E. (2021). Quantum technology and its impact on security in mobile networks. *Ericsson Review*, *2021*(12), 2–12. https://doi.org/10.23919/etr.2021.9904724

[17] Jonsson, P., & Ericsson. (2022). Ericsson Mobility Report (Review. Part II). *Synchroinfo Journal*, *8*(4), 33–41. https://doi.org/10.36724/2664-066x-2022-8-4-33-41

[18] Ganapathi, A. (2025). Zero trust evolution: Advanced architectures for resilient cloud-native IAM systems. *International Journal of Scientific Research in Computer Science, Engineering and Information Technology*, *11*(1), 679–685. https://doi.org/10.32628/cseit25111261

[19] Sadiq, K. A., Favour-Bethy Thompson, A., Ayeni, O. A., & Arome, G. J. (2023). A real-time intrusion detection system for service availability in cloud computing environments. In *Security and Privacy Schemes for Dense 6G Wireless Communication Networks* (pp. 261–290). Institution of Engineering and Technology.

[20] Siriwardhana, Y., Porambage, P., Liyanage, M., & Ylianttila, M. (2021). AI and 6G Security: Opportunities and challenges. In *2021 Joint European Conference on Networks and Communications & 6G Summit (EuCNC/6G* Summit*)*, Porto, Portugal.

[21] Fang, H., Qi, A., & Wang, X. (2020). Fast authentication and progressive authorization in large-scale IoT: How to leverage AI for security enhancement. *IEEE Network*, *34*(3), 24–29. https://doi.org/10.1109/mnet.011.1900276

[22] Wang, M., Zhu, T., Zhang, T., Zhang, J., Yu, S., & Zhou, W. (2020). Security and privacy in 6G networks: New areas and new challenges. *Digital Communications and Networks*, *6*(3), 281–291. https://doi.org/10.1016/j.dcan.2020.07.003

17 Next-Gen Communication Security

Bridging 6G and IoT

Neetu Rani, Narinder Yadav, Ishdeep Singla, and Sujoy Ray

17.1 INTRODUCTION

The advancement of wireless communication technologies has given rise to the next-generation network, known as 6G. 6G is expected to be available in the 2030s and aims to provide unprecedented connectivity, ultra-low latency, higher data rates, and smart automation [1]. With AI, edge computing, and even quantum communication, 6G will likely outperform 5G by providing seamless and intelligent communication between a wide range of devices. The IoT has enhanced the digital world by providing an enormous array of devices that can communicate on their own. IoT devices encompass home automation systems, wearables, industrial sensors, and even robotic cars. The union of IoT and 6G will lead to an over-connected ecosystem in which trillions of devices will efficiently transfer data to one another [2]. Key enablers of 6G and IoT include:

- Terahertz (THz) communication for speedy data transfer.
- AI-powered network management for prompt decision making.
- Secure data transmission through quantum communication.
- Edge and fog computing, which reduce latency and improve efficiency.
- Supported by THz communication, which enables ultra-high-speed data transfer and ultra-fast data processing.
- AI-powered algorithmic network management and smart optimization of networks operating with AIs for low latency and high performance.
- Holographic & XR will include a future-proof immersive experience while leading forwarding Key Technologies (KFT).
- Quantum-based communication security (QCS) (due to cryptography sensitive), meaning quantum-secured communication with cells.
- Contextual edge computing & blockchain – for real-time precision, distributed data management.
- Massive scale integrated IoT connectivity – MaCConnect standards provide seamless integration with trillions of IoT devices along with frictionless support.
- Energy-efficient networks are cost-controlled, green, eco-friendly networking solutions because operating energy is below set levels.

DOI: 10.1201/9781003586456-17

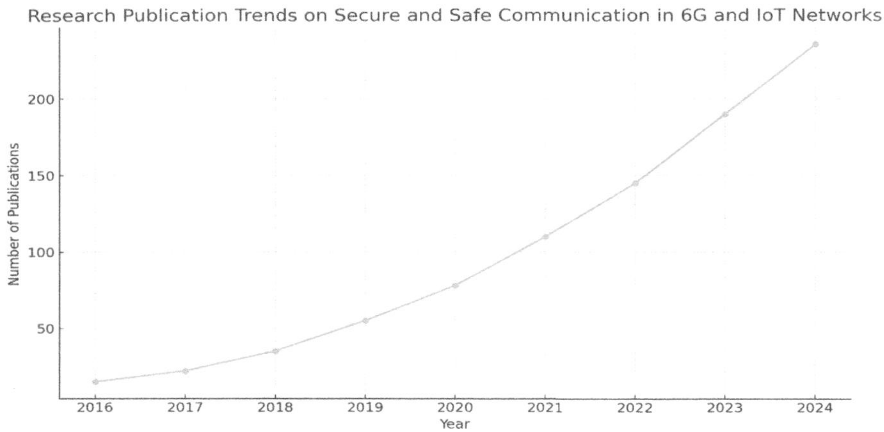

FIGURE 17.1 Research publications trends in 6G and IoT networks.

The large-scale implementation of these technologies, along with their enormous potential, also creates extraordinary security challenges requiring robust actions in order to ensure secure communication.

Figure 17.1 illustrates the trend of research papers on safe and secure communication in 6G and IoT networks between 2016 and 2024. There has been a consistent and high rise in interest, particularly since 2020, reflecting the worldwide momentum toward next-generation communication technologies and secure IoT systems.

17.2 IMPORTANCE AND NEED OF SECURE AND SAFE COMMUNICATION

With the advancements of 6G and IoT networks, there is a need for safety and security mechanisms to transfer the data over the networks. Security in communication systems focuses on protecting data integrity, confidentiality, and availability in order to keep unauthorized users out of the network. Safe communication guarantees that the data that is transmitted is reliable and uncorrupted, undetectable, and not able to be accessed without authorization [3,4]. There is a need for secure and safe communication in 6G and IoT networks for the following aspects:

17.2.1 Protection from Cyber Threats

The IoT devices suffer from many attacks such as distributed denial-of-service (DDoS), Man-in-the-Middle (MitM) attacks, and many more. Such threats need advanced security mechanisms to be incorporated into 6G infrastructure systems.

17.2.2 Preservation of Privacy

The gathering of data through IoT devices creates a wide range of privacy-invasive breaches. Secure communication protects data privacy, legislations such as general data protection regulation (GDPR), and others.

17.2.3 Trustworthy Data Transfer

In critical applications such as healthcare and autonomous vehicles, information security is crucial. Secured protocols make sure that transmitted data is accurate, coherent, and untampered.

17.2.4 Trust and Authentication

Analyzing approved user devices before allowing connections to the network eliminates unwanted access breaches. IoT and 6G networks face security threats. Non-trivial privacy and national security concerns arise without selecting strong cybersecurity measures (Figure 17.2).

The remaining parts of this study are organized as follows. The threats of IoT security are explained in Section 17.2. Privacy issues in IoT and edge devices are presented in Section 17.3. Analyses of 6G and IoT network security incidents are presented in the fourth section. 6G and IoT security architectures are presented in Section 17.5, and cryptographic practices for protection of confidential information are illustrated in Section 17.6, and instructive conclusions along with limitations and further research are defined in Section 17.7.

17.3 KEY CHALLENGES IN 6G AND IOT SECURITY

The integration of new technology has not solved the multitude of issues surrounding security for 6G and IoT networks [5,6]. The following are some key challenges in 6G and IoT.

17.3.1 Issues of Connectivity and Scalability

AI-driven robots and advanced infrastructures integrate additional devices into the 6G network, exceeding the current limits of modern technologies. This creates a

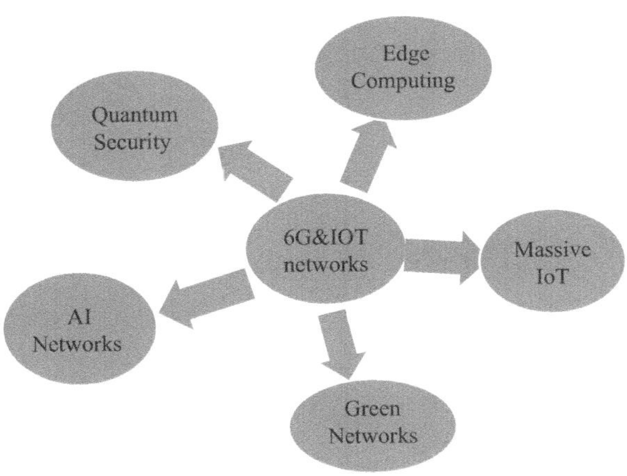

FIGURE 17.2 6G and IoT trends [3].

need for dynamic security management that can address the diverse and extensive definitional borders of ecosystems.

17.3.2 Resource Limits in IoT Devices

The absence of sufficient processing, memory, or battery life in most IoT devices hinders strong encryption implementation. Enhancing device security without overburdening the device necessitates creating lightweight cryptography.

17.3.3 Cyber Threats Powered by AI

Aside from several attacks, AI is also employed by criminals to execute automated, sophisticated attacks. There is a need for proactive threat detection to defend against malicious threats. AI models are trained to bypass conventional security countermeasures.

17.3.4 Risks Related to Collecting Data and Privacy Sharing

With the use of IoT devices, user information is captured perpetually, which, if misused, can lead to privacy concerns. A few vital problems in 6G and IoT security involve data anonymization, differential privacy, and safe data storage.

17.3.5 Risks of Quantum Computing

With the development of quantum computing, traditional methods of encryption will no longer be useful. 6G networks need to integrate PQC to defend against quantum attacks and ensure long-term data security (Figure 17.3).

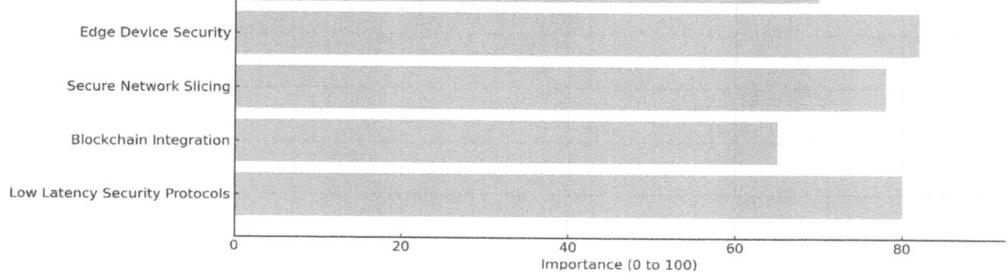

FIGURE 17.3 Challenges in 6G and IoT security [5].

17.3.6 Secure Communication between Devices (D2D)

6G will facilitate vehicles to communicate directly with one another without going through a central system network. Authentication and data transfer in a non-centralized network still remains one of the most difficult problems [7].

17.4 ISSUES RELATING TO REGULATIONS AND COMPLIANCE

The security standards for 6G and IoT are still under development. The existence of different laws from different nations makes it hard to develop universal security standards. A solution requires collaboration with multiple disciplines, such as cryptography, AI, blockchain technology, and resilient network protocols to create secure, robust communication frameworks. Table 17.1 presents some of the main security challenges in relation to 6G and IoT networks alongside their implications and some possible approaches to mitigate them. It also discusses other issues such as AI-assisted cyber warfare, threats from quantum computing, and the regulatory dilemma known as colossal connectivity. Addressing these problems involves employing AI-driven cyber defenses, PQC, and Zero Trust frameworks.

17.5 SCOPE AND AIM OF THE CHAPTER

This chapter centers on the scrutiny of 6G and IoT networks' security and safety challenges and provides modern advances in solutions to communicational security. The key considerations of this chapter include:

TABLE 17.1
Security Challenges in 6G and IoT Networks [8]

Challenge	Description	Impact
Connectivity and scalability	6G will accommodate a record number of devices, which will be difficult to secure and scale effectively.	Greater attack surface
Resource constraints among IoT	Several IoT devices have inadequate processing power and memory, so it is difficult to apply robust encryption and security mechanisms.	Devices are susceptible to attacks
Threats from AI-driven cyberattacks	AI-driven attacks have the capability to take advantage of network security vulnerabilities by leveraging advanced cyber threats like adversarial ML.	AI-based cyberattacks can
Privacy risks in data collection and sharing	IoT devices continuously gather sensitive user information, which raises the threat of unauthorized access, identity theft, and data breaches.	Inability to have mechanisms for data protection can lead to financial loss, identity theft, and damage to reputation.
Quantum computing threats	Quantum computers have the ability to compromise conventional cryptography techniques, leading to the necessity of implementing post-quantum cryptographic solutions.	Current encryption algorithms will be obsolete, leaving confidential data vulnerable to decryption.

1. Understanding the importance, needs, and key challenges of 6G and IoT ecosystems.
2. Identification of various threats that are vulnerable to 6G IoT security.
3. Privacy issues in IoT and edge devices.
4. Analysis of various real-life case studies along with 6G and IoT security architectures.
5. Cryptographic practices for the protection of confidential information and various AI and machine learning (ML) techniques in this field.

17.6 THE THREATS OF IoT SECURITY

As more and more industries are integrating IoT devices, there is an increase in the need for IoT devices to be protected from emerging dangers. IoT devices are susceptible to multiple security issues such as malware infections, DDoS assaults, and leaks of confidential data [9–12]. Below are some important concerns regarding security in IoT technology.

17.6.1 Ransomware and Malware Attack

The lack of protective protocols leaves an IoT ecosystem susceptible to malware and ransomware attacks. Within the IoT, the devices that are interconnected to each other are particularly susceptible due to the lack of protective measures. Ransomware attacks targeting IoT devices can capture and ransom critical infrastructures, demanding a payment to restore access to vital information.

17.6.2 DDoS Attacks

In terms of DDoS attacks, the focus is on crippling services by flooding the network with extremely high amounts of traffic. Since there is usually a large volume of IoT devices, the attacker can use compromised IoT nodes to launch botnet-driven DDoS attacks that affect the entire network.

17.6.3 MitM Attacks

MitM attacks occur when an unauthorized entity eavesdrops on communications between two devices with the intent of altering information and masquerading as a legitimate device. Since modern IoT networks are primarily wireless in nature, such networks can fall prey to MitM attacks with the absence of proper cryptography and authentication mechanisms.

17.6.4 Lack of Proper Configuration and Authentication

Brute force attacks become easier to execute because most of the passwords set in IoT devices are either default or comparatively weak. With no strong protective authentication frameworks such as multi-factor authentication (MFA) or even biometrics, unauthorized access and data theft is bound to happen on IoT devices.

17.6.5 Firmware Exploitation and Supply Chain Attacks

Attacks can target the firmware and software supplied by third parties to IoT devices due to inadequate security measures. Supply chain attacks involve the insertion of malware into equipment during manufacturing or distribution, sabotaging security before deployment.

17.7 PRIVACY ISSUES IN IoT AND EDGE DEVICES

The 6G networks paired with IoT technology create opportunities to capture and process vast amounts of user data [13,14]. However, the methods of data collection, storage, and subsequent sharing are of concern. Some of the privacy concerns are illustrated below:

17.7.1 Unwarranted Surveillance and Data Accumulation

The vast array of IoT devices picks up and sends data regarding users, their activities, mobile locations, and a host of signifying personal information. If this data is made accessible, it may result in surveillance and subsequent violation of personal privacy.

17.7.2 Security Vulnerabilities of Edge Computing

Processing closer to the data source instead of relying on remote cloud servers is referred to as edge computing and is a prerequisite to IoT and 6G networks. While this reduces latency, it introduces new security issues such as a lack of protection for edge nodes, data leakage possibilities, and exposed edges.

17.7.3 Data Theft and Breach of Information

An IoT system that lacks adequate protection will certainly expose key details to registered users. This can lead to identity theft, financial damage, and loss of reputation. The exposed cloud storage and data transfer systems provide an opportunity for hostile agents to breach and acquire sensitive materials.

17.7.4 Absence of Uniform Privacy Standards

The presence of IoT devices in different regions with differing privacy laws makes it challenging to apply security measures. Data privacy law requires compliance with certain standards like the GDPR and the California Consumer Privacy Act (CCPA). Table 17.2 discusses privacy issues in IoT and edge devices, their impacts, and mitigation strategies.

17.8 ANALYSIS OF 6G AND IOT NETWORK SECURITY INCIDENTS

The increasing adoption of 6G technology interconnected with IoT systems has improved automation and connectivity, but has also introduced significant security risks. Many real-life security breaches have demonstrated how vulnerabilities

TABLE 17.2
Privacy Concerns in IoT and Edge Devices [13]

Privacy Concern	Description	Potential Impact	Possible Mitigation Strategies
Unauthorized data collection and surveillance	IoT devices collect personal data, location, and behavioral patterns, raising risks of unauthorized surveillance and misuse.	Loss of user privacy, unauthorized profiling, and increased mass surveillance risks.	Implement strong data encryption, user consent mechanisms, and access control policies.
Edge computing security risks	Edge computing processes data closer to the source, reducing latency but increasing exposure to cyber threats and data leaks.	Sensitive data exposure, increased attack surface, and lack of centralized control over security.	Secure edge nodes with AI-driven threat detection, real-time monitoring, and authentication measures.
Data breaches and identity theft	Compromised IoT networks can expose sensitive user data, leading to identity theft, financial fraud, and reputation loss.	Financial loss, identity fraud, and unauthorized access to critical personal information.	Use end-to-end encryption, secure cloud storage, and advanced identity protection techniques.
Lack of standardized privacy regulations	Different privacy regulations across regions make enforcing consistent security standards difficult, leading to non-compliance risks.	Inconsistent data protection practices, legal challenges, and risks of cross-border data exploitation.	Establish global data privacy frameworks, align regional regulations, and enforce GDPR/CCPA compliance.

within IoT ecosystems and their integration can be exploited by hackers to launch large-scale attacks. Such attacks and the severe consequences are depicted in the case studies below to emphasize the need for improved security in modern communication infrastructures.

17.8.1 The Mirai Botnet Attack (2016)

Hacking of the IoT networks for security cameras, routers, and DVRs like the Mirai botnet became known in 2016. The particular problem was security cameras, various routers, and DVRs, which all utilized some form of default and weak passwords. Upon infecting a device, the attacker converts it into a botnet. The botnet included one of the largest DDoS attacks, which brought down very popular services such as Twitter, Reddit, GitHub, and Netflix. Instead of focusing on the damage done, this incident highlighted the gaping holes that lie in the security systems of various unprotected IoT devices. It further depicts the need to shift the responsibility of IoT security from only users to the network level, where bounded anomaly detection systems and robust multi-layered firewalls with ML are essential [15,16].

17.8.2 The Stuxnet Worm (2010)

Although the Stuxnet worm was not considered an attack on the IoT at the time, it stands as one of the sophisticated forms of cyberattacks on industrial control systems (ICS). Discovered in 2010, the Stuxnet virus attacked Siemens programmable logic controllers (PLCs) that interfaced with industrial units. Stuxnet's primary goals include the Iranian nuclear facility at Natanz, where it was able to sabotage uranium enrichment centrifuges, causing them to fail and, subsequently, destroy. The worm infected air-gapped networks with USB devices and spread through Windows networks, instantly modifying operational configuration parameters. Stuxnet caused industrial processes to be suspended while generating false normal status outputs. By doing so, Stuxnet exploited the vulnerabilities of critical infrastructure systems that are automated and networked due to feedback control systems. Simultaneously, it laid the groundwork for what are termed cyber-physical assaults, showing the need for better interruption of critical infrastructure, real-time monitoring for deviation from established baselines, and reliance on isolated security systems to protect against sophisticated malware [16].

17.8.3 The Breach of Tesla's IoT Security Systems

The smart features in Tesla vehicles, such as feature unlocking, navigation, over-the-air (OTA) updates, and intelligent system integration, rely on Internet-based connectivity. Advanced OTA (AoT) updates require top-of-the-line security features as tampering with system access could lead to the loss of functional control of the automobile and its important mechanics. Keen Security Lab Researchers demonstrated the presence of a multitude of critical flaws in Tesla software's security, proving the possibility of remote access to vital system controls. Through the car's Bluetooth and Wi-Fi vulnerability, control over braking and door unlocking functions, as well as infotainment system disabling and enabling, could be achieved without the driver's consent. The relative damage of such exploits is great; passengers' safety, vehicle integrity, and the security of surrounding vehicles are at stake. The breaches from this research further exacerbate the problem as they decrypt communications, remove the periodic update requirement, put backup channels, and the need for detection systems for vehicles and communications [17]. The rapid measures Tesla enforced in security updates respond to constant monitoring challenges of the software controlling the vehicle's network system and incessant IoT connections. These measures reiterate the continuous need for systematic oversight and management of vehicle software security.

17.8.4 Verkada Security Camera Leak (2021)

In 2021, hackers dubbed "Terrible Herd" gained access to Verkada's security camera systems and cloud architecture, showcasing over 150,000 surveillance streams in real time. They went beyond simply hacking into the cameras, breached advanced security systems, and bypassed controls. Cameras hacked included those placed in hospitals, corporate offices, factories, and even prisons. The hackers made use of stolen administrator credentials that led them to security loopholes that allowed access to live feeds from Tesla, Cloudflare, and numerous other healthcare facilities. This particular incident brought to light critical risks posed to the security of surveillance systems that are hosted over the cloud, exposing the damage caused due to a single credentials exposure. As a result, this incident shed light on the steps that need to be taken by IoT-based security frameworks in order to prevent unauthorized surveillance and information breaches, such as MFA, tighter control policies over accessibility, and utilizing encryption from one end to the other [18].

17.8.5 The SolarWinds Supply Chain Attack (2020)

The SolarWinds cyberattack is the most devastating attack in history. The attack was a supply chain attack that had infected companies globally, both private and public, including government institutions. The attackers put malware in the SolarWinds Orion software updates, which were later disguised as real updates. After installation, the malware that was planted opened back doors to allow frequent access to the corporate networks. These allowed them to collect sensitive data, change system settings, and conduct other subsequent attacks. This specific attack was difficult to detect, which is why it remained unnoticed for quite a long period of time. Countless companies were compromised for months on end. The attack itself exploited the insecurities that were present in the supply chains, especially concerning the software code integrity audit, vendor security audit, and other trustless security frameworks that were in place, aimed to mitigate such incursions. Such violations of security within IoT and connected networks bring forth real-world implications that need to be dealt with. The Mirai, Stuxnet, Tesla, Verkada, and SolarWinds breaches highlight the key areas these malicious actors focus on default passwords, software vulnerabilities, network hardware, and supply chain defenses [19].

To prevent such attacks, industries need to implement strong controls for authentication, real-time monitoring, firmware updates, and administrative requirements. The advancement of 6G and IoT technologies will ensure trust and reliability in the next-generation communication networks based on the security frameworks established in the present.

17.9 6G AND IOT SECURITY ARCHITECTURES

As 6G and IoT networks evolve, security across all tiers of the network architecture must be taken into consideration. Advanced methods are critical for securing data, preventing cyberattacks, and ensuring reliable communication. The subsequent section discusses the four basic security frameworks for IoT and 6G networks along with their interconnections.

17.9.1 For 6G: Secure Network Slicing

Network slicing is one of the key features of 6G, since it enables numerous virtual networks to coexist on a single physical network. Each one of them needs to cater to a specific use case like self-driving cars, smart healthcare, or industrial automation, and has to be configured with certain policies, especially security ones. Some security challenges of network slicing include:

- Slices can be singularly exploited because of incorrect security policies. An intruder from one slice can breach other slices [20].
- Needless to say, cross-slice attacks have to be prevented, and data isolation between slices is necessary.
- When security is to be controlled over several service providers and over many industries, it becomes a complicated issue to deal with.

17.9.2 AI-Driven Security Frameworks

In the case of 6G and IoT networks, security is reinforced with the help of ML and AI by automating threat identification, anomaly detection, and securing measures [21]. Underlying Responsibilities of AI in Security:

- **Intrusion Detection and Prevention**: AI so far has been successful in correctly identifying intrusions into specific systems and ensuring intervention in real time.
- **Behavioral Analysis**: ML systems are used to detect anomalies by observing the basic actions of a device.
- **Automated Threat Mitigation**: AI will act swiftly to neutralize cyberattacks by isolating contaminated devices or shutting down other networks.

17.9.3 Blockchain for IoT and 6G Security

Trust, transparency, and security in 6G and IoT networks can be created using blockchain technology, which provides a decentralized and immutable ledger [22]. Applications of Blockchain in IoT and 6G Security:

- **Decentralized Authentication**: Authentication among IoT devices is enhanced by removing centralized control points with blockchain.
- **Immutable Data Storage**: Modifying transactions on the blockchain is impossible, thus ensuring the integrity of stored data (Table 17.3).

17.9.4 Zero Trust Architecture (ZTA) for IoT Networks

ZTA has a motto of "Never Trust, Always Verify". In 6G and IoT, where dynamically billions of devices are connected, Zero Trust ensures that every device, user, and application authenticate themselves before every access [23]. The most important features of ZTA in IoT Security include:

- **Micro-Segmentation:** Dividing networks into smaller, distinct units prevents attackers from moving laterally within the network.
- **Continuous Authentication:** Each user and device must validate themselves at every interaction, thus reducing insider risk.
- **Least Privilege Access Control:** Access rights for devices and users are given only for the bare minimum needed for their roles (Table 17.4).

TABLE 17.3
Blockchain in IoT and 6G Security [22]

Challenges	Blockchain Security Solutions
Blockchain consumes high computing power	Use lightweight blockchain solutions optimized for IoT
Limited scalability for large IoT networks	Implement off-chain processing and sharding techniques
Data privacy concerns in public blockchains	Use private or permissioned blockchain networks for IoT security

TABLE 17.4
Implementation Challenges and Solutions in ZTA [23]

Challenges	Zero Trust Security Solutions
High computational overhead in verifying each device	Use AI-based real-time authentication to optimize verification speed
Complexity in enforcing Zero Trust across large-scale IoT networks	Implement automated policy enforcement and monitoring tools
User experience trade-offs due to frequent authentication	Optimize authentication workflows using behavioral biometrics

The security architectures presented here, Secure Network Slicing, AI-Driven Security, Blockchain, and Zero Trust, are all useful in the protection of 6G and IoT networks. With these technologies evolving, there will be a need for a multi-layered security approach based on AI and cryptographic methods with decentralized security frameworks to ensure integrity against cyberattacks and maintain data integrity.

17.10 CRYPTOGRAPHIC PRACTICES FOR PROTECTION OF CONFIDENTIAL INFORMATION

The subsequent section illustrates the various cryptographic techniques used in 6G and IoT security:

17.10.1 Cryptography Techniques for Safe Exchange of Information

In today's world of digital networks, confidential communication is one of the most fundamental requirements where information should be private, verified, and authenticated. With the emergence of new technologies like 6G and the mass proliferation of IoT devices, novel security mechanisms will need to be created to respond to new threats [24]. Conventional methods of encryption may be inadequate to defend against quantum computing attacks or may not be appropriate for IoT-constrained environments. Therefore, to enhance security measures, there is a need for new mechanisms such as PQC, lightweight encryption, end-to-end encryption, and secure key management.

17.10.2 PQC in 6G

The security of 6G networks is susceptible to attacks from quantum computers capable of decoding traditional cryptographic algorithms such as RSA and Elliptic Curve Cryptography (ECC). Algorithms that will remain secure under quantum assaults are termed PQC. These lattice, hash, and code-based algorithms provide alternatives to well-established cryptographic methods and enable secure key distribution, authentication, and information encryption for 6G networks [25]. Due to 6G's ultra-low latency and high data speed requirements, PQC cannot compromise performance in optimization efforts made on security. Research and standardization bodies like NIST are currently investigating the development and standardization of PQC in order for seamless integration into future networks.

17.10.3 Encryption Techniques with Low-Resource Consumption for IoT Devices

Unlike other devices, IoT devices operate in resource-constrained environments. This means the less powerful devices may find traditional encryption techniques inefficient in terms of computation and power usage. Lightweight cryptography makes it possible to provide strong security at low memory and computational overhead. Algorithms tailored to protect the confidentiality and integrity of data with low delay and power consumption include present, simon, speck, and hight. These methods must manage the trade-off between security and performance due to the variety of IoT applications, from health care to industrial automation. In addition, low-resource cryptography needs to withstand assault from side channel and differential cryptanalysis attacks, making the design an important consideration for IoT security [26].

17.10.4 IoT Communication and End-to-End Encryption

Every IoT network comprises numerous devices where every device interacts and shares confidential information over the Internet, putting them at risk of data modification, interception, and unauthorized access. E2EE secures data from the moment it is sent until it reaches the intended recipient. E2EE guarantees that no intermediaries to the communication are capable of viewing the data in plaintext and that any intermediaries are solely privy to the data in encrypted form. Unlike previously used encryption schemes, E2EE has more complex requirements for key management and encryption techniques tailored to low-power IoT nodes. In IoT communication, TLS/DTLS and encrypted MQTT or AES-based implementations are widely used. E2EE can be fortified further through the application of homomorphic encryption and Zero Trust frameworks that restrict data access not only to clouds but also to service middlemen and sub-providers. This is most needed for IoT solutions in the healthcare sector, smart city systems, and industrial IoT [27].

17.10.5 Secure Key Management in Large-Scale IoT Deployments

Managing secret keys in a secure manner within the context of large-scale IoT systems is particularly difficult due to the number of devices that are interconnected. The compromise of a single key could invite multiple security threats to the entire network. Therefore, key management systems must be capable of resolving these issues. Secure distribution, storage, and revocation of keys are fundamental processes necessary for safeguarding the integrity and confidentiality of IoT networks. Public Key Infrastructure (PKI) technology, blockchain-based key management, and group key agreements aid in the distributed management of cryptographic keys in IoT systems and also being developed to provide additional security to the keys are the techniques of generating ephemeral keys and physically unclonable functions (PUFs). Automated or programmed lifecycle management, including zero-touch provisioning and dynamic key rotation, also increases the resiliency of vast IoT networks [28–30].

17.11 THE INTEGRATION OF AI AND ML IN SAFEGUARDING THE 6G AND IOT NETWORKS

17.11.1 Security AI Algorithms for Anti-Cyber Attack Strategies

Cybersecurity is one of the major challenges facing today's organizations because of rapid technological advances. IoT systems, as well as cloud infrastructure integrated with networks, give rise to sensitive data that is constantly being monitored and accessed in real time. Thus, having proper security measures in place is more important than ever. The implementation of older procedures causes gaps that expose systems to complex cyber threats. As such, employing AI techniques in threat identification and developing adaptive security approaches is critical. Real-time automated data controls and analytics are vital for proactively yielding AI-secured systems. AI-powered security solutions actively analyze data, detect patterns, recognize and block cyber threats, and mitigate any risks instantly [31].

17.11.2 AI and ML Threat Detection

AI and ML are extremely important in cybersecurity because they scan a large database for dissimilar information and outlier prospective threats with much greater efficiency than traditional methods dependent on rule-based approaches. Computers could learn to recognize pieces of malware, phishing attempts, insider threats, and even network intrusions using supervised, unsupervised, or deep learning algorithms. AI-powered solutions utilize anomaly detection and behavioral analytics to correlate cyberattacks with patterns using predictive modeling to identify threats. Popular methods include Support Vector Machines (SVM), Random Forest, and Neural Networks. Threat detection is also enabled by deep learning models such as Convolutional Neural Networks (CNN) and Recurrent Neural Networks (RNN), as they analyze past attacks to predict future ones. AI security systems provide automated threat intelligence to enhance reaction time and fortitude against attacks [32].

17.11.3 Monitoring and Managing Intrusions: The Intrusion Detection and Prevention Systems (IDPS) Procedure

IDPS monitors and examines data from a particular organization, looking out for violations related to policies and damaging actions. IDPS can be divided into two categories: Network-Based Intrusion Detection Systems (NIDS) and Host-Based Intrusion Detection Systems (HIDS). Advanced AI-powered IDPS employs ML models to identify irregular patterns and classify and prevent breaches. Traditional IDPS would rely on outdated signature-based detection methods that scan for known attack signatures, but newer AI-based systems employ behavioral monitoring techniques that scan networks for activity pattern deviations. With the application of AI, IDPS can adapt to evolving risks, detect novel threats, and autonomously initiate countermeasures. These systems are now part of protective infrastructures used by enterprises, cloud-based, and even IoT service providers [33–35].

17.11.4 AI – Enhancing Adaptive Security Automation

Every day, organizations face new cyber-related challenges that force them to adapt their cybersecurity policies. In order to use AI for automation, policies need to be adjusted on real-time telemetry data (either through behavior analytics or contextual interpretation) to take action against actively existing threats. Organizations can manage security-associated incidents and change policies simultaneously using AI-powered SIEM solutions. For instance, MFA can be enforced, or access to sensitive data can be blocked through user monitoring AI. Multi-faceted AI authentication creates more adaptive security policies that learn from breaches and attack patterns, and significantly increases resilience against cyber threats while reducing false alarms and unnecessary restrictions [36–39].

17.11.5 IoT Security through the Prisms of Federated Learning (FL) and Privacy Preservation

FL is a novel subdivision of AI, which allows collaborative ML model training on multiple devices without disclosing their raw data, hence ensuring privacy and security. FL is particularly relevant in IoT security, where the data is scattered across remote edge devices and cloud servers. With traditional centralized ML models, data has to be sent to a central server, which increases the risk of data leaks and unauthorized access [40–42]. In contrast, FL allows for IoT devices to learn from the data locally and share only model updates, thus preserving user privacy. FL enhances security through the decentralized detection of IoT threats, anomaly detection, and fraud prevention while protecting sensitive data. It is widely used in healthcare and finance, and in smart city applications to protect user data while applying AI for security purposes. The adoption of FL combined with blockchain and differential privacy technologies enables organizations to strengthen security, making it easier to comply with data protection regulations [43–45].

17.12 CONCLUSION AND FUTURE SCOPE

The development of the 6G networks, which are characterized by ultra-high speed, low latency, and AI-based automation, is a consequence of advancements in global communications technology. The proposed research highlights the growing demand for 6G and IoT networks for secure communications and the protection against cyberattacks, privacy violations, quantum computing risks, and other emerging threats. The current research also highlights primary concerns like the massive scale of the connectivity, low-power consuming IoT devices, AI-powered cyberattacks, and regulatory fragmentation. These technologies enhance privacy and authentication as well as strengthen the ability to detect threats in real time. Firms must adopt solutions such as two-way encryption, ZTA, and incorporate AI for autonomous intrusion detection scanning to secure communications. Adherence to international security regulations such as GDPR and NIST is essential in mitigating compliance exposure. Appropriate measures to maintain the integrity of the data include secure key management and the use of low-power encryption technologies for IoT devices. As 6G technology and IoT systems proliferate, multidisciplinary approaches from government, industry, and academic institutions are needed to ensure ethical, continual advancement of research associated with security challenges. A strong infrastructure of privacy-preserving and efficient next-generation cyber defense systems of ever-evolving cyber threats aims to sustain the balance between safeguarding user privacy and enhancing operational efficiency.

REFERENCES

1. V. Ziegler, P. Schneider, H. Viswanathan, M. Montag, S. Kanugovi, and A. Rezaki, "Security and trust in the 6G era," *IEEE Access*, vol. 9, pp. 142314–142327, 2021, doi: 10.1109/ACCESS.2021.3119691
2. A. Kaushik, R. Singh, M. Li, H. Luo, S. Dayarathna, R. Senanayake, X. An, R. A. StirlingGallacher, W. Shin, and M. Di Renzo, "Integrated Sensing and Communications for IoT: Synergies with Key 6G Technology Enablers," *IEEE Internet of Things Magazine*, vol. 7, no. 5, pp. 136–143, Sep. 2024, doi: 10.1109/IOTM.001.2400052.
3. D. D. Kumar, J. D. Mukharzee, C. V. D. Reddy, and S. M. Rajagopal, "Safe and secure communication using SSL/TLS," in *Proc. 2024 Int. Conf. Emerging Smart Comput. Informatics (ESCI)*, Pune, India, Mar. 5–7 2024, pp. 1–6.
4. Parker, S.K., Axtell, C.M. and Turner, N., 2001. Designing a safer workplace: Importance of job autonomy, communication quality, and supportive supervisors. *Journal of Occupational Health Psychology*, 6(3), pp. 211–218.
5. Abdel Hakeem, S.A., Hussein, H.H. and Kim, H., 2022. Security requirements and challenges of 6G technologies and applications. *Sensors*, 22(5), pp. 1–43.
6. Nguyen, V.L., Lin, P.C., Cheng, B.C., Hwang, R.H. and Lin, Y.D., 2021. Security and privacy for 6G: A survey on prospective technologies and challenges. *IEEE Communications Surveys & Tutorials*, 23(4), pp.2384–2428.
7. P. Porambage, G. Gür, D. P. Moya Osorio, M. Livanage, and M. Ylianttila, "6G security challenges and potential solutions," in *Proc. 2021 Joint Europ. Conf. Networks Commun. & 6G Summit (EuCNC/6G Summit)*, Porto, Portugal, Jun. 8–11, 2021, pp. 622–627, doi: 10.1109/EuCNC/6GSummit51104.2021.9482609.
8. Jahid, A., Alsharif, M.H. and Hall, T.J., 2023. The convergence of blockchain, IoT and 6G: Potential, opportunities, challenges and research roadmap. *Journal of Network and Computer Applications*, 217, pp. 1–18.

9. Schiller, E., Aidoo, A., Fuhrer, J., Stahl, J., Ziörjen, M. and Stiller, B., 2022. Landscape of IoT security. *Computer Science Review*, 44, pp. 1–18.
10. A. Ankita and S. Rani, "Machine learning and deep learning for malware and ransomware attacks in 6G network," in *Proc. 2021 4th Int. Conf. Comput. Intell. Commun. Technol. (CCICT)*, Sonepat, India, Jul. 3, 2021, pp. 39–44, doi: 10.1109/CCICT53244.2021.00016.
11. Benlloch-Caballero, P., Wang, Q. and Calero, J.M.A., 2023. Distributed dual-layer autonomous closed loops for self-protection of 5G/6G IoT networks from distributed denial of service attacks. *Computer Networks*, 222, pp. 1–16.
12. N. Mohamed and A. A. Ahmed, "AI in combatting man-in-the-middle attacks: A comprehensive review," in *Proc. 2024 15th Int. Conf. Comput., Commun. Netw. Technol. (ICCCNT)*, Kamand, India, Jun. 24-28 2024, pp. 1–6.
13. Alwarafy, A., Al-Thelaya, K.A., Abdallah, M., Schneider, J. and Hamdi, M., 2020. A survey on security and privacy issues in edge-computing-assisted internet of things. *IEEE Internet of Things Journal*, 8(6), pp.4004–4022.
14. Rehman, S.U., Khan, I.U., Moiz, M. and Hasan, S., 2016. Security and privacy issues in IoT. *International Journal of Communication Networks and Information Security*, 8(3), p.147.
15. M. Antonakakis, T. April, M. Bailey, M. Bernhard, E. Bursztein, J. Cochran, Z. Durumeric, J. A. Halderman, L. Invernizzi, M. Kallitsis, and D. Kumar, "Understanding the Mirai Botnet," in *Proc. 26th USENIX Security Symposium (USENIX Security '17)*, Vancouver, BC, Canada, Aug. 16–18, 2017, pp. 1093–1110.
16. J. Huddleston, P. Ji, S. Bhunia, and J. Cogan, "How VMware exploits contributed to SolarWinds supply-chain attack," in *Proc. 2021 Int. Conf. Comput. Sci. Comput. Intell. (CSCI)*, Las Vegas, NV, USA, Dec. 15–17 2021, pp. 760–765, doi: 10.1109/CSCI54926.2021.00143.
17. A. P. Shende, B. Shiragpur, G. Raj, and P. Tamhankar, "Securing the future: Comprehensive strategies for IoT security in Industry 4.0 and beyond," in *Modelling of Virtual Worlds Using the Internet of Things*, S. P. Singh and A. Solanki, Eds. Boca Raton, FL, USA: CRC Press, 2024, pp. 19–73.
18. A. Pawlicka, D. Puchalski, M. Pawlicki, and M. Choraś, "How to secure the IoTbased surveillance systems in an elegant way," in *Proc. 2023 IEEE Int. Conf. Cyber Security Resilience (CSR)*, Isola San Servolo, Venice, Italy, Jul. 31–Aug. 2 2023, pp. 636–640, doi: 10.1109/CSR57506.2023.10224938.
19. J. Huddleston, P. Ji, S. Bhunia, and J. Cogan, "How VMware exploits contributed to SolarWinds supply-chain attack," in *Proc. 2021 Int. Conf. Comput. Sci. Comput. Intell. (CSCI)*, Las Vegas, NV, USA, Dec. 2021, pp. 760–765, doi: 10.1109/CSCI54926.2021.00143.
20. Ren, Z., Li, X., Jiang, Q., Wang, Y., Ma, J. and Miao, C., 2022. Network slicing in 6G: An authentication framework for unattended terminals. *IEEE Network*, 37(1), pp.78–86.
21. Sarker, I.H., Furhad, M.H. and Nowrozy, R., 2021. Ai-driven cybersecurity: An overview, security intelligence modeling and research directions. *SN Computer Science*, 2(3), pp. 1–20.
22. Pajooh, H.H., Demidenko, S., Aslam, S. and Harris, M., 2022. Blockchain and 6G-enabled IoT. *Inventions*, 7(4), pp.1–18.
23. Syed, N.F., Shah, S.W., Shaghaghi, A., Anwar, A., Baig, Z. and Doss, R., 2022. Zero trust architecture (ZTA): A comprehensive survey. *IEEE Access*, 10(1), pp. 57143–57179.
24. Kol, G. and Naor, M., 2008, March. Cryptography and game theory: Designing protocols for exchanging information. In *Theory of Cryptography Conference* (pp. 320–339). Berlin, Heidelberg: Springer.
25. Partala, J. (2021). Post-quantum cryptography in 6G. In Wu, Y., Singh, S., Taleb, T., Roy, A., Dhillon, H. S., Kanagarathinam, M. R., & De, A. (Eds.), *6G Mobile Wireless Networks* (pp. 431–448). Springer, Cham, Switzerland.

26. Singh, S., Sharma, P.K., Moon, S.Y. and Park, J.H., 2024. Advanced lightweight encryption algorithms for IoT devices: Survey, challenges and solutions. *Journal of Ambient Intelligence and Humanized Computing*, *15*(2), 1625–1642.
27. Alluhaidan, A.S.D. and Prabu, P., 2023. End-to-end encryption in resource-constrained IoT device. *IEEE Access*, 11, pp.70040–70051.
28. Rao, P.M. and Deebak, B.D., 2023. A comprehensive survey on authentication and secure key management in internet of things: Challenges, countermeasures, and future directions. *Ad Hoc Networks*, 146, p.103159.
29. Raza, S., Seitz, L., Sitenkov, D. and Selander, G., 2016. S3K: Scalable security with symmetric keys—DTLS key establishment for the Internet of Things. *IEEE Transactions on Automation Science and Engineering*, *13*(3), pp.1270–1280.
30. Overmars, A. and Venkatraman, S., 2020. Towards a secure and scalable IoT infrastructure: A pilot deployment for a smart water monitoring system. *Technologies*, *8*(4), p.50.
31. Liu, L., Sajid, Z., Kravaris, C. and Khan, F., 2024. Detection and analysis of cybersecurity challenges for processing systems. *Process Safety and Environmental Protection*, 185, pp.1061–1071.
32. Lekkala, S., Avula, R. and Gurijala, P., 2022. Big data and AI/ML in threat detection: A new era of cybersecurity. *Journal of Artificial Intelligence and Big Data*, 2(1), pp.32–48.
33. Srinivasan, M. and Senthilkumar, N.C., 2025. Intrusion detection and prevention system (IDPS) model for IIoT environments using hybridized framework. *IEEE Access*, *13*, 26608–26621
34. Mudzingwa, D. and Agrawal, R., 2012, March. A study of methodologies used in intrusion detection and prevention systems (IDPS). In *2012 Proceedings of IEEE Southeastcon* (pp. 1–6). IEEE.
35. Scarfone, K. and Mell, P., 2010. Intrusion detection and prevention systems. In *Handbook of Information and Communication Security* (pp. 177–192). Berlin, Heidelberg: Springer.
36. Wen, S.F., Shukla, A. and Katt, B., 2025. Artificial intelligence for system security assurance: A systematic literature review. *International Journal of Information Security*, 24(1), pp.1–42.
37. Rangaraju, S., 2023. Secure by intelligence: Enhancing products with AI-driven security measures. *EPH-International Journal of Science and Engineering*, 9(3), pp.36–41.
38. Akinade, A.O., Adepoju, P.A., Ige, A.B., Afolabi, A.I. and Amoo, O.O., 2021. A conceptual model for network security automation: Leveraging AI-driven frameworks to enhance multi-vendor infrastructure resilience. *International Journal of Science and Technology Research Archive*, *1*(1), pp.39–59.
39. Shen, S., Zhu, T., Wu, D., Wang, W. and Zhou, W., 2022. From distributed machine learning to federated learning: In the view of data privacy and security. *Concurrency and Computation: Practice and Experience*, *34*(16), p.e6002.
40. Kasula, V.K., Yadulla, A.R., Konda, B. and Yenugula, M., 2024. Fortifying cloud environments against data breaches: A novel AI-driven security framework. *World Journal of Advanced Research and Reviews*, *24*, pp. 1613–1626.
41. Ouchani, S., Boudouaia, M.A. and Tasidou, A., 2025. PRISM: A PRIvacy safeguard for smart multiservices. *IEEE Software*, *42*(5), pp. 107–115.
42. Ali, M., Naeem, F., Tariq, M. and Kaddoum, G., 2022. Federated learning for privacy preservation in smart healthcare systems: A comprehensive survey. *IEEE Journal of Biomedical and Health Informatics*, 27(2), pp.778–789.

43. Awan, K.A., Din, I.U., Almogren, A. and Rodrigues, J.J., 2023. Privacy-preserving big data security for IoT with federated learning and cryptography. *IEEE Access*, 11, pp.120918–120934.
44. Abdel-Basset, M., Hawash, H., Moustafa, N., Razzak, I. and Abd Elfattah, M., 2024. Privacy-preserved learning from non-i.i.d. data in fog-assisted IoT: A federated learning approach. *Digital Communications and Networks*, 10(2), pp.404–415.
45. Prabowo, S., Putrada, A. G., Oktaviani, I. D., Abdurohman, M., Janssen, M., Nuha, H. H., & Sutikno, S. (2025). Privacypreserving tools and technologies: Government adoption and challenges. *IEEE Access*, 13, 33904–33934.

18 Usability Perspectives and Applications of 6G and IoT Networks

A Case Study Approach

Anand Singh and Gaganpreet Kaur

18.1 INTRODUCTION

The development of the tight-knit connection between the physical and virtual worlds requires constant dependence on high-form sensors [1]. For this, there is a need for ultra-high peak rate and low latency rate of 6G, which facilitates IoT devices for maximum coverage. The constant use of sensor networks requires high energy and the development of a strict system that may cause 6G to combat data security concerns and pave the way for green communication [2].

The sensor-based speedy communication requires fog intelligence [2]. In the next decade or two, scientists will create fog intelligence and offer IoT services with ultra-low latency by integrating 6G with Artificial Intelligence (AI). Such devices can be further used in collecting data under the neural network framework for data communication.

The growing 6G technology will result in edge intelligence, in which there will be AI on chips to upgrade the IoT systems [3]. Edge intelligence is reliant on edge nodes for fast communication, especially for categorization and prediction-making system building. For such systems, there is a need for a secure solution that is possible with 6G-based IoT networks [4]. Maximum coverage of 6G in space, air, ground, and underground is essential to effectively integrate IoT to expand innovative activities and develop industries [5].

There will be the construction of communication pathways on the above-mentioned four major grounds for 6G-IoT connectivity. For terrestrial communication, the scientists are required to use the THz to enhance bandwidth efficiency and data rates, while for underwater communication, submarines working in broad and deep-sea operations carry communication channels between the hub and the water bodies [6].

The coverage of 6G-IoT is possible with the integration of satellite communication networks in the current wireless systems, also known as Satellite IoT (SIoT) [7]. These satellites are divided into three networks in terrestrial communication: LEO, MEO, and GEO. With the emergence of 6G, scientists can launch hundreds of satellites to let the LEO network maximize efficiency and coverage. Scientists are in the

DOI: 10.1201/9781003586456-18

process of developing an inter-satellite communication network by using THz bands for operational excellence and to accommodate more satellites.

With the growing dependence on digital devices and the advancement of computation techniques, IoT devices are predicted to increase to 32.1 billion in 2030 and by 2033, with the largest number of devices in China [8]. Its major use case will be consumer media devices like the internet. Other than this, there will be more than a billion IoT devices in asset tracking, Information Technology, and vehicles.

6G is the technology of 2030, but the current experiments with existing digitized tools paved the way to bring the upcoming generation. There are multiple use cases for it.

18.1.1 Mixed Reality (MR)

Pico-second communication channels merge the digital and real world, leading to the advancement in AR technology [9]. Products like Vision Pro, Orion, and neural wristbands provide a glimpse of upcoming 6G technology and new use cases of it.

Such products demand a consistent and hyped user experience, due to which developers are unable to connect the tool with the network. In the case of multiplayer MR gaming, they need a network that can expose APIs to compute offload, position the system, and sense minute things [10]. These applications demand technicalities related to the upstream of the environment of the players, the downstream of the virtual zone, and game logic.

Regarding latency, encoding on the AR glasses defines the limits at a range of ~100 ms. High frame rates result in a low latency rate in the range of ~20 ms [11]. For MR, the bandwidth of around 20 Mbps is required, which is likely to increase by 50 points more for each player. MR needs 10 Mbps per player for spatial computing. It is also estimated to increase with the resolution according to the compression ratio.

18.1.2 Smart City

Smart cities are sustainable, aligning with the goals of the United Nations, due to which their market size is expected to increase to $12 trillion in 2033 [12]. 6G is likely to leverage the development of such cities because of its data handling capabilities and its wide applications in smart grids, vehicles, healthcare, and management systems in the form of holographic communication, VR, and gaming experience [13].

To build a smart city, the foremost requirement of 6G is to support a wide range of IoT devices that have high efficiency and low latency rates [14]. Besides, there is a need to use AI for resolving network concerns to optimize the data handling process.

AI has played a keen role in improving city governance with its personalized policymaking tools and decisions by utilizing IoT-collected data [15]. IoT utilizes multiple sources such as security cameras, sensors, and GPS, and home devices for data collection, which they transfer to the cloud to compile them later. The compiled data is later used during the decision-making process. Smart cities consist of automated industries and virtual and real-life connected manufacturing.

Hexa-X is an example of it [16]. It is a flagship project of the European Union that combines the digital, physical, and human worlds to form the global Internet of

Everything (IoE). In this project, humans are the core part whose virtual replicas will be made. For this, the digital twin (DT) technology will be utilized to make virtual replicas of different physical objects and for problem resolution [17].

6G is a prime need for the development of smart cities; it will promote sustainable mobility with smart and green parking, pave the way for the Internet of Bio-Nano Things to build human–machine connections, and offer personalized solutions. 6G leverages an IoT-based optimal routing path protocol to optimize waste management systems and offer smart gridding services to promote local living.

18.1.3 Tactile Internet

Tactile internet goes beyond real-time interaction by giving access to a sense of emotions and touch through communication [18]. Because of the involvement of kinesthetic communication and haptic interaction, it heavily relies on 6G and its integration with AI, due to which the users have to wear sensitive devices like a stylus, gloves, or any other device that is connected to the surface of the human body. Such devices stimulate the touch sensations and send their feedback to the actuator system connected to it. In return, the system would replicate the movement of users via tactile sensors to leverage remote touch with ultra-low latency communication (Table 18.1).

The major challenge in the development of the tactile internet is lowering the latency rate as per the human senses via reliable communication [19]. It has to be designed in the same way as the hand and brain of a typical human. When a human touches anything, the organs and neurons send feedback to the brain in a fraction of a second [20]. To fasten the system, there is a need to utilize AI in network architecture and optimization, and edge and transmission technologies [21].

Edge technology enables AI to build predictive models before the occurrence of actual action [22]. In this way, users can avoid delays by anticipating the movement of others and transmitting to them earlier. Yet, limited values can be predicted via it as humans move in four ways: push and pull.

TABLE 18.1
Characteristics of Tactile Internet

Application Features	Augmented Reality	Gaming	Smart Cities	Brain–Machine Interface
Latency	H	H	M	H
Reliability	H	M	H	H
Data rate	H	H	M	H
Security	H	M	H	H
Edge computing capability	H	H	H	H
AI integration	H	H	H	H
Multi-sensory integration	H	M	M	H
Interactivity	H	H	M	H
Scalability	M	M	H	M
Localization accuracy	H	M	H	H

Tactile internet, if AI and latency concerns are sorted, will expand remote education, handle material handling, and revolutionize telemedicine, which will let users have skills and services while being away. Tactile internet aims to develop a bilateral human-to-machine communication channel to deploy real-time data by utilizing network protocols for reducing the latency rate [23]. AI-enabled models maximize fault detection prior to problems to ensure reliability. Besides, AI improves haptic communication to drive accuracy [24]. For this, developers refine the network by testing different scenarios, aiming to enhance human–machine interaction. The team develops an adaptive system and interface to leverage user experience [25].

18.1.4 Holographic Communication

6G communication offers energy-efficient solutions to fulfill the demands of consumers with the development of reconfigurable intelligent surfaces (RIS) [26]. RIS consists of a large number of antennas to control electromagnetic waves on the radio surface. It leverages holographic communication as well, but those devices are called reconfigurable holographic surfaces (RHS).

RHS is made of metamaterial radiation elements that are arranged in a holographic manner for beam generation in the expected direction [27]. Its transceiver is developed on the conventional hologram principles; based on this, it does not require an extra control unit to shift and convert hologram signals. Scientists can obtain well-directed radiation cost-effectively by setting up a MIMO antenna.

Scientists can apply the RHS configuration from sub-6 GHz to terahertz (THz) on the radio spectrum [28]. They can also reuse hardware of the whole network system by integrating localization, sensing, and communications in it and enhancing configurations with a vehicle to everything (V2X), internet of Intelligent Things (IoIT), IoE, internet of space things (IoST), and IoT [29]. Scientists integrate rate splitting multiple access (RSMA) and non-orthogonal multiple access (NOMA) like schemes with RHS configurations to develop more novel schemes than before for constant improvement in communication networks, cost-effectively.

However, RHS poses security concerns; illegitimate configurations can result in jamming attacks, signal leakage, and pilot contamination [30]. Unfortunately, due to its passive nature, detecting contamination is difficult, which makes deterioration of the security layer very likely. Yet, the current projects on the security characteristics of 6G communication could solve the issue.

Besides RHS, holographic communication has made teleconferences more convenient as participants can remotely collaborate by showcasing their presence in 3D as holograms [31]. It happens because whenever a person enters the meeting, the sensors capture the information and replicate it to make it 3D as a hologram on the other side of the person in the meeting, to make it appear 3D [32]. This new technology reduces the gap between the physical and digital worlds, mixing them as they are the same. In the case of users, they can grasp haptic information for a better experience because of the use of the sensors. They can sense touch due to haptic sensors.

Holographic technology is going to be the need of the coming generation, and 6G makes it digitally secure and enhances user experience with its latency rate and data-holding characteristics.

18.1.5 Telemedicine

The ubiquitous speed and low latency of 6G modernize healthcare by making it more accessible for patients and healthcare workers, especially in telemedicine, remote diagnosis, and surgery [33]. This accessibility allows both parties to communicate with no requirement for commuting and cost-effectively. Besides, people from underserved areas and rural places would have access to health facilities.

6G will let patients have access to real-time video consultations. The doctors can send high-resolution imaging seamlessly, which would enhance accuracy [33]. Wearable devices like smartwatches and sensors make it possible for doctors to consistently monitor the health of their patients while they are away [34–35]. The devices record data in real time and transmit it to the healthcare professional for timely analysis. The use of AI and machine learning with 6G results in systems that enable doctors to identify health concerns in an early stage before they become critical [36]. For such diagnosis, the professionals use different tools such as smartwatches, health apps, and others.

Remote surgery benefits many because of 6G; the surgeons would be able to operate while being away by using robotic arms or video [37]. They lead and direct robots on what to do next. However, there are some challenges that need to be fixed by 6G, such as imprecision and delays. 6G is assured to reduce the cost of healthcare for patients with telemedicine and remote diagnostics, along with offering high-quality care from anywhere globally.

18.1.6 Smart Transport

The Intelligent Transport System (ITS) evolved into Autonomous Intelligent Transport System (6G-AITS) because of artificial intelligence, cloud computing, blockchain, edge, and data computing [38]. The key to 6G-AITS is a human-centric three-dimensional transport system that would be service-oriented and based on collaboration, sharing, intelligence, and data. It will have an updated global navigation satellite system (GNSS) positioning system [39]. Unlike the 5G network, the upcoming generational network utilizes IoT tools to guide accurately, maximizing safety.

6G-AITS is part of a smart city; therefore, its other name is smart transport. The key part of smart transport is the traffic management system that aims to reduce accidents, pollution, energy consumption, and traffic rate [40]. Other important characteristics are intelligent parking system (IPS), transportation management system (TMS), and electronic toll collection system (ETC).

These characteristics were established by using a 5G network. Yet, to upgrade them to 6G-AITS, there is a need to achieve a data rate of 1Tb/s, as it will be based on hyper-connectivity to offer personalized services [41] or whose core component will be Integrated Space–Air–Ground–Underwater Network (ISAGUN). According to scientists, in the process of making 6G-based commute successful, there will be the formation of the traffic brain; it will have a long-term memory that will function like a human hippocampus, leading computers to learn the ability to think.

However, this won't be achieved in the near future because of the unavailability of the required technology and the funds required to bridge the gap to turn this aim into reality. Yet, the scientists are working to make the transportation seamless by utilizing present resources and working on upgrading to 6G.

There are multiple use cases of 6G, but many of them are the dreams that are in the process of becoming reality because of the lack of required technology and the applicable framework to make systems that offer bespoke services to humans.

18.2 USE CASES OF IOT

18.2.1 Process Automation

Around 58% of organizations have automated their operations by integrating IoT tools such as IPCs and PLCs in their process [42]. It streamlines the task and minimizes the requirement of human intervention and errors [43]. Moreover, it enhances security and enables companies to collect real-time data, which can be used to optimize the workflow and implement predictive analysis. Automating workflows is seen to have improved work efficiency by approximately 8%.

There is an example of an Australian craft brewery, Burnley Brewing, that decided to expand its operations as the demand for its product grew. For this, they partnered with Bon Industry, an engineering firm, to automate their operations. The firm utilized its previous experience of working with Rockwell Automation to set up FactoryTalk Craft Brew; it's an automation process mainly developed for the brewing industry. Bon Industry placed PLCs, IoT sensors, and action-driving hardware throughout the workflow area and connected them to FactoryTalk Craft Brew to streamline the data collection process. Consequently, the labor work was reduced from the beginning to the end of the process, and efficiency increased, which enabled the company to expand with preparation.

18.2.2 Energy Monitoring

The growing energy concerns and need for sustainable solutions led companies to adopt energy monitoring methods by 55%. The organizations place smart grids and sensors to monitor the energy consumption of every individual. The studies proved that such methods have reduced the energy cost by 8.1% in companies.

Rolls-Royce benefited from monitoring energy consumption via IoT tools [44]. The company had 35 kinds of commercial craft and 13,000 engines. They used DTs for monitoring and maintenance. The sensors in the DTs are used to create replicas of engines to collect data on their real-time health. In response, the engines became more efficient; the need for maintenance was halved, and money spent on energy was reduced as well.

18.2.3 Quality Control and Management

Adopting IoT-based quality management practices gives a quick payoff to the companies, due to which 55% of companies have brought it to their places. While 53% of them grew within 2 years.

Quality management practices enable companies to collect real-time data on major factors like pressure, temperature, and humidity during operations, which allows them to detect deviation immediately prior to it causing defects. Integrating machine learning with IoT tools helps in analyzing the collected data prior to the concerns being shown. Therefore, studies prove that such practices have improved product quality by 4.6%.

There is an example of a robotics manufacturer, Fanuc, that adopted IoT sensors for quality control to reduce downtime [45]. The data collected from sensors enabled them to predict when equipment would malfunction, which ultimately enhanced the quality of their products, leading them to successfully achieve zero downtime.

18.2.4 Real-Time Inventory Management

Smart supply chain solutions keep companies from overstocking and stockouts. Therefore, 54% of the companies shared that they have either fully adopted real-time inventory management or are adopting it. Its adoption rate has increased by 35% from 2021 to 2024. Implementation of cost-effective and IoT-based supply chain practices has improved return on equity by approximately 8%.

Tyson Foods, a food company in the US, adopted real-time inventory management to automate manual work such as inventory counting [46]. The team wanted to integrate machine learning tools to improve performance. With SageMaker Services, they integrated the same solution to automate the counting of the chicken tray. Ultimately, the solution saved around 15,000 labor hours and prevented under- and overproduction.

18.2.5 Remote Asset Monitoring

Unlike other IoT systems, remote asset monitoring is easy to set up. In 2021, with 34% of companies adopting it, it became the most adopted use case. Currently, its adoption has increased to 48%. Companies with great assets are required to have this tool as it reduces the cost by 6% to 7% on average.

There is an example of Hindustan Coca-Cola; they implemented remote asset monitoring to keep an eye on their can-bottling lines and PET. The team was facilitated to monitor major factors, such as can-rinse pressure. The company used the collected data with their customer experience to update them about the quality.

Fortunately, the data collection process was streamlined; its integration with IoT allowed employers to stay informed about the factory. Besides, it has improved the product and ensured the assets were not stolen.

18.2.6 Asset Performance Optimization

IoT devices for optimizing the assets are updated versions of asset performance management, which integrate, visualize, and analyze the data. These tools combine software and IoT gateways to manage them at optimum levels, due to which the analyzers get well-prepared data that ultimately improves overall equipment effectiveness by 9.5%.

There is an example of Longroad Energy, an American company that is affiliated with WindESCo to improve the yield of their turbines. The company installed the Swarm System to get real-time turbine data, which would be stored on the cloud and fed into an analytics platform for operational excellence. Consequently, the production increased by 2.5%, having an increment of $430,000 in revenue.

18.2.7 Location Tracking

IoT tools for location tracking are the most utilized use case. These tools are used at all kinds of places, such as factories, warehouses, offices, schools, and others. To optimize asset utilization and ensure safety, they gather data from the devices and locate them. Location tracking tools are beneficial for both vendors and users. Its usage improves the revenue usually by 1.6%.

IoT for location is usually used in the e-commerce industry. Amazon uses Radio Frequency Identification (RFID) tags and IoT sensors to track their products in their warehouses [47]. The company has employed sensors in delivery operations to track the movement from the warehouse to customers' homes. These tools reorder the products that are running low once they detect their need [48]. Therefore, the warehouses and delivery of Amazon are highly automated.

18.2.8 Supply Chain Track and Trace

IoT for supply chain management results in fast growth in companies, due to which 54% of companies have employed it, and 60% of them grew within two years. It includes satellite technologies, GPS systems, sensors, and LPWAN. They track goods throughout the shipment and delivery operations and provide real-time visibility [49]. The IoT tracking system offers an advanced analytics system that minimizes losses and maximizes delivery accuracy by predicting disruptions. Therefore, it has a record of improving on-time delivery by 8.5%.

There is an example of Maersk that moved to edge computing in 2022 as they needed real-time instructions for the equipment handling, and the cloud was resisting time sensitivity [50].

As part of edge computing, Maersk employed IoT sensors and RFIDs to examine the contents in the container upon arrival at the terminal [51]. The team has a plan to install 6G networks in the near future to improve connectivity and data analytics operations with the aim to enhance global supply chain management.

IoT tools offer solutions in multiple spectrums that automate the process and reduce the cost, leading to various advantages for the organization that implements them [52]. With the advantage of 6G, IoT use cases will be adopted at higher percentages to secure more benefits.

18.3 CHALLENGES

6G network and IoT are intertwined with each other. There is a need to improve the technology to bring a 6G network, and IoT devices rely on 6G to offer better solutions to the users (Akbar et al., 2024). Scientists faced numerous challenges in IoT during the 5G era, and there are many ahead for 6G and IoT for the next decade to enter into a new generation.

Allocation of bandwidth in the THz spectrum between 100GHz to 10 THz is the biggest challenge [53]. Scientists have to ensure the presence of required spectrum recourses to manage the data accumulation process effectively. This spectrum is ideal for high data rates, but oxygen and water vapor can easily absorb THz, which ultimately limits its transmission. To combat this challenge, there is a need for a highly developed network infrastructure to support the 6G network. The hardware must have efficient network components, antennas, sensors, and signal processing capabilities to handle ultra-low latency and data rates. The infrastructure needs to be energy efficient as it will handle lots of data at an affordable cost. For this, it is required to integrate edge technology, heterogeneous networks, and satellite communication, which is a challenge, but it is a prime requirement.

The second biggest challenge is security. The transmission of large amounts of data, which increases the chances of threats, due to which there is a need to have detection systems, encryption methods, and authentication protocols to keep user data from cyber threats. There is a need to make 6G private and secure because of its use in user-driven industries such as healthcare and finance. In the case of security, the other sub-challenge is the need to address social and ethical considerations. There is a need to have a legal body that sets laws regarding algorithm bias, access to private data, intellectual property, and others. Such laws would add a layer to the security system of 6G, making people utilize the network with ease.

Deployment of 6G is the third main challenge because it requires hefty investment to build architecture, network systems, and create standards in technology. There is a need to make protocols as they will combine different networks, which is possible in the presence of algorithms.

These challenges cannot be addressed soon, as developing 6G is primarily dependent on the availability of technology that can support it. The availability of such a structure would be followed by the development of ethical codes and a deployment strategy to bring it into daily life with ease.

18.4 CONCLUSION

6G and IoT are intertwined with each other. They both offer multiple use cases to a wide range of industries. 6G is important for developing smart cities, combining physical and virtual worlds, holographic communication, tactile internet, and telemedicine. IoT offers remote asset management, process automation, quality control and management, real-time inventory management, and energy monitoring. However, the complete deployment of 6G is not possible yet because of the numerous prime challenges such as bandwidth spectrum, security concerns, and deployment strategy. Currently, computer scientists are in the research phase to develop and improve IoT tools to supplement it.

REFERENCES

1. Ali, A., Mujtiba, S., & Begh, G. R. (2023). 6G and IoT Use Cases. In 6G Wireless (pp. 315–340). https://www.researchgate.net/publication/370739610_6G_and_IOT_Use_Cases
2. Fernando, X., & Lăzăroiu, G. (2024, September 23). Energy-Efficient Industrial Internet of Things in Green 6G Networks. Applied Sciences, 14(18). https://doi.org/10.3390/app14188558

3. Liu, Y., Wang, Q., Liu, H., Zong, J., & Yang, F. (2022, November 15). *Edge Intelligence-Based RAN Architecture for 6G Internet of Things* (B. Rang, Ed.). Forward Series. https://doi.org/10.1155/2022/4955498
4. Duy La, Q., Ngo, M. V., Quang Dinh, T., Quek, T. Q.S., & Shin, H. (2019, February 5). Enabling Intelligence in Fog Computing to Achieve Energy and Latency Reduction. Digital Communications and Networks, 5(1), 3–9. https://doi.org/10.1016/j.dcan.2018.10.008
5. Liao, Y., Liu, S., Hong, X., Shi, J., & Cheng, L. (2023, October 31). Integration of Communication and Navigation Technologies toward LEO-Enabled 6G Networks: A Survey. Space: Science and Technology, 3. 10.34133/space.0092
6. Kumar Padhi, P., & Charrua-Santos, F. (2021, February 11). 6G Enabled Industrial Internet of Everything: Towards a Theoretical Framework. Applied System Innovation, 4(1). https://doi.org/10.3390/asi4010011
7. Wikström, G., & Thorson, S. (2024, December 19). Explore the Impact of 6G: Top Use Cases You Need to Know. https://www.ericsson.com/en/blog/2024/12/explore-the-impact-of-6g-top-use-cases-you-need-to-know
8. Lin, Y. (2024, January). Satellite-MEC Integration for 6G Internet of Things: Minimal Structures, Advances, and Prospects. IEEE Open Journal of the Communications Society. 10.1109/OJCOMS.2024.3418860
9. Sujay Vailshery, L. (2024, September 11). Number of IoT Connections Worldwide 2022–2033. IoT Connected Devices Worldwide. https://www.statista.com/statistics/1183457/iot-connected-devices-worldwide/
10. Hatami, M. (2024, October 18). Survey of the Real-Time Metaverse: Challenges and Opportunities. Future Internet, 16(10), 1–52. https://doi.org/10.3390/fi16100379
11. Hazarika, A. (2023, April 02). Towards an Evolved Immersive Experience: Exploring 5G- and Beyond-Enabled Ultra-Low-Latency Communications for Augmented and Virtual Reality. *Sensors*, 23(7). https://doi.org/10.3390/s23073682
12. Qadir, Z., Le, K. N., Saeed, N., & Munawar, H. S. (2023, June 23). Towards 6G Internet of Things: Recent Advances, Use Cases, and Open Challenges. *ICT Express*, 9(3), 296–312. https://doi.org/10.1016/j.icte.2022.06.006
13. Saeed, M. M., Saeed, R. A., Hasan, M. K., Sayed Ali, E., & Mazha, T. (2025, March 18). A Comprehensive Survey on 6G-Security: Physical Connection and Service Layers. Discover Internet Of Things, 5. https://link.springer.com/article/10.1007/s43926-025-00123-7
14. Murroni, M. (2023, August 30). 6G—Enabling the New Smart City: A Survey. Sensors, 23(17). https://doi.org/10.3390/s23177528
15. Nawaz, N. A., Abid, A., Rasheed, S., Farooq, M. S., Shahzadi, A., & Mubarik, I. (2022, April 21). Impact of Telecommunication Network on Future of Telemedicine in Healthcare: A Systematic Literature Review. International Journal of Advanced and Applied Sciences, 9(7), 122–138. https://science-gate.com/IJAAS/Articles/2022/2022-9-7/1021833ijaas202207013.pdf
16. Wolniak, R., & Stecuła, K. (2024, June 10). Artificial Intelligence in Smart Cities—Applications, Barriers, and Future Directions: A Review. Smart Cities, 7(3), 1346–1389. https://doi.org/10.3390/smartcities7030057
17. Sharma, S., Popli, R., Singh, S., Chhabra, G., Saini, G. S., Singh, M., Sandhu, A., Sharma, A., & Kumar, R. (2024, August 16). The Role of 6G Technologies in Advancing Smart City Applications: Opportunities and Challenges. Sustainability, 16(16). https://doi.org/10.3390/su16167039
18. Sharma, N. (2025, March 10). 5G and IoT: Emerging Technologies with Endless Use Cases. 5G and IoT: Benefits, Real World Applications and More. Retrieved March 10, 2025, from https://appinventiv.com/blog/5g-iot-technology/

19. Chaudhari, B. S. (2025, March 11). Enabling Tactile Internet via 6G: Application Characteristics, Requirements, and Design Considerations Top of Form. Future Internet, 17(3). https://www.researchgate.net/publication/389768113_Enabling_Tactile_Internet_via_6G_Application_Characteristics_Requirements_and_Design_Considerations_Top_of_Form
20. Ali Yahiya, T., & Kirci, P. (2019, June 1). Issues and Challenges Facing Low Latency in Tactile Internet. UKH Journal of Science and Engineering, 3(1), 47–58. https://www.researchgate.net/publication/334013390_Issues_and_Challenges_Facing_Low_Latency_in_Tactile_Internet
21. Abraira, V. E., & Ginty, D. D. (2013, August 21). The Sensory Neurons of Touch. Neuron, 79(4). https://doi.org/10.1016/j.neuron.2013.07.051
22. Singh, R., & Singh Gill, S. (2023, March 13). Edge AI: A Survey. Internet of Things and Cyber-Physical Systems, 3. https://doi.org/10.1016/j.iotcps.2023.02.004
23. Tychola, K. A., Voulgaridis, K., & Lagkas, T. (2023, September 15). Tactile IoT and 5G & Beyond Schemes as Key Enabling Technologies for the Future Metaverse. Telecommunication Systems, 84, 363–385. https://link.springer.com/article/10.1007/s11235-023-01052-y
24. El-Hajj, M. (2025, January 06). Enhancing Communication Networks in the New Era with Artificial Intelligence: Techniques, Applications, and Future Directions. Network, 5(1). https://doi.org/10.3390/network5010001
25. WindESCo. (2021, Decemeber 07). Longroad Deploys WindESCo Swarm Tech on 306MW Operational Utah Site. Longroad Deploys Swarm. https://www.windesco.com/news/longroad-deploys-swarm
26. Xu, Y. (2024, April 04). Reconfigurable Holographic Surface-Assisted Wireless Secrecy Communication System. Electronics, 13(7), 1359.
27. Dhar, P. (2021, June 24). Study: 6G's Haptic, Holographic Future? Possibilities and Challenges for Future 6G Communications Networks. Retrieved June 24, 2021, from https://spectrum.ieee.org/6g-haptic-holography
28. Chen, R. (2025, February 21). A Review of Circuits and Systems for Advanced Sub-THz Transceivers in Wireless Communication. Electronics, 14(5). https://doi.org/10.3390/electronics14050861
29. Niu, M., Huang, X., & Wang, H. (2022). Vehicle-To-Anything: The Trend of Internet of Vehicles in Future Smart Cities. In Intelligent Electronics and Circuits - Terahertz, ITS, and Beyond. Intech Open. https://doi.org/10.5772/intechopen.105043
30. Ahmed, A., Zia, M., & Bhatti, N. (2022, January 1). Higher Secrecy Capacity by Successive Pilot Contamination and Jamming Cancellation. IEEE Access. 10.1109/ACCESS.2022.3229615
31. Shvetsov, A. V., & Hamood Alsamshi, S. (2024, December 11). Holographic Communication has Made Teleconferences More Convenient as Participants can Remotely Collaborate by Showcasing their Presence in 3D as Holograms. IEEE Access, 12. https://ieeexplore.ieee.org/stamp/stamp.jsp?arnumber=10788695
32. Bron, D. (2023, April 24). The Emergence of Holographic Telepresence: Disrupting Communication and Collaboration in the Digital Age. Medium. https://medium.com/chain-reaction/the-emergence-of-holographic-telepresence-disrupting-communication-and-collaboration-in-the-cd3461811e
33. Kumar, A., & Masud, M. (2025, April 04). Integrating 6G Technology in Smart Hospitals: Challenges and Opportunities for Enhanced Healthcare Services. Frontiers in Medicine, 12. https://www.frontiersin.org/journals/medicine/articles/10.3389/fmed.2025.1534551/full

34. Hosseini, M. M. (2023, November 03). Smartwatches in Healthcare Medicine: Assistance and Monitoring; A Scoping Review. BMC Medical Informatics and Decision Making. https://pmc.ncbi.nlm.nih.gov/articles/PMC10625201/
35. Alowais, S. A. (2023, September 22). Revolutionizing Healthcare: The Role of Artificial Intelligence in Clinical Practice. BMC Medical Education, 23. https://bmcmededuc.biomedcentral.com/articles/10.1186/s12909-023-04698-z
36. Handa, A. (2024). Role of Robotic-Assisted Surgery in Public Health: Its Advantages and Challenges. Cureus, 16(6). https://doi.org/10.7759/cureus.62958
37. Deng, X., Wang, L., Jiang, P., & Gui, J. (2023, June 1). A Review of 6G Autonomous Intelligent Transportation Systems: Mechanisms, Applications and Challenges. Journal of Systems Architecture, 142(3). 10.1016/j.sysarc.2023.102929
38. Othman, W. M. (2025, March 17). Key Enabling Technologies for 6G: The Role of UAVs, Terahertz Communication, and Intelligent Reconfigurable Surfaces in Shaping the Future of Wireless Networks. Sensor and Actuator Network, 14(2). https://doi.org/10.3390/jsan14020030
39. Mazur, S. (2020, December 09). An Introduction to Smart Transportation: Benefits and Examples. Digi. https://www.digi.com/blog/post/introduction-to-smart-transportation-benefits
40. Singh, S. (2024, July 20). Unlocking the Potential of 6G: How Ultra-Fast Connectivity is Set to drive Business Innovation. Appinventiv. https://appinventiv.com/blog/6g-technology-transforming-enterprises/
41. Paraskevopoulos, D. (2024, September 11). The Top 10 IoT Use Cases. Retrieved September 11, 2024, from https://iot-analytics.com/top-10-iot-use-cases/
42. Alahi, M. E. E., Sukkuea, A., Wazed Tina, F., & Nag, A. (2023, May 30). Integration of IoT-Enabled Technologies and Artificial Intelligence (AI) for Smart City Scenario: Recent Advancements and Future Trends. Sensors, 23(11). https://doi.org/10.3390/s23115206
43. Goldenberg, B. (n.d.). Driving Innovation: Rolls Royce's Success with Digital Twins. In Rolls Royce Use of Digital Twin Technology Case Study. https://ismguide.com/rolls-royce-use-of-digital-twin-technology-case-study/
44. Zelinska, J. (2023, August 18). Top 10 Industrial IoT Applications with Real-Life Examples. https://nix-united.com/blog/top-10-industrial-iot-applications-with-real-life-examples/
45. Render, C. (2022, July 27). Tyson Foods Boosts Efficiency with Computer Vision and Machine Learning from AWS. https://aws.amazon.com/blogs/industries/tyson-foods-boosts-efficiency-with-computer-vision-and-machine-learning-from-aws/
46. Swedberg, C. (2023, September 25). Amazon Employs RFID For Apparel Automated Purchasing. https://www.rfidjournal.com/news/amazon-employs-rfid-for-apparel-automated-purchasing/200454/
47. Andriy. (2024, September 04). IoT in Transportation: 8 Examples of How It Can Improve the Future. Vakoms. https://vakoms.com/blog/iot-in-transportation-8-examples-of-how-it-can-improve-the-future/
48. Rooney, P. (2022, December 12). Maersk Embraces Edge Computing to Revolutionize Supply Chain. https://www.cio.com/article/415822/maersk-embraces-edge-computing-to-revolutionize-supply-chain.html
49. Karim, M. R., Rodgers, D. C., & Hossain, M. A. (2024, November 12). The Role of Internet of Things (IoT) in Real-Time Supply Chain Monitoring. IJRISS, 1800–1816. https://dx. doi.org/10.47772/IJRISS.2024.8100157
50. Akbar, M. S., Hussain, Z., Ikram, M., Sheng, Q. Z., & Mukhopadhyay, S. (2024, October 17). On Challenges of Sixth-Generation (6G) Wireless Networks: A Comprehensive Survey of Requirements, Applications, and Security Issues. Journal of Network and Computer Applications. https://www.sciencedirect.com/science/article/pii/S1084804524002170

51. Ma, J. (2025, March 28). Terahertz Channels in Atmospheric Conditions: Propagation Characteristics and Security Performance. Fundamental Research, 5(2). https://doi.org/10.1016/j.fmre.2024.09.012
52. Gallego-Madrid, J., Sanchez-Iborra, R., Ortiz, J., & Santa, J. (2023, March 27). The Role of Vehicular Applications in the Design of Future 6G Infrastructures. ICT Express, 9(4), 556–570. https://doi.org/10.1016/j.icte.2023.03.011
53. Abdel Hakeem, S. A., Hussein, H. H., & Kim, H. (2022, March 02). Security Requirements and Challenges of 6G Technologies and Applications. Sensors, 22(5). https://doi.org/10.3390/s22051969

19 Future Networks and Services
The 6G-IoT Paradigm Shift

Anand Singh and Gaganpreet Kaur

19.1 INTRODUCTION

5G has leveraged Internet of Things (IoT) with its ultra-low latency and large capacity [1]. Their convergence has upgraded and fastened the way the data is exchanged by seamlessly connecting the sensors, gadgets, and machines for intercommunication externally and within the centralized system. The characteristics of 5G, which include gigabytes data rate, network reliability, and low latency, paved the way for IoT.

5G technology maximizes operational efficiency by connecting a vast number of devices without compromising productivity, due to which the deployment of IoT becomes seamless. It eventually enhances the decision-making process and offers multiple benefits, which makes the integration of 5G into IoT smooth.

19.1.1 Data Transfer Speeds

IoT systems are installed with software, sensors, and other devices to share data internally and externally via the internet. The speed at which the data is transferred determines the effectiveness and efficiency of IoT systems. 5G significantly increases the data transfer speed through higher frequency bands, which allows seamless transmission of a large amount of data. Rapid transfer of data allows IoT applications to process and act upon received data fast.

Fast transmission is beneficial for industries that require real-time data to make critical decisions in a short time, especially in healthcare. Nowadays, in hospitals, the medical professionals use wearable IoT devices to monitor vitals. They collect data and send real-time information to the professional who can suggest and recommend medicines at the time of need.

19.1.2 Energy Efficiency

Energy efficiency determines the quality of battery-operated IoT devices because users expect each device to run on a single battery charge for a long time. Energy-efficient IoT systems need less power and less battery disposal, which eventually makes them effective and useful for a long time.

DOI: 10.1201/9781003586456-19

5G technology maximizes the energy efficiency of IoT devices by enhancing the data transfer speed, which consequently reduces their energy consumption and operation time. The combined features improve the battery life of IoT devices and reduce maintenance costs, battery waste, and power usage.

5G handles small data packets more efficiently, which are significantly used in IoT, enabling the operations on minimal energy consumption with constant connectivity. Because of its capability to maintain connectivity in packet-based energy systems, 5G allows IoT applications to be used in agricultural sensors and environmental monitoring as well.

19.1.3 Latency

Latency is defined as the response time, which starts from the moment the data is sent from an IoT device to the moment the data is received, processed, and acted upon. Its rate has an impact on the efficiency and performance of an IoT device, the transmission of real-time data, and the decision-making process. In the case of industries, there is a need for Industrial Internet of Things (IIoT) that minimizes the latency rate to make operations efficient and reduce safety concerns.

5G reduces the latency rate to a few milliseconds, enabling IoT systems in industries to provide real-time data and instantaneous actions, which consequently enhances the reliability and performance of the IoT systems and applications.

19.1.4 Network Reliability

Network reliability determines the consistency in the data transfer process in IoT systems. It ensures that an IoT device can monitor real-time data and control other devices connected to it.

Network reliability is a critical aspect in IoT applications that are related to infrastructure management and healthcare monitoring, where interruptions can result in critical situations. 5G enhances the network reliability because of its ability to handle a large number of connected devices at a time. This feature consequently strengthens connections in complex settings, which means that via the 5G network, IoT systems can connect to other devices with poor signals as well.

Integration of 5G into IoT systems enhances its capabilities by reducing the response time, which results in an improvement in efficiency and network reliability. These features together increase the connectivity of IoT systems with other connected devices even in settings that have poor signals or are complex, leading to the provision of real-time data that hones the decision-making process.

19.2 CHALLENGES OF 5G IN IoT

The integration of 5G and IoT poses a set of challenges that need to be solved to embrace transformative technology. These hurdles include bandwidth issues, security and privacy concerns, and deployment strategies. Exploring these issues could pave the way for a solution to unlock the full potential of the current generational network.

19.2.1 Bandwidth

The growing data traffic and the rise in connected devices to 40 billion have led 5G and IoT to offer a fast network and more capacity to support data transmission. To offer a better bandwidth range, there is a need to enhance the IoT infrastructure. For this, the network operators need to connect IoT devices via the cloud, as it will help them to improve scalability, but it will raise security concerns over the sensitive data transferred via the cloud. Addressing these issues is essential, or it could affect businesses and internet users.

Increasing the bandwidth could lower the network reliability and speed because the increased size may affect internet exchange points. However, there are security concerns because of the growing number of connected devices, as well as the vulnerabilities they bring to the IoT system. To combat such issues, it's important to install a gateway-based access management and authentication system. Besides, there is a need to introduce protective measures, such as threat detection and encryption, to ensure IoT security.

19.2.2 Interdependence and Digital Infrastructure

IoT infrastructure determines its strength. To address the challenges related to interdependence and infrastructure of 5G and IoT integration, network operators should assess potential concerns of IoT systems and deliver user-based service. While offering a solution, they should consider the following points:

- **Interconnected Networks:** The 5G-IoT integration is possible in the presence of an interconnected network to offer connected devices. To make this network, the specialists should maintain network reliability and stability by ensuring that the network can handle the bandwidth requirements of IoT systems.
- **Infrastructural Risks:** Poor implementation of IoT applications poses a risk to the infrastructure of the device. The vulnerability in one sensor can potentially create service gaps by affecting the production lines. The network operators are required to follow effective strategies to mitigate risks by prioritizing security measures.
- **IoT Project Assessment:** IoT specialists, infrastructure leaders, and network operators should assess the projects thoroughly to determine service gaps because of present issues. Consequently, the deep assessment will enable network operators to offer the required resources to carry out IoT projects successfully.

19.2.3 5G Big Data Management

Big data management is the summation of data handling and data processing that IoT systems have generated efficiently. The growing demand for data exchange due to the increasing number of connected devices has revealed weak points in the 5G-IoT infrastructure, which cannot handle increased bandwidth.

The key challenge that network operators will face is sustaining network stability because disruption in the network infrastructure can significantly affect the consumers. To sort out this challenge, there is a need to implement applicable data management strategies that process and analyze the data efficiently to extract meaningful insights and outcomes from it.

- **Privacy and Security:** The convergence of 5G and IoT necessitates the implementation of data protection measures because of rising privacy and security challenges. The following are key methods to ensure privacy.
- **Data Protection:** To protect data during data transmission, it is important to update software regularly, implement encryption tools, and ensure the data is transmitted safely.
- **Authentication Methods:** An effective IoT security measure includes the implementation of unique authentication tools that only allow authorized users access to the devices, aiming to reduce data breaches.
- **Network Integrity:** To safeguard against security threats, there is a need to maintain network integrity by monitoring traffic, updating protocols, and implementing threat detection tools.

19.3 IoT CAPABILITIES AND SECURITY

- The challenges of 5G-IoT integration can be addressed by strategically utilizing the IoT capabilities to ensure a resilient network infrastructure. It can be done by implementing network auditing, network management, and IoT gateway management solutions.
- **Network Auditing:** The practice of auditing the network regularly can help a company, a user, or a network operator identify potential security threats, which eventually allows them to assess the risk and mitigate to maintain network integrity and security.
- **Network Management:** Network management tools and resources are significant for enterprises as they allow the companies to maintain, manage, and monitor their IoT devices and systems in complex settings effectively by ensuring their stability and security. However, these solutions should be made available for medium and small businesses and individual consumers to provide them with the means to ensure their data security.
- **IoT Gateway Management:** IoT gateway management resources ensure the protection of an IoT device of a user, which usually includes the installation of firewalls and utilizing different preventive means to keep the device from cyberattacks by inhibiting unauthorized access. Regularly updating the device is important to utilize the gateway management solutions effectively.

The capabilities of IoT significantly address the security concerns that arise because of the convergence of 5G and IoT technology via network management and gateway management solutions. Meanwhile, network auditing ensures the threats are identified earlier than they become alarming. It is expected that these capabilities will pave the way to make the 6G network secure in the near future.

19.4 5G USE CASES IN DIFFERENT INDUSTRIES

Currently, there are around 76 million 5G connections [2], leading to 44 use cases, with a few top of the charts because of their high-end dependence on 5G networks to enhance operational excellence with the provision of real-time data to streamline the decision-making process.

19.4.1 Automobiles and Transportation

The transport industry is the top sector that has adopted the convergence of 5G-IoT. Being the most mature "IoT" user with a hundred thousand connected cars, the sector utilizes numerous telematics applications for user-based insurance, vehicle diagnostics, and location tracking. Because of the need to transfer minimum non-real-time data in these applications, they can be used on generational networks that are lower than 5G without compromising the performance.

However, 5G enhanced the operations of these applications by allowing them to manage more data in real time to offer more complicated and personalized services, such as car dealerships and pricing plans of insurance companies. In these applications, the network operators utilize Long-Term Evolution (LTE)technology to connect with the vehicles. It is expected that in the coming years, the specialists will make alterations in the telematics to enable them to support 4G and 5G technology, aiming to make them work with all networks. The finest example of this is the telematics solution Harman has introduced for the 5G car.

19.4.2 In-Vehicle Infotainment

In-vehicle infotainment is another relatively mature transportation application that is expected to benefit a lot from 5G connectivity. This use case involves the delivery of information and entertainment services to drivers and passengers, such as updates on the current traffic situation or weather forecasts, but also more data-intensive video and music streaming [3]. 5G networks upgraded the in-vehicle infotainment applications by offering and adding new services to them, such as AR-based navigation systems, and in-car retail.

It is expected that in the near future, there will be 5G connectivity via tethered connections, and connected cars can use these applications on both 4G and 5G. This will eventually lead to the provision of more high-end services to companies and individual users.

19.4.3 Smart Grid Automation

5G technology is highly beneficial to automate the smart electricity grid by detecting faults to respond to them effectively and optimize the performance. To enhance the operations, 5G integrates Distributed Energy Resources (DER) with the power grid in adherence to laws and safeguards against cybersecurity threats. However, integrating DER into the electricity system is difficult because the energy production varies a lot. Besides, it needs an automated feeder system and real-time control to maintain reliability and keep from overloading the grids. These requirements can be met via wired communication methods as it reduces the response time and ensures network reliability.

Therefore, most of the utilities have a wire-communication-based power grid monitoring system. These systems at some utilities are also based on either copper or optical fiber. These grids are also available in wireless connections that offer less critical services. In the industry, the adoption of 5G relies on its flexibility and cost effectiveness when compared to wired communication. The example of automated smart grids is the demonstration of URLLC based on 5G that was made by Kalmar, ABB, and Nokia.

19.4.4 Robotics

With the presence of sensors, the growing technology has sharpened robots and enabled them to perform clerical tasks on automation, such as repairing and moving products. Low latency of 5G has paved the way for robots as it enables them to process a large amount of data. There is an example of cloud-based mobile robots that perform tasks in a warehouse as part of the 5G transformer project. In industries, 5G has mobilized the robots, resulting in their range between static collaborative robots and autonomous guided vehicles (AGVs) [4].

5G is able to add more processing solutions into these robots with its low latency rate and significant reliability score, which makes offloading easier via cloud servers and reduces complexities. The example of the use of 5G in robotics is a cloud mobile robot that performs warehouse operations with better mobility.

19.4.5 Cooperative Intelligent Mobility

Cooperative mobility intelligence offers public and private transportation via 5G-based Vehicle-to-Everything (V2X) infrastructure. It is made of sensors, requiring V2X base stations.

Such systems assist drivers and passengers with their service. Thy warn drivers about calamities, road conditions, traffic, and weather forecasts, to maximize efficiency and follow road safety rules. Currently, V2X is made on LTE as part of 3GPP Rel-14 and 15, whereas 5G-based V2X was introduced as 3GPP Rel-16.

19.4.6 Video Surveillance

5G-based IoT devices play an integral role in enhancing security and surveillance systems by adopting wireless communication for video surveillance [5]. It costs less than the wired connection and is easier to set up than the latter. Nowadays, makers use LTE for video surveillance, which includes cameras installed on cars, transport, and drones. 5G has enhanced the capabilities of LTE by offering services that assist in analyzing real-time data collected from multiple cameras. Currently, there are cameras from KDDI and Samsung that transmit Ultra-High Definition (UHD) surveillance video in real time via 5G [6].

Currently, there are two kinds of surveillance in general: static and mobile. In the case of static surveillance, LTE is commonly used in public events like concerts and construction sites. However, in mobile video surveillance, cameras are installed on vehicles. It also includes drones and body-worn cameras. LTE in video surveillance systems records HD videos that can be analyzed with the provision of real-time data to make effective decisions. 5G has advanced IoT technology; it is estimated that by the end of 2025, there will be 1.2B devices connected to 5G, which includes both non-IoT and IoT. Therefore, it is expected that 6G will revolutionize the services to the next level.

19.5 ROLE OF 6G IN ADVANCING IoT

The low latency and large bandwidth of 5G have already enabled IoT to transform industries by upgrading with more digitized and sophisticated systems [7]. However, 6G is expected to address the limitations of the current generational network by lowering latency further, and increasing bandwidth that would lead to developing smart cities, healthcare, and industries with maximum automation. There are the following ways in which 6G will enhance IoT.

19.5.1 High Bandwidth and Fast Transmission

6G will increase the data rate, leading to having a large enough bandwidth to transmit a massive amount of data and handle massive data streams. There will be larger THz spectrum bands and mid-bands, due to which data transmission will become faster as they will have more capacity to hold and store data [8].

Such technology will fulfill the need for speedy communication of manufacturing industries as they would be able to get real-time data that would improve efficiency rate and control the cost during the decision-making process. 6G will optimize the performance in data-oriented workplaces and environments, as data can flow fast between IoT devices in it.

19.5.2 Low Latency Communication (URLLC)

6G will reduce the response time, leading to ultra-low latency communication, which will be beneficial for the healthcare and automation industry, minimizing the problems that arise due to a minute delay in the communication [9]. It is expected that the response time will be less than a millisecond, which means that machines and devices can make decisions in a split second. 6G ensures to minimize the downtime and maximize the reliability to make them applicable for critical digital situations.

19.5.3 Sustainability and Energy Efficiency

6G will enhance the energy efficiency to solve the current significant issue and allow IoT tools and machines to carry on the operations for a longer time with a minimum requirement of power.

Via AI and machine learning, 6G machines will optimize the resource allocation to make the power usage system efficient with minimum wastage, making it more sustainable [10]. Such systems will be required in the development of smart cities in which thousands of sensors will monitor the environment and require low latency for the timely transmission of data.

19.5.4 Edge Computing

Edge computing is the key way in which 6G will enhance IoT devices. By edge computing, it will process data near IoT devices instead of sending it to a centralized cloud. This local method will ultimately fasten the data flow and transmission [11].

To understand it further, take an example of traffic. Because of edge computing, in the 6G era, the systems will process the data received from the connected vehicles at their places without sending it to a data center. Edge computing will enhance the efficiency of the industrial sector and maximize the success rate in the healthcare sector with the provision of timely services.

19.5.5 AI and Machine Learning

The integration of AI and machine learning with 6G will upgrade the IoT devices, enabling them to predict the future based on the information they learned from the environment themselves and make decisions without human interference [12].

Such technology would be beneficial in the healthcare sector, where AI-based wearables could monitor vitals and alert professionals to the potential risks. This will ultimately make IoT machines more reliable and enable them to adapt to the real-time surroundings.

6G will enhance IoT systems further, taking the charge from where 5G has left off to make the environment more digitized. Both 6G and IoT depend on each other; IoT creates a system that supports 6G, while 6G improves the performance by reducing the latency rate and data transfer speed. By converging with edge computing and machine learning, it will enhance the bandwidth and energy efficiency of IoT, making the combination of the two useful for the user.

19.6 QUALITY SERVICE AND EXPERIENCE IN 6G

People have high expectations from 6G technology, due to which there are two broader metrics to judge its capability: QoS and QoE [13]. QoS is to provide a reliable service in connecting people and things. QoE is the qualitative measurement of QoS from the perception of the end-user; it is more about user experience with the technology. This metric enables the developers and scientists to make it more personalized via a network design approach (Figure 19.1).

In order to maximize the efficiency rate, there is a need to maintain and develop a fine balance between QoS and QoE.

19.6.1 QoS

There are multiple opportunities in QoS management, but those opportunities have become important for 6G to meet the requirements [14]. It is very important for QoS mechanisms to adapt to the upcoming IoT technology being developed for 6G. There is a set of QoS features for which there is a need for data packets. Those packets are classified according to the QoS class. These classes are defined via Deep Packet Inspection (DPI) technology or Operations Support Systems (OSS).

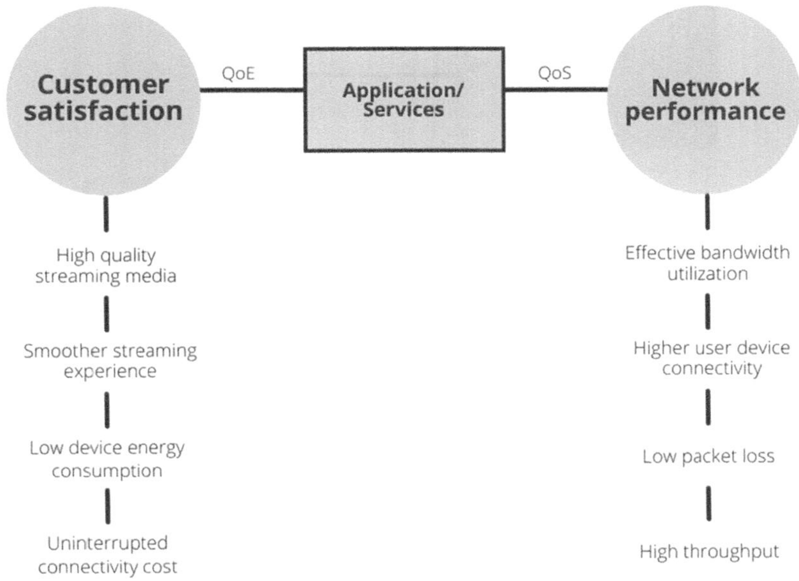

FIGURE 19.1 Relationship between QoS and QoE.

After implementing QoS features, it is important not to have traffic more than the available capacity. In a fixed network, it is easy to determine this thing because of the limited ports on one cable, due to which it is easy to calculate the remaining space. Via DPI, traffic becomes controllable, but maintaining SLA remains a challenge.

To sort this issue, scientists are developing solutions to optimize network capacity that allow profile changes.

19.6.2 QoE

QoE is the satisfaction of the user with the service [15]. The intensity and percentage of the fulfillment of their expectation with regard to the service are directly dependent on their current state and personality (Figure 19.2).

Currently, there is a lack of understanding of the user experience with regard to 6G and IoT in research. It is expected that 6G would offer personalized mobile experiences via AI-based network management, but it is important to address specific considerations with regard to 6G to offer a quality experience.

The connotation of QoE has evolved with regard to telecommunication. In the beginning, it was understood as the acceptability of service by the end-user, but it includes end-to-end effects and the influence of the expectation of the users [16]. This evolution has improved aggregated scores that include the mean opinion score (MOS). Yet, this change is the by-product of the upgradation in-network delivery after the rise of IP-based services and smartphones. With current work on 6G, QoE is discussed and worked upon in new dimensions that are related to enriching cyber-physical experiences in AR/VR.

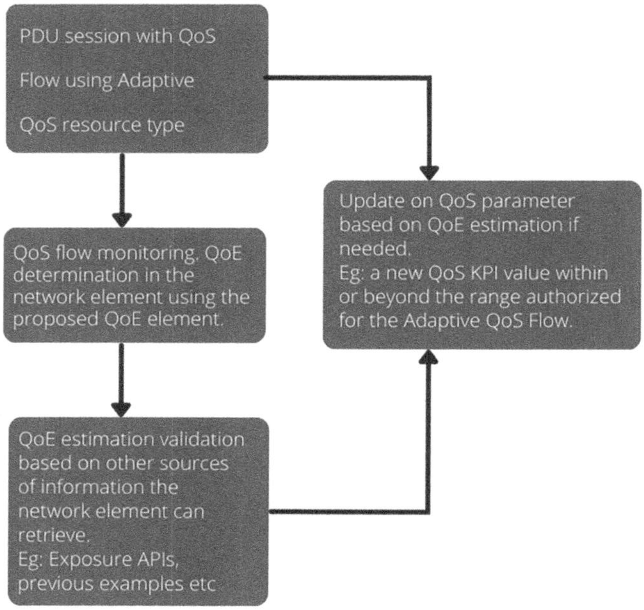

FIGURE 19.2 QoE cycle based on a sophisticated flowchart.

19.7 CURRENT TRENDS AND BREAKTHROUGHS IN IoT

IoT develops with every generational network, leading to the current state where there is the availability of wearables and sensors to monitor real-time data [17]. With further development, it can prepare the environment for 6G, during which its installation in different fields has led to numerous following trends.

19.7.1 Blockchain

Blockchain is more than half based on IoT because of its crucial role in data security and communication between network nodes [18]. It assures that the records will be kept safe. Therefore, it is assumed that blockchain is ideal for IoT machines, and its market would grow to 2409 million by 2026. Because of IoT, blockchain has made its way into agriculture, banking, transportation, finance, and other professions.

19.7.2 Smart Cities

IoT is paving the way to develop data-driven cities that upgrade the residential facilities, decrease deviance, and reduce the mortality rate and assault [19]. However, it will take time to become a reality. The governments of several countries have started working on IoT projects to implement digital solutions regarding traffic management, sustainability, energy consumption, and citizen safety.

19.7.3 Traffic Management

IoT is capable of combating traffic issues worldwide, due to which several organizations are going with IoT-installed systems to develop smart traffic networks.

According to the report of McKinsey, smart technologies can reduce commute time by 10% to 15% [20]. This smart technology is only possible with IoT systems on which the companies are working.

19.7.4 Metaverse

IoT is a key component in the metaverse that enhances its features, due to which their interplay develops new ways to expand in the tech sector [21]. IoT is assumed to grow further by the application of metaverse in different industries, such as e-commerce, media, games, entertainment, engineering, manufacturing, training, retail, and others.

The two are expected to complement each other, leading modernization in all industries. With the work expansion in these fields, IoT is growing and becoming smarter with the installation of AI, paving the way for 6G to come.

19.8 BREAKTHROUGHS IN 6G

6G is under research by many organizations, due to which breakthroughs in research and tech are making their way. Here are two key breakthroughs in 6G that are expected to make 6G possible in 2030.

19.8.1 In-Network Service Acceleration Platform (ISAP)

Recently, NTT DOCOMO (DOCOMO) and NTT Corporation (NTT) developed and demonstrated the ISAP, an integral part of the NTT- and DOCOMO-built Inclusive Core mobile network architecture, to meet the data processing needs of the 6G era (NTT, 2025). During the demonstration, Nokia and the two companies showed how ISAP reduced the response time during the data exchange process and improved AI analysis to 90%.

The current work has enabled ISAP to develop the Inclusive Core architecture. It is developed in the lens of In-network Computing (INC). INC, enabled by ISAP, has the ability to transfer processing function from the application layer to the data function within the same network while reducing response time and energy consumption at the terminal. In this architecture, ISAP is used to mediate the data processing on terminals and clouds that synchronize the data on both sides of the terminal quickly.

Because of this, ISAP is capable of advancing numerous services that can help technology reach a variety of users. Therefore, NTT and DOCOMO are intending to work on INC further to provide mobile-network-linked computing services in the 6G era. They are collaborating with Nokia to work on cloud and telecommunication to pave the way for the next-generation network.

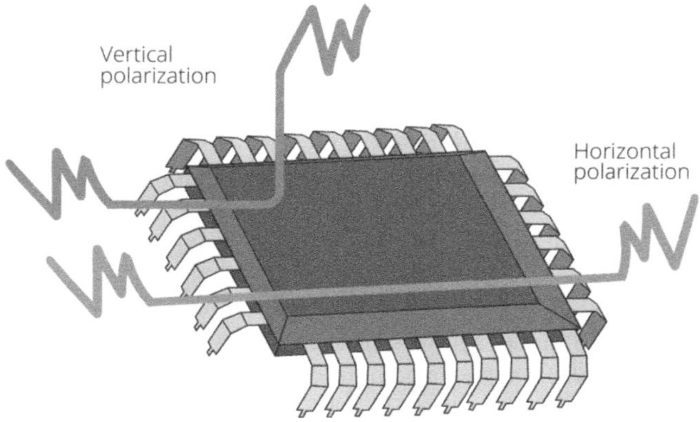

FIGURE 19.3 Terahertz polarization multiplexer of a silicon chip.

19.8.2 Silicon Chip

The team of researchers at the University of Adelaide, under the leadership of Withawat Withayachumnankul, has developed a silicon chip that enables 6G communications [22].

This chip can process the data at 220 to 330 GHz in the sub-terahertz J-band, which assures that it could be used for the next-generation network [23] (Figure 19.3).

This silicon chip has a terahertz polarization (de)multiplexer on the silicon base. The multiplexer on it picks one signal out of multiple signals and transmits it to the output line. Via polarization, the chip can divide waves according to the direction of oscillation. This technology, as researchers expect, would be utilized in the near future to support the 6G era with IoT and AI integration in it.

6G needs the support of multiple heavy technologies, and such breakthroughs are paving the way for it, hoping researchers will make a future reality soon. However, this going-to-be reality is incomplete without ethics.

19.9 ETHICAL IMPLEMENTATION OF 6G AND IoT TECHNOLOGY

Increasing technology is enhancing security on the one hand, but also growing room for hackers to utilize the same technology for their mischievous tasks with the wrong use of data. Therefore, it has become important to have an ethical framework for the development and deployment of IoT and 6G.

19.9.1 Core Components of Ethical IoT and 6G

The integral pointers drive the ethical framework in technological components [24].

- **Public Trust:** It is important that available technology is adopted by the masses to drive the potential of improving society morally and economically.

- **Transparency:** Manufacturers are required to be honest about what they have and how it will be used, so the users would be aware of where their provided data is going. This will ultimately help them make informed decisions.
- **Accountability:** The companies are required to store and use the data of users responsibly. They must receive consent from the users before sending it to anyone. If it involves financial and personal data, it should be kept confidential.
- **Security:** There should be an ecosystem in every company and government that ensures the protection of the data of users to keep them from cyber threats. The ecosystem should give importance to the user's consent before giving information to anybody.

Currently, numerous initiatives are being taken to draft an ethical framework and ethical technologies, which include The Institute of Electrical and Electronics Engineers (IEEE) Global Initiative on Ethics of Autonomous and Intelligent Systems, The Centre for Cybersecurity del World Economic Forum, and the European Internet of Things policy [25].

Besides, many tech companies are making an effort to implement transparency and ethics to protect the user data, which is generated via IoT sensors and devices, to empower people with their digital rights. However, there is a need for a global framework to keep nations and people from malicious attacks [26]. To make a legal framework, there is a need to give importance to moral principles and make companies develop a product that aligns with them. The companies must be bound to address the social implications associated with the recent systems to keep people safe. The framework is required to ensure that algorithms are unbiased to avoid discrimination. Based on a constitution, there needs to be a system that ensures all recent technologies do not cause harm and violate moral principles. There needs to be a body that monitors whether technology is affecting the employment rate, socio-economic disparities, and inequality or not. The framework should ensure that technology promotes inclusivity and ensures affordability among the masses to reduce class differences because the motive of technology is to enhance the life quality of the common man, and it is only possible when such things are included in the framework.

19.10 INTERNATIONAL STANDARDS FOR IOT TECHNOLOGY

Four main international organizations have made some standards on a universal level for individuals and organizations to follow one path to ensure that IoT is being utilized ethically. Those four organizations are the International Electrotechnical Commission (IEC), the IEEE, the International Organization for Standardization (ISO), and the International Telecommunication Union (ITU). They have drafted numerous standards, but some of them are significant in ensuring compliance with data privacy laws. Some of them are following.

- **ISO/IEC 21823-1:2019:** This standard outlines peer-to-peer interoperability because of its applications to the different independent IoT frameworks.

- **ISO/IEC 23093-1:2020:** This standard overviews the architecture of the IoT framework to determine the data flow and application programming interfaces (APIs).
- **ISO/IEC 21823-2:2020:** This standard expounds on the distributed network, constant interaction, and data exchange between different IoT frameworks.
- **ISO/IEC TR 30164:2020:** This standard describes use cases, terminologies, and upgradation in edge computing to determine regions of standardization in adherence to IoT.
- **ISO/IEC TR 30166:2020:** This standard discusses new advancements and future standards in IIoT. It covers risk analysis and utilitarian components of the same technology as well.
- **ISO/IEC 27030:** This standard outlines the security of IoT devices by expounding on its controlling measures, risk assessment, and principles.
- **ISO/IEC AWI 30147:** This standard is under development; it determines lifecycle processes of IoT frameworks.
- **ISO/IEC AWI 30149:** This standard is under process; it sheds light on the trust framework of IoT technology.

These four organizations finalized these standards after taking current issues into consideration. However, these standards generated a gap because of a lack of standards that deal with security issues of hardware and software in the realm of IoT. It is expected that this will be addressed in the near future by making standards that ensure security

19.11 CONCLUSION

IoT is paving the way for 6G, but the quality of the upcoming generational network is determined by two factors: QoS and QoE. To strike a fine balance between the two, scientists are making an effort to enhance IoT technology, leading to its use in different fields, such as blockchain, traffic management, smart cities, and the metaverse. Their work has resulted in collaboration with big tech to make mobile computing platforms and a chip that can be used in the 6G era. This breakthrough signals that 6G is near, prompting social scientists to understand the need to have an ethical framework to use the technology responsibly, which minimizes its negative impact on society.

REFERENCES

1. Sharma, N. (2025, March 10). *5G and IoT: Emerging Technologies with Endless Use Cases*. 5G and IoT: Benefits, Real World Applications and More. Retrieved March 10, 2025, from https://appinventiv.com/blog/5g-iot-technology/
2. Pasqua, E. (2019, September 09). *The Leading 5G IoT Use Cases*. Retrieved September 09, 2019, from https://iot-analytics.com/the-leading-5g-iot-use-cases-2019

3. SBD. (2012, February 01). *2025 Every Car Connected: Forecasting the Growth and Opportunity* [A whitepaper written by SBD for the GSMA]. GSMA. Retrieved February 01, 2012, from https://www.gsma.com/solutions-and-impact/technologies/internet-of-things/wp-content/uploads/2012/03/gsma2025everycarconnected.pdf
4. Stój, J., Kampen, A.-L., Cupek, R., & Drewniak, M. (2022, December 23). Industrial Shared Wireless Communication Systems—Use Case of Autonomous Guided Vehicles with Collaborative Robot. *Sensors*, *23*(1). https://doi.org/10.3390/s23010158
5. Gieske, J. (2024, October 02). *5G IoT: The Transformative Impact of 5G on IoT.* 5G IoT. https://nybsys.com/5g-iot/
6. Korea. (2019, January 10). *Samsung and KDDI Demonstrate Real Time 4K Video Communication Powered by 5G at a Japanese Train Station.* Samsung and KDDI Demonstrate Real Time 4K Video Communication Powered by 5G at a Japanese Train Station. https://news.samsung.com/global/samsung-and-kddi-demonstrate-real-time-4k-video-communication-powered-by-5g-at-a-japanese-train-station
7. Taoglas. (2024, September 20). *How 6G Technology Will Enhance IoT Applications.* How 6G Technology will Enhance IoT Applications - Taoglas. Retrieved September 20, 2024, from https://www.taoglas.com/blogs/how-6g-technology-will-enhance-iot-applications/
8. Alwis, C. D., Kumar, P., Pham, Q.-V., Dev, K., Kalla, A., Liyanage, M., & Hwang, W.-J. (2022, October 25). Towards 6G: Key Technological Directions. *ICT Express*, *9*(4). https://doi.org/10.1016/j.icte.2022.10.005
9. Kamath, S. (2024, January 17). A Review of Recent Developments in 6G Communications Systems. *Engineering Proceedings*, *59*(1). https://www.mdpi.com/2673-4591/59/1/167
10. Kumar, R. (2023, November 28). From Efficiency to Sustainability: Exploring the Potential of 6G for a Greener Future. *Sustainability*, *15*(23). https://www.mdpi.com/2071-1050/15/23/16387
11. Ergen, M., & Saoud, B. (2024, October 14). Edge Computing in Future Wireless Networks: A Comprehensive Evaluation and Vision for 6G and Beyond. *ICT Express*, *10*(5), 1151–1173. https://www.sciencedirect.com/science/article/pii/S2405959524000948
12. Alhammadi, A., Shayea, I., El-Saleh, A. A., & Hadri Azmi, M. (2024, March 25). Artificial Intelligence in 6G Wireless Networks: Opportunities, Applications, and Challenges. *International Journal of Intelligent Systems.* https://onlinelibrary.wiley.com/doi/10.1155/2024/8845070
13. García-Torres, M., Pinto-Roa, D. P., & Núñez-Castillo, C. (2024, March 01). Feature Selection Applied to QoS/QoE Modeling on Video and Web-Based Mobile Data Services: An Ordinal Approach. *Computer Communications*, *217*(C), 230–245. https://doi.org/10.1016/j.comcom.2024.02.004
14. Skarp, M., & Costa-Requena, J. (2025). *Quality of Service (QoS): In Sliced Network.* IntechOpen. https://www.intechopen.com/chapters/1195123
15. Tondwalkar, A., Andres-Maldonado, P., Chandramouli, D., Liebhart, R., Sanchez Moya, F., & Kolding, T. (2024, January 01). Provisioning Quality of Experience in 6G Networks. *IEEE Access*, *12*, 127007–127017. https://doi.org/10.1109/ACCESS.2024.3455938
16. Varela, M. (2012, June 03). *Qualinet White Paper on Definitions of Quality of Experience* (1.1) [The Qualinet White Paper Aims to Establish a Working Definition of Quality of Experience (QoE) in Multimedia Communication Systems to Enhance Interdisciplinary Collaboration and Communication.]. Academia.
17. Verma, D., Singh, K. R., & Yadav, A. K. (2022, May 11). Internet of Things (IoT) in Nano-Integrated Wearable Biosensor Devices for Healthcare Applications. *Biosensors and Bioelectronics: X*, *11*. https://doi.org/10.1016/j.biosx.2022.100153
18. Antino. (2025, March 03). *Top IoT Trends in 2025 and What IoT Holds for the Future? Top 9 IoT Trends.* https://www.antino.com/blog/top-9-iot-trends

19. Sharma, N. (2025, March 28). *The Role of IoT in Building Smart Cities – 10 Applications and Use Cases*. IoT for Smart Cities: Transforming Urban Infrastructure. https://appinventiv.com/blog/role-of-iot-in-smart-cities-applications/
20. Woetzel, L., Remes, J., Boland, B., Lv, K., Sinha, S., Strube, G., Means, J., Law, J., Cadena, A., & Tann, V. (2018, June 05). *Smart Cities: Digital Solutions for a more Livable Future*. Smart City Technology for a More Liveable Future | McKinsey. https://www.mckinsey.com/capabilities/operations/our-insights/smart-cities-digital-solutions-for-a-more-livable-future
21. Asif, R. (2023, September 12). Exploring the Confluence of IoT and Metaverse: Future Opportunities and Challenges. *IoT*, *4*(3), 412–429. https://doi.org/10.3390/iot4030018
22. Cointelegraph. (2024, August 30). *Chip Breakthrough Could Bring Ultra-Fast 6G Network 'Within a Decade'*. Retrieved August 30, 2024, from https://www.binance.com/en-KZ/square/post/12875433193025
23. Yazgin, E. (2024, August 30). *Silicon Chip Could Pave the Way for 6G Communications*. Silicon Chip Could Pave the Way for 6G Communications. https://cosmosmagazine.com/technology/internet/silicon-chip-adelaide-6g/
24. Telefonica Tech. (2022, December 20). *Ethical IoT: Principles for an Implementation that Respects People's Rights*. https://telefonicatech.com/en/blog/ethical-iot-principles-for-an-implementation-that-respects-peoples-digital-rights
25. The Future Society. (2018, May 28). *The IEEE Global Initiative on Ethics of Autonomous and Intelligent Systems*. https://thefuturesociety.org/the-ieee-global-initiative-on-ethics-of-autonomous-and-intelligent-systems/
26. ApekshaTelecom. (2024, February 03). *Ethical Considerations in the Development of 6G*. https://www.telecomgurukul.com/post/ethical-considerations-in-the-development-of-6g

Index

For Product Safety Concerns and Information please contact our EU representative GPSR@taylorandfrancis.com Taylor & Francis Verlag GmbH, Kaufingerstraße 24, 80331 München, Germany